Experimental Methods and Instrumentation for Chemical Engineers

Experimental Methods and Instrumentation for Chemical Engineers

Gregory S. Patience

AMSTERDAM • BOSTON • HEIDELBERG • LONDON
NEW YORK • OXFORD • PARIS • SAN DIEGO
SAN FRANCISCO • SINGAPORE • SYDNEY • TOKYO

ELSEVIER

Elsevier
225 Wyman Street, Waltham, MA 02451, USA
The Boulevard, Langford Lane, Kidlington, Oxford OX5 1GB, UK
Radarweg 29, PO Box 211, 1000 AE Amsterdam, The Netherlands

First edition 2013

Notices
Knowledge and best practice in this field are constantly changing. As new research and experience broaden our understanding, changes in research methods, professional practices, or medical treatment may become necessary.

Practitioners and researchers must always rely on their own experience and knowledge in evaluating and using any information, methods, compounds, or experiments described herein. In using such information or methods they should be mindful of their own safety and the safety of others, including parties for whom they have a professional responsibility.

To the fullest extent of the law, neither the Publisher nor the authors, contributors, or editors, assume any liability for any injury and/or damage to persons or property as a matter of products liability, negligence or otherwise, or from any use or operation of any methods, products, instructions, or ideas contained in the material herein.

British Library Cataloguing-in-Publication Data
A catalogue record for this book is available from the British Library

Library of Congress Cataloging-in-Publication Data
A catalog record for this book is available from the Library of Congress

ISBN: 978-0-444-53804-8

For information on all Elsevier publications
visit our website at store.elsevier.com

This book has been manufactured using Print On Demand technology. Each copy is produced to order and is limited to black ink. the online version of this book will show color figures where appropriate.

Working together
to grow libraries in
developing countries

www.elsevier.com • www.bookaid.org

Contents

Preface

Throughout the day, we constantly make use of experimental methods, whether or not we are aware of it: we estimate physical properties like time, distance, and weight, as well as probability and expectation. For example, what is the probability that I will be late if I sleep another five minutes? (What is the probability that I will only sleep an additional five minutes?) Many of us look at the weather forecast to gauge what clothes to wear. Following a recipe to bake or prepare a meal is an example of an experimental procedure that includes the classic engineering quantities of temperature, time, mass (volume) and length.

The principles of chemistry and chemical engineering were perhaps first formulated in the kitchen.

The undergraduate course on Experimental Methods offered in my department was, in the beginning, primarily based on the textbook written by J.P. Holman entitled "Experimental Methods for Engineers." This is an excellent textbook and is particularly suited for mechanical (and perhaps electrical) engineers, but elements particular to Chemical Engineering are lacking. For this reason, we embarked on the daunting task of compiling a version suited to the needs of Chemical Engineers.

The chapters often begin with a historical perspective to recognize the work of early pioneers but also to stimulate the imagination of the students. For example, 10 000 years ago, man created plaster from limestone. Plaster requires temperatures nearing 900 °C, which is well over 100 °C hotter than an open pit fire. This technology required considerable resources: 1 t of wood (chopped by stone axes), 500 kg of limestone, a pit 2 m in diameter and 0.7 m deep, rocks to insulate, and two days to burn. Modern manufacturing errors are costly and a nuisance; in prehistoric times, errors would have been considerably more than just an inconvenience.

In Chapter 1, the rules of nomenclature are reviewed—units of physical quantities, abbreviations, conversion between SI and British Units—and the various national and international standards bureaus are mentioned. Chapter 2 introduces significant figures and concepts of accuracy, precision and error analysis. Experimental planning is discussed in some detail in Chapter 3. This subject is enormous and we try to distil the essential elements to be able to use the techniques. Chapters 4 and 5 cover many aspects of measuring pressure and temperature. The industrial context is often cited to provide the student with a picture of the importance of these measurements and some of the issues with making adequate measurements. Flow measurement instrumentation is the subject of Chapter 6. A detailed list of the pros and cons of most commercial flow meters is listed. Example calculations are detailed throughout the book to help the students grasp the mechanics of solving problems but also to underline pitfalls in making these calculations. Chapter 7 deals with the three major physicochemical properties in Chemical Engineering, including thermal conductivity, viscosity, and diffusion. Measuring gas and liquid concentration is the subject of Chapter 8—many analytical instruments are mentioned but chromatography is primarily described. Finally, in Chapter 9 we discuss powder and solids analysis—physical characterization as well as practical examples in Chemical Engineering.

This manuscript has been a collaborative effort from the beginning. I would particularly wish to recognize the contributions of Melina Hamdine who early on in the project drafted several chapters in French including Physicochemical Properties, Analysis of Powders and Solids, and Design of Experiments. Much of the material on DOE was based on the contribution of Prof. Bala Srinivasan. Katia Senécal was "instrumental" in gathering the essential elements for the chapters including Measurement Analysis, Pressure, Temperature and Flow Rate. Prof. Bruno Detuncq collaborated in the revision of these chapters. Danielle Béland led the redaction of the chapter on chromatography to determine concentration with some assistance from Cristian Neagoe. He also wrote the section concerning spectroscopy. Amina Benamer contributed extensively to this project, including preparing solutions to the problems after each chapter, writing sections related to refractometry and X-ray and translating. Second-year students from the Department also participated by proposing original problems that were added at the end of each chapter (together with the name of the author of the problem). Ariane Bérard wa devout at identifying errors and proposing additional problems. I would particularly like to recognize Paul Patience for his tremendous contribution throughout the creative process of preparing this manuscript. The depth of his reflection has been appreciated tremendously (LATEX). He also co-authored the section on pyrometry. Christian Patience prepared many of the drawings and Nicolas Patience helped with translating from French to English, as did Nadine Aboussouan.

Chapter 1

Introduction

1.1 OVERVIEW

Experimental methods and instrumentation—for the purpose of systematic, quantifiable measurements—have been a driving force for human development and civilization. Anthropologists recognize tool making, together with language and complex social organizations, as a prime distinguishing feature of *Homo sapiens* from other primates and animals. However, the animal kingdom shares many concepts characteristic of experimentation and instrumentation. Most animals make measurements: cheetahs, for example, gauge distance between themselves and their prey before giving chase. Many species are known to use tools: large arboreal primates use branches as levers for displacement from one tree to another; chimpanzees modify sticks as implements to extract grubs from logs; spiders build webs from silk to trap their prey; beavers cut down trees and use mud and stones to build dams and lodges. Adapting objects for a defined task is common between man and other animals. If the act of modifying a twig to extract grubs is considered "tool making" then a more precise differentiating factor is required. Man uses tools to make tools and a methodology is adapted to improve an outcome or function. One of the earliest examples of applying methodology is in the manufacture of chopping and core tools—axes and fist hatchets—that have been used since before the Lower Paleolithic period (from 650 000 to 170 000 BC): blades and implements were produced through cleaving rocks with a certain force at a specific angle to produce sharp edges. The raw material—a rock—is modified through the use of an implement—a different rock—to produce an object with an unrelated function (cutting, scraping, digging, piercing, etc.). Striking rocks (flint) together led to sparks and presumably to the discovery of how to make fire.

Experimental Methods and Instrumentation for Chemical Engineers. http://dx.doi.org/10.1016/B978-0-444-53804-8.00001-0

Throughout the day, we make measurements and employ instrumentation. The clothes that we wear, the food that we eat, the objects that we manipulate have all been developed and optimized through the use of standardized procedures and advanced instrumentation. The transportation sector is an example where instrumentation and sensors are commonplace: gauges in the car assess speed, engine temperature, oil level, fuel level, and even whether or not the seat belt is engaged. One of the key factors in homes is maintaining the correct temperature either in rooms, refrigerators, hot water heaters, ovens, or elements on the stove. Advanced scales now display not only body weight but also percent fat and percent water!

Development is the recognition and application of unrelated or non-obvious phenomena to a new or improved application—like making fire. Optimization of innovations and technology can be achieved through accidents, trial-and-error testing, or systematic approaches. Observation is the fundamental basis for measuring devices and it was the main technique employed by man to understand the environment in which we lived as interpreted by our senses: sight, sound, smell, touch, hearing, time, nociception, equilibrioception, thermoception, etc.

The manufacture of primitive stone tools and fire required a qualitative appreciation for the most common measures of mass, time, number, and length. The concept of time has been appreciated for millennia. In comparative terms it is qualified by longer and shorter, sooner and later, more or less. Quantitatively, it has been measured in seconds, hours, days, lunar months, and years. Calendars have existed for well over 6000 yr and clocks—instruments to measure time intervals of less than a day—were common as long as 6000 yr ago. Chronometers are devices that have higher accuracy and laboratory models have a precision of 0.01 s.

One of the first 24-h clocks was invented by the Egyptians with 10 h during the day, 12 h at night, and 1 h at dawn and dusk—the shadow hours. The night time was measured by the position of the stars in the sky. Sun dials were used at the same time by Babylonians, Chinese, Greeks, and Romans. The water clock (clepsydra) was developed by Egyptians to replace the stars as a means of telling time during the night: Prince Amenemhet filled a graduated vessel with water and pierced a hole in the bottom to allow the water to drain (Barnett, 1998). Records of the hourglass date back to the early 13th century but other means to "accurately" measure time included burning candles and incense sticks.

Recording time required a numbering system and a means of detecting a change in quantity. In the simplest form of a water clock, time was read based on the liquid level in the vessels as indicated by a notch on the side. The system of using notches on bones, wood, stone, and ivory as a means of record-keeping dates before the Upper Paleolithic (30 000 BC). Notch marks on elongated objects are referred to as tally sticks. Medieval Europe relied on this system to record trades, exchanges, and even debt, but it was mainly used for the illiterate. It was accepted in courts as legal proof of a transaction. Western civilization continues to use tally marks as a means of updating intermediate results.

This unary numeral system is written as a group of five lines: the first four run vertically and the fifth runs horizontally through the four.

Perhaps one of the driving forces throughout the ancient civilizations for numbering systems was for taxation, lending, land surveying, and irrigation. The earliest written records of metrology come from Sumerian clay tablets dated 3000 BC. Multiplication tables, division problems, and geometry were subjects of these tablets. The first abacus—an ancient calculator used to perform simple arithmetic functions—appeared around 2700–2300 BC. Later tablets— 1800–1600 BC—included algebra, reciprocal pairs, and quadratic equations. The basis for 60 s in a minute, 60 min in an hour, and 360° in a circle comes from the sexagesimal numeral system of the Sumerians (Mastin, 2010). Moreover, unlike the Greeks, Romans, and Egyptians, they also had a decimal system. The Pythagorean doctrine was that mathematics ruled the universe and their motto was "all is number."

1.2 UNITS OF PHYSICAL QUANTITIES

The notion of weight, or mass, emerged during the same period as counting. Throughout history, systems have been developed for weights, measures, and time. Often these systems were defined by local authorities and were based on practical measures—the length of an arm, a foot, or a thumb. In the late 18th century the French National Assembly and Louis XVI commissioned the French Academy of Science to conceive a rational system of measures. The basis for the modern standards of mass and length was adopted by the National Convention in 1793.

Originally, the meter was to be defined as the length of a pendulum for which the half cycle was equal to 1 s:

$$t = \pi \sqrt{\frac{L}{g}}, \tag{1.1}$$

where L is the length of the pendulum and g is the gravitational constant. Eventually, the Assemblée Constituante defined the meter as one ten-millionth of the distance between the equator and the North Pole. In 1795, the gram was defined as the mass of melting ice occupying a cube whose sides equal 0.01 m the reference temperature was changed to 4 °C in 1799. At the Metre Convention of 1875, the Système international (SI) was formally established and a new standard for measuring mass was created: an alloy composed of 90% Pt and 10% Ir that was machined into a cylinder of height and diameter equal to 39.17 mm. Iridium was included in the new "International Prototype Kilogram" to increase hardness. The kilogram is the only unit based on a physical artifact and not a property of nature as well as the only base unit with a prefix.

The definition of the meter and the techniques used to assess it have evolved with technological advances. In 1799, a prototype meter bar was fabricated to represent the standard. (It was later established that this bar was too short by 0.2 mm since the curvature of the Earth had been miscalculated.) In 1889, the standard Pt bar was replaced with a Pt(90%)-Ir(10%) bar in the form of an **X**. One meter was defined as the distance between two lines on the bar measured at 0 °C. In 1960, the standard was changed to represent the number of wavelengths of a line in the electromagnetic emission of ^{86}Kr under vacuum. Finally, in 1983, the standard was defined as the distance that light travels in a vacuum in $1/299\,792\,458$ s.

The standard to measure the base unit of time—the second—has evolved as much as the standard to measure distance. During the 17–19th centuries, the second was based on the Earth's rotation and was set equal to 1/86 400 of a mean solar day. In 1956, recognizing that the rotation of the earth slows with time as the Moon moves further away (about 4 cm yr^{-1}), Ephemeris Time became the SI standard: 1/31556925.9747 the length of the tropical year of 1900. In 1967, the second was based on the number of periods of vibration radiation emitted by a specific wavelength of ^{133}Cs.

The International System of Units (Système international d'unités or SI) recognizes seven base properties as summarized in Table 1.1—time, length, mass, thermodynamic temperature, amount of matter, electrical current, and luminous intensity. Other measures include the plane angle, solid angle, sound intensity, seismic magnitude, and intensity. The standard changed from the cgs—centimeter, gram, second—standard to the present one in 1960. In 1875 at the Convention du Mètre, three international organizations were formed to oversee the maintenance and evolution of the metric standard:

- General Conference on Weights and Measures (Conférence générale des poids et mesures—CGPM).
- International Bureau of Weights and Measures (Bureau international des poids et mesures—BIPM).
- International Committee for Weights and Measures (Comité international des poids et mesures—CIPM).

TABLE 1.1 SI Base Units

Property	Quantity	Measure	Unit	Symbol
Time	t	T	second	s
Length	l, x, y, z, r	L	meter	m
Mass	m	M	kilogram	kg
Amount of matter	n	N	mole	mol
Temperature	T	θ	kelvin	K
Luminous intensity	I_v	J	candela	cd
Electrical current	I, i	I	ampere	A

1.3 WRITING CONVENTIONS

Table 1.1 lists not only the seven standard properties recognized by the International System of Quantities (SIQ) but also the symbols representing each property and its dimension as well as the base unit and its symbol. All other quantities may be derived from these base properties by multiplication and division (Bureau International des Poids et Mesures, 2006). For example, speed equals distance (or length) divided by time and is expressed as L/T. Several forms of energy have now been defined—kinetic, potential, thermal, etc.—but energy was originally defined by Leibniz as the product of the mass of an object and its velocity squared. Thus, energy is expressed as ML^2/T^2 and the units are kg m^2 s^{-2}. The kg m^2 s^{-2} has also been designated as the Joule (J) in honor of the contributions of the 19th century English physicist. Pressure is defined as the force (ML/T^2) exercised on a unit area and has units of $ML^{-1}T^{-2}$. The standard unit for pressure is the Pascal (Pa) after the French physicist who demonstrated the change in atmospheric pressure with elevation.

Quantities or properties may either be extensive—properties that are additive for subsystems, for example mass and distance—or intensive, in which case the value is independent of the system, like temperature and pressure. Prefixes are added to some properties to further qualify their meaning, for example "specific" and "molar." Specific heat capacity is the heat, or energy, required to raise the temperature of a given mass by an increment. The SI units for specific heat capacity are J kg^{-1} s^{-1}. The units of molar heat capacity are J mol^{-1} s^{-1}. The volume occupied by 1 mol of a substance is referred to as the molar volume. Several derived quantities together with their SI derived unit, symbol, and SI base units are shown in Table 1.2. Those units that are a combination of the first four derived units will have their name omitted for reasons of space.

Other symbols that are recognized as part of the SI unit system but fall outside the standardized nomenclature are shown in Table 1.3.

Units with multiple symbols should be separated by a space or a half-high dot: the viscosity of water at 0 °C equals 0.001 Pa s. Negative exponents, a solidus, or a horizontal line may be used for the case of derived units formed by division. Only one solidus should be used, thus atmospheric pressure may be expressed as 101 325 $\frac{\text{kg}}{\text{m s}^2}$ or 101 325 kg m^{-1} s^{-2}. As in the case for the symbol for pressure "Pa," symbols of units named after a person are capitalized ("N"—Newton, "Hz"—Hertz, "W"—Watt, "F"—Faraday). Note that since symbols are considered as mathematical entities, it is incorrect to append a period after the symbol—"min." is unacceptable (except at the end of a sentence). Moreover, symbols do not take an "s" to indicate the plural. Regardless of the font used, unit symbols should be expressed in roman upright type.

The CGPM has made several recommendations and standards to represent a quantity including the numerical value, spacing, symbol, and combinations of symbols. A space should follow the numerical value before the unit symbol: 454 kg. In the case of exponential notation, the multiplication symbol should be preceded and followed by a space: 4.54×10^3 kg. The plane angular symbols

TABLE 1.2 SI Coherent Derived Units

Quantity	Unit	Symbol	SI Base Units
Force	Newton	N	$kg\ m\ s^{-2}$
Pressure	Pascal	Pa	$kg\ m^{-1}\ s^{-2}$
Energy	Joule	J	$kg\ m^2\ s^{-2}$
Power	Watt	W	$kg\ m^2\ s^{-3}$
Moment of force	–	N m	$kg\ m^2\ s^{-2}$
Surface tension	–	$N\ m^{-1}$	$kg\ s^{-2}$
Dynamic viscosity	–	Pa s	$kg\ m^{-1}\ s^{-1}$
Heat flux density, irradiance	–	$W\ m^{-2}$	$kg\ s^{-3}$
Entropy	–	$J\ K^{-1}$	$kg\ m^2\ s^{-2}\ K^{-1}$
Specific entropy, heat capacity	–	$J\ kg^{-1}\ K^{-1}$	$kg\ m^2\ s^{-2}\ K^{-1}$
Specific energy	–	$J\ kg^{-1}$	$m^2\ s^{-2}\ K^{-1}$
Molar energy	–	$J\ mol^{-1}$	$kg\ m^2\ s^{-2}\ mol^{-1}$
Energy density	–	$J\ m^{-3}$	$kg\ m^{-1}\ s^{-2}$
Molar entropy	–	$J\ mol^{-1}\ K^{-1}$	$kg\ m^2\ s^{-2}\ K^{-1}\ mol^{-1}$
Thermal conductivity	–	$W\ m^{-1}\ K^{-1}$	$kg\ m\ s^{-3}\ K^{-1}$

TABLE 1.3 SI Recognized Units

Unit	Symbol	SI
minute	min	60 s
hour	h	3600 s
day	d	86 400 s
hectare	ha	10 000 m^2
liter	l or L	0.001 m^3
tonne	t	1000 kg
decibel	dB	–
electronvolt	eV	$1.60217653 \times 10^{-19}$ J
knot	kn	1852 m h^{-1}
fathom	ftm	1.82880 m
nautical mile	M	1852 m

representing degrees, minutes, and seconds are exceptions and should be placed immediately after the numerical value without a space. Note that for temperature expressed in degrees Celsius, a space is required between the numerical value and the symbol—25.0 °C. In 2003, both the comma and period were recognized

TABLE 1.4 SI Prefixes

Multiples			Fractions		
Name	Symbol	Factor	Name	Symbol	Factor
deca	da	10^1	deci	d	10^{-1}
hecto	h	10^2	centi	c	10^{-2}
kilo	k	10^3	milli	m	10^{-3}
mega	M	10^6	micro	μ	10^{-6}
giga	G	10^9	nano	n	10^{-9}
tera	T	10^{12}	pico	p	10^{-12}
peta	P	10^{15}	femto	f	10^{-15}
exa	E	10^{18}	atto	a	10^{-18}
zetta	Z	10^{21}	zepto	z	10^{-21}
yotta	Y	10^{24}	yocto	y	10^{-24}

as decimal markers. In practice, English-speaking countries and most Asian countries have adopted a period while other nations typically use a comma. To avoid confusion in writing large quantities, it is recommended that a space can be used as the thousand separator ($c = 299\,792\,458$ m s^{-1}). For numbers from -1 to 1, the decimal marker is preceded by zero: $R = 0.008314$ kJ mol^{-1} K^{-1}.

Prefixes may be added to the unit as a matter of practice and, in many cases, they are standardized conventions of particular fields. For instance, the unit MW is common to the power industry. The unit nm is used in crystallography for the physicochemical characterization of solids—pore diameter is an example. All prefixes are multiples of ten and are summarized in Table 1.4. The symbol is capitalized for multiple factors greater than 10^3. The symbols for 10^{21} and 10^{24} are the same as for 10^{-21} and 10^{-24}, the only difference being the former are capitalized and the fraction factors are in lower case. The only Greek letter used as a symbol is for 10^{-6}—micro—and the only two-letter symbol is da (which is rarely used in practice, especially combined with meters).

1.4 UNIT CONVERSION

Together with the SI system, two other unit systems commonly used are the cgs (centimeter-gram-second) and the fps (foot-pound-second). While the cgs system was essentially supplanted by SI units (also termed mks), the fps system is still in use in different parts of the world and most notably in the United States. Conversion between the cgs and SI systems is generally straightforward—usually a factor of 10 or 1000 is involved. Conversion between fps (also known as the Imperial system of units) and SI is more complicated.

In the cgs system, the standard of mass is the gram and thus the conversion between cgs and mks is a factor of 1000. In the fps system the standard unit of mass is the avoirdupois (which means "to have weight" in French) with the abbreviation lb (or lb_m—pound-mass), which is derived from the Latin word *libra* (meaning scale or balance). The factor to convert from pounds to kilograms, by definition, is:

$$1 \text{ lb} = 0.45359327 \text{ kg.}$$

The length standard is the centimeter for the cgs system and the foot for the fps system, with the abbreviation ft:

$$1 \text{ ft} = 0.3048 \text{ m} = 30.48 \text{ cm.}$$

Other commonly used length measures in the fps system include the inch (12 in. ft^{-1}), the yard (3 ft yd^{-1}), and the mile (5280 ft mi^{-1}).

Volume measures in both cgs and SI are derived readily from units of length. The most common measure of volume in the fps system is the gallon (gal) of which there are two standards: the US gallon is approximately equal to 3.79 l while the imperial gallon equals 4.54 l. A barrel of oil equals 0.159 m^3.

The time standard is the same for all three systems. The cgs and SI systems share the standards for temperature and for quantity of matter (mol). The standard for thermodynamic temperature in fps is the Rankine:

$$1.8 \,^\circ\text{R} = 1 \text{ K.}$$

The Fahrenheit scale is the equivalent of the Celsius scale and the two are related as shown below:

$$T_{\text{Fahrenheit}} = 32 \,^\circ\text{F} + 1.8 \,^\circ\text{F} \,^\circ\text{C}^{-1} \times T_{\text{Celsius}}.$$

At 0 °C, the temperature in the Fahrenheit scale is 32 °F. The boiling point of water is 212 °F and absolute zero (0 K) equals −459.67 °F (which is equal to 0 °R).

In most practical applications, the mol is too small and thus chemical engineers often adopt the unit kg-mol (also written kmol), which is 10^3 mol. To minimize ambiguity, often the mol will be written as g-mol. In the fps system, the lb-mol is the standard:

$$1 \text{ lb-mol} = 453.59237 \text{ g-mol} = 0.453592378 \text{ kg-mol.}$$

Mixed units are often used in chemistry: molar concentration should be expressed as mol m^{-3} but almost all chemical literature employs mol dm^{-3} or more commonly mol l^{-1}. Industrially, the unit kmol m^{-3} is also used. These units are referred to as molar with the designation of M. Prefixes may be added to M for low values. Thus, μM represents μmol l^{-1} and nM refers to nmol l^{-1}.

As with SI units, important derived units have been assigned independent symbols such as force, energy, and power. The unit for force in SI is the Newton (N), which is equal to the product of mass and acceleration:

$$1 \text{ N} = 1 \text{ kg m s}^{-2}.$$

The dyne is the unit for force in the cgs system:

$$1 \text{ dyn} = 1 \text{ g cm}^2 \text{ s}^{-1},$$
$$1 \text{ N} = 10^5 \text{ dyn}.$$

The dyne is most commonly used to measure surface tension: the surface tension of distilled water is 72 dyn cm^{-1} (at 25 °C), which equals 72 mN m^{-1}.

In the fps system, the pound force (lb$_f$) is the quantity measured by an avoirdupois pound at the surface of the earth and is equal to 4.448 N. The lb$_f$ and lb$_m$ are related through the gravitational constant:

$$1 \text{ lb}_f = 1 \text{ lb}_m \cdot g_c = 1 \text{ lb}_m \cdot 32.174 \text{ ft}^2 \text{ s}^{-1}.$$

Pressure is a derived unit that represents the force applied to an area perpendicular to the force. The SI derived unit is the Pascal with the symbol Pa:

$$1 \text{ N m}^{-2} = 1 \text{ kg m}^{-1} \text{ s}^{-2} = 1 \text{ Pa}.$$

Atmospheric pressure equals 101 325 Pa at sea level and other derived units are common, such as bar and atm:

$$1 \text{ bar} = 100\,000 \text{ Pa},$$
$$1 \text{ atm} = 101\,325 \text{ Pa}.$$

The common unit for pressure in the fps system is the lb$_f$ in^{-2} and the commonly accepted symbol is psi. One atmosphere of pressure equals 14.696 psi.

The Joule (J) has been adopted to represent energy in the SI system whereas the erg is used in the cgs system:

$$1 \text{ J} = 1 \text{ kg m}^2 \text{ s}^{-2} = 10^7 \text{ erg} = 10^7 \text{ g cm}^2 \text{ s}^{-2}.$$

A more common unit of energy in the cgs system is the calorie, defined as the energy necessary to raise the temperature of 1 g of water by 1 K. The following conversion factors are required to convert to ergs and Joules:

$$1 \text{ cal} = 4.184 \times 10^7 \text{ erg} = 4.184 \text{ J}.$$

The unit for energy in the fps system is the British thermal unit (Btu):

$$1 \text{ Btu} = 1055.06 \text{ J}.$$

Finally, power is another common derived unit in SI that is represented by the Watt (W), which is the rate of change of energy conversion:

$$1 \text{ W} = 1 \text{ J s}^{-1} = 1 \text{ kg m}^2 \text{ s}^{-3}.$$

In the fps system, power is represented by horse power (hp):

$$1 \text{ hp} = 745.7 \text{ W}.$$

1.5 METROLOGY

Metrology is the science of measurement and is derived from the Greek words *metron* (measure) and *logos* (logic, study, calculation, reason, etc.). It has been defined by the International Bureau of Weight and Measures as a science that encompasses theoretical and experimental measures at any level of uncertainty in the fields of science and engineering. It comprises not only the instruments applied to quantify the magnitude of a physical phenomenon but also standards, procedures, quality control, training, documentation, etc. Analysis and quantification of uncertainty is a core element, as is traceability—which relates to an instrument's measurements to known standards as well as the documented accreditations to national and international standards.

Together with the physical aspects of recording data accurately and repeatedly, metrology requires the verification and validation of the data collected by the test equipment and may also include its enforcement. Enforcement is a critical aspect not only for consumer goods—baby carriages, helmets, and the like—but also for industrial equipment such as vessel design (pressure vessels), materials of construction (quality of steel), and safety procedures.

Along with international organizations that maintain standards for the basic measures of distance, weight, etc. most countries also maintain their own system of metrology (Table 1.5). For example, the National Institute of Standards and

TABLE 1.5 International Standards Organizations

Organization	Founded
ASTM (American Society for Testing and Materials)	1898
BSI (British Standards Institute—BS)	1901
SAE (Society of Automotive Engineers)	1905
DIN (Deutsches Institut für Normung)	1917
JIS (Japanese Industrial Standard)	1921
ISO (International Organization for Standards)	1926
NF (Norme Française)	1926
CEN (European Committee for Standardization)	1961

Technology (NIST), formerly the National Bureau of Standards founded in 1918, is responsible for maintaining both scientific and commercial metrology in the United States. Its mission is to promote American innovation and competitiveness and supplies industry, academia and government with certified standard reference materials, including documentation for procedures, quality control, and materials for calibration. The German Institute for Standards (DIN) was founded in 1917 while in the United Kingdom the BSI was formed in 1901.

Further to national standards, many industries have promoted and maintained their own standards. One of the most well-known and oldest non-governmental standards organizations is the American Society for Testing and Materials (ASTM), which was established in 1898. It collects and maintains over 12 000 standards that are available to the public and include 82 volumes (at a price of $9700 in 2010). The origin of the organization was the desire to improve the quality of the rail system that had been plagued by breaks.

Although the International Organization for Standards—ISO—is a non-governmental organization, it has the ability to set standards that become law through treaties or through the national standards organizations that are represented in the organization. Of the 203 nations in the world, 163 are members of ISO. The making of a standard in ISO follows a ten-step procedure:

1. Preliminary work item.
2. New work item proposal.
3. Approved new work item.
4. Working draft.
5. Committee draft.
6. Final committee draft.
7. International standard draft.
8. Final international standard draft.
9. Proof of a new international standard.
10. International standard.

Three common standards include:

- ISO 5725: Accuracy of Measurement Methods and Results Package.
- ISO 9001:2008: Quality Systems Management—Requirements.
- ISO 17025:2005: General Requirements for Calibration Laboratories.

The ISO 9001 standard was originally based on BS 5750. A primary objective of this standard is to ensure the commitment of management to quality with respect to the business as well as to customer needs. The Quality Systems Management standard recognizes that employees require measurable objectives. In addition to a detailed record system that shows the origin of raw materials and how the products were processed, it includes auditing (both internal and external, in the form of certification) at defined intervals to check and ensure conformity and effectiveness.

The standard for calibration laboratories (ISO 17025) is closely aligned with the ISO 9001 standard but includes the concept of competency. Moreover, continual improvement of the management system itself is explicitly required as well as keeping up to date on technological advances related to the laboratory.

1.6 INDUSTRIAL QUALITY CONTROL

Industrial metrology concerns accuracy as much in the laboratory as in the field but it is more constrained in that measurements must often be made in hostile environments including high temperature, dust, vibration, and other factors. Moreover, the factor of time and financial cost must also be considered. Quality control systems have been implemented to take these aspects into account. The ability to measure accurately and consistently and then interpret the results correctly to make coherent decisions is the basis of modern manufacturing. In advanced commercial chemical installations, thousands of independent measurements are collected at frequencies greater than 1 Hz and stored in massive databases. Operators read data in real time through consoles in a central location (control room) through distributive control systems (DCS). Modern databases and information management systems can easily be interrogated for both offline analysis and online monitoring. They serve to control the plant, troubleshoot, detect deviations from normal operation, analyze tests designed for process optimization, and are also a historical record in the case of accidents. Additionally, the databases may be used for environmental reporting to the authorities. The most common measurements are temperature and pressure. Many flow rate measurements are calculated based on pressure and temperature readings. Online analytical devices are less common but increase the level of confidence in operations and allow for mass balance and process performance calculations in real time—this greatly enhances production tracking as well as one's ability to troubleshoot.

Duplicate and triplicate measurements of pressure and temperature of critical pieces of equipment are common in order to ensure safe operation. When a reading exceeds a certain threshold value, an alarm might sound or a reading may appear on the console for the operator to take action. Generally, there are different levels of alarms as well as interlocks. Alarms generally require operator intervention while an interlock condition usually will shut the process or equipment down automatically.

In addition to pressure and temperature measurements, duplicates of some pumps and control valves are installed in parallel. This allows the equipment to be bypassed and serviced during operation of the plant, which avoids the costly necessity to shut down the process. The choice of installing spares depends on many factors but it is generally recommended for major feed and product streams and for lines that are considered critical for safe operation of the plant.

Although redundant measurements and equipment such as fail-safe devices and the like are often mandatory, accidents still happen. The 2010 Macondo well disaster in the Gulf of Mexico is an example where instrumentation was insufficient to warn operators of an impending blowout. Accidents may be attributed to human error, instrument error, mechanical failure, or a combination of these factors. At times, a process may be operated near design limits and thus alarms become a nuisance and are ignored. Shutting down a process to fix instrumentation or equipment outside the normal maintenance cycle is very expensive and can represent millions of dollars of lost production. Engineers and managers may choose unorthodox methods to keep a plant running. In one example, a vessel operating over 600 °C lost the refractory lined bricks that insulated the metal wall from the high temperature. To avoid shutting down the plant, cold water was sprayed on the outer wall. This operation is clearly non-standard and introduced a potentially hazardous situation—if the water spray were inadvertently shut off, the wall temperature could increase sufficiently high to cause a perforation and result in an explosion. The chemical industry has made tremendous efforts in producing goods and services in such a way as not to impact the health and well being of society. Before commissioning a new plant or equipment, detailed operating procedures are written and all aspects are considered to minimize hazards. Different methodologies are may be followed to assess the risks and include a What-if, Checklist (Human Factor Checklist or General Hazards Idenitification Checklist, for example), Hazard and Operability Study (HAZOP), Failure Mode and Effect Analysis (FMEA) or a Fault Tree Analysis. Together with general safety, other aspects that are assessed include occupational health, ergonomics, fire safety, process safety, product stewardship. Instrumentation is a cornerstone to process safety management.

1.7 EXERCISES

1.1 (a) Derive the value of the gas constant R (8.314 J mol^{-1} K^{-1}) in British units (ft^3 psi lb-mol^{-1} °R^{-1}).

  (b) What is the value of R in units of atm l mol^{-1} K^{-1}?

1.2 The operating temperature of a reactor is approximately 50.00 °C and the effluent stream is theoretically at the same temperature. Ten measurements of the effluent (in °C) are: 50.12, 50.03, 49.97, 50.09, 60.2, 50.05, 50.00, 49.99, 49.98, and 50.13. The range of the instrument is 0–60 °C and its precision is to within 0.1%FS (full scale).

  (a) Represent the data graphically neglecting measurement error (in K).

  (b) Calculate the absolute and relative error of each measurement.

  (c) Is it possible that all these values are reliable because they were measured electronically?

  (d) List three sources of error and how to reduce them.

1.3 The pressure gauge of a distillation column indicates 1500 mbar at the exit. The pressure differential across the column is 150 inH$_2$O. What is the absolute pressure in atm at the bottom of the column?

1.4 Calculate the temperature at which the numerical value on the Celsius scale coincides with that of the Fahrenheit scale.

1.5 The standard unit for vacuum is the Torr and 1 Torr is equivalent to 1 mmHg pressure. Convert 5 mTorr to kPa.

1.6 In the development of a new mosquito repellent, you are required to estimate if a 100 ml spray bottle is sufficient for an individual for 3 months of standard use. Make a detailed list of assumptions and the experimental equipment necessary to make such an estimate.

1.7 Sieving is a standard operation for separating powders according to particle size using woven wire screens. The Tyler mesh size represents the number of openings per inch or the number of parallel wires that form the opening.

(a) What is the diameter of a spherical particle that can pass through a 200 Tyler mesh with a 0.0021 in diameter wire?

(b) Calculate the minimum diameter retained by a 60 Tyler mesh screen with a 0.0070 in diameter metal wire.

1.8 How many seconds have we lost in the last 2000 yr since the adoption of the modern definition of the second compared to the one used before 1956?

1.9 A scale records your weight as 160 lb on Earth.

(a) How much do you weigh on the Moon, in SI units, where the force of gravity is one-sixth that of Earth?

(b) What is your mass on Uranus, its mass being 14 times that of Earth?

1.10 A brewer racks beer from an 800 l fermentation tank into 7.93 gal (US) conditioning tanks. How many tanks are filled if 0.200 ft^3 are lost for each tank and they are filled to 98% of their capacity? *M. Bourassa-Bédard*

REFERENCES

Barnett, J.E., 1998. Time's Pendulum: From Sundials to Atomic Clocks, the Fascinating History of Timekeeping and How Our Discoveries Changed the World. Plenum Press, NY. ISBN: 0-15-600649-9.

Boyer, C.B., 1991. A History of Mathematics. John Wiley & Sons, Inc.

Bureau International des Poids et Mesures, 2006. The International System of Units (SI), eighth ed. <http://www.bipm.org/utils/common/pdf/si_brochure_8_en.pdf>.

ISO 17025, 2005. General Requirements for Calibration Laboratories.

ISO 5725, 1998–2005. Accuracy of Measurement Methods and Results Package.

ISO 9001, 2008. Quality Management Systems—Requirements.

Mastin, L., 2010. Sumerian/Babylonian Mathematics. Retrieved 2011, from The Story of Mathematics: <http://www.storyofmathematics.com/sumerian.html>.

Chapter 2

Measurement and Analysis

2.1 OVERVIEW

The paradigm shift from qualitative observation to quantitative measurement is the cornerstone of science. This shift was advocated by Sir Francis Bacon in the 17th century who insisted that repeated observations of a phenomenon were the first step in a scientific methodology that included experimentation (Jardine 2011). The objective of an experiment might be to test a hypothesis, maximize or minimize a process operation, improve a formulation, reduce variability, compare options, etc. Experiments may be carefully controlled measurements in a laboratory in which extraneous factors are minimized or they may be observational studies of nature, such as an epidemiological study, or field studies.

2.2 SIGNIFICANT FIGURES

Regardless of the care taken in measuring a variable, the measurement is subject to uncertainty or errors. The reliability of an instrument is a critical factor in its selection and in the interpretation of collected data. There are few quantities that are known or can be measured precisely. One exception is the speed of light in a vacuum that is exactly $299\,792\,458\,\text{ms}^{-1}$. For other quantities, the uncertainty in the value is expressed in two ways: the first is to define the value with a limited number of digits (or figures) and the second is to include a second number indicating the probability that the measured value lies within a defined interval—its uncertainty, which is discussed in detail later in this chapter. It represents an interval in which the true value of a measurement lies. If the uncertainty of the measured volume in a graduated cylinder equals ± 1 ml,

Experimental Methods and Instrumentation for Chemical Engineers. http://dx.doi.org/10.1016/B978-0-444-53804-8.00002-2

the volume of a one-quarter liter graduated cylinder would be expressed as:

$$V = (250. \pm 1) \text{ ml}.$$

Significant figures include digits—1, 2, 3, 4, 5, 6, 7, 8, 9, and 0—in a number. The number of significant figures for the graduated cylinder is three when the volume measured is greater than 100 ml but it is only two for volumes less than 100 ml. The period after 250 in the example above indicates that the zero is significant. The distance 0.0254 m has three significant figures, as does 25.4 mm and both represent the value of 1 inch. To represent the three significant figures of 1 in, we should write 1.00 in. For a process vessel that costs 100 000 $, only one significant figure is expressed. To express three significant figures for this vessel, we could write 100. k$ or 1.00×10^5 $.

A single convention has yet to be universally accepted for the number of significant figures of the uncertainty. The simplest convention is that the uncertainty has one significant figure and has the same magnitude as the rightmost digit of the measurand. Uncertainty may be either derived from statistical analysis of repeated measurements (Type A) or a rectangular probability distribution is adopted (Type B) for which the interval is derived from the instrument resolution. This convention is respected in the example of the graduated cylinder. A standard 250 ml Erlenmeyer flask has graduation marks at 25 ml intervals and its uncertainty equals 12 ml (Type B uncertainty). (The resolution for physical instruments is at least half of the value of the demarcation and can be as low as 10%.) The uncertainty of the Erlenmeyer flask is to the tens digit, while the graduation marks go down to the ones digit—five.

$$V = (250 \pm 25/2) \text{ ml}.$$

The choice for the uncertainty could be ± 12.5 ml, ± 20 ml, or ± 10 ml. While two significant digits may be acceptable in similar small-scale circumstances, for most engineering applications one significant figure is adequate. Adding significant figures to the uncertainty may lead to a false appearance of precision. In this case, we could choose either ± 10 ml or ± 20 ml.

Rounding is an operation that reduces the precision of a numerical value. For example, π is an irrational number that cannot be expressed as a ratio of two integers and its decimal representation never repeats. Often it will be represented by six significant figures: 3.14159. To round it to a lower precision involves discarding the digits to the nth level of significance. Rounding π to five, four, and three significant figures gives 3.1416, 3.142, and 3.14, respectively. Another approximation to π is the fraction 22/7, which equals 3.1429.

To minimize rounding errors, the digit of the nth level of significance is increased by 1 when the digit of the $n(\text{th}+1)$ level of significance is greater than 5 and it remains constant if it is less than 5. Thus, when π is rounded to five digits, 3.14159 becomes 3.1416: the 5 in the fifth digit increases to 6. When rounding π to four digits, 3.14159 changes to 3.142: the 1 in the fourth digit becomes 2. However, the value of the third digit remains unchanged when it is rounded to three significant figures. In the case that the digit preceding the nth level of significance equals 5, no general rule has been accepted. Several rules have been proposed such as the "round half to even" convention that stipulates the digit should be increased by 1 when the resulting digit is an even number and to leave it unchanged if the number is already even. According to this rule, 24.5 would be rounded down to 24 while 25.5 would be rounded up to 26.

When summing a series of numbers, the resulting value will generally have the same number of significant figures as the number with the greatest absolute uncertainty. In the following example, liquid from two graduated cylinders, one of 250 ml and one of 10 ml, are added together with a volume of liquid from a graduated flask:

$$(250. \pm 1)\ ml + (5.6 \pm 0.4)\ ml + (125 \pm 13)\ ml = (380.6 \pm 14.4)\ ml$$
$$= (381 \pm 13)\ ml$$
$$\cong (380 \pm 15)\ ml.$$

Theoretically, the final volume of liquid equals 380.6 ml but since the uncertainty in the value of the graduated flask is so high, it is unreasonable to carry four or even three significant figures. Thus, the best estimate of the combined volume is (380 ± 13) ml. (Note that the uncertainty in the total volume is the square root of the sum of the squares—the variance—of each measure.)

Subtracting a series of numbers leads to a loss in accuracy. Compared to the preceding example, the calculated absolute error is the same but the relative error increases substantially:

$$(250. \pm 1)\ ml - (5.6 \pm 0.4)\ ml - (125 \pm 13)\ ml = (119.4 \pm 13)\ ml$$
$$= (119 \pm 13)\ ml$$
$$\cong (120 \pm 15)\ ml.$$

As with addition and subtraction, the result of a multiplication or division should have as many significant figures as contained in the least accurate number. Often the proper use of significant figures is neglected in scientific and engineering literature. Too many significant figures are carried in models developed to characterize experimental data. A mass balance for even a simple reaction is seldom within a relative error of 1%. However, rate expressions and activation energies derived from this data often have five significant digits. Correlations involving exponents should rarely have more than two significant figures. Calculating equipment and investment costs is an example where engineers frequently carry too many significant figures: prices of equipment costing over $1000 and certainly over $1 000 000 should never be reported to the cent! In typical engineering economic calculations, three significant figures are sufficient to communicate the cost of an item or process. The uncertainties in installation, commissioning, etc. are sufficiently large that most times even two significant figures are enough to represent cost.

Empirical models are used to correlate thermodynamic properties, unit operations, heat transfer, fluid dynamics, etc. but often the parameters in these equations also carry too many significant figures. Heat capacity, for example, is approximated by polynomials of up to three degrees and the coefficients in some models carry as many as seven significant figures (Reid et al. 1977, Himmelblau 1962, Howell and Buckius 1992, Spencer 1948 the NIST[1] correlations, for example):

- 273–1800 K, Kyle (1984)

$$C_{p,N_2} = 28.90 - 0.001571T + 0.8081 \times 10^{-5}T^2 + 2.873 \times 10^{-9}T^3,$$

- 298–3000 K, Kelley (1960)

$$C_{p,N_2} = 28.58 - 3.77 \times 10^{-3}T - 0.50 \times 10^5 T^{-2},$$

- 300–3500 K, Van Wylen and Sonntag (1978)

$$C_{p,N_2} = 39.060 - 512.79T^{-1.5} + 1072.7T^{-2} - 820.4T^{-3},$$

- 100–500 K, NIST

$$C_{p,N_2} = 28.98641 + 1.853978T - 9.647459T^2 \\ + 16.63537T^3 + 0.000117T^{-2},$$

- 500–2000 K, NIST

$$C_{p,N_2} = 19.50583 + 19.88705T - 8.598535T^2 \\ + 1.369784T^3 + 0.527601T^{-2},$$

where T is the temperature (K) and C_{p,N_2} is the molar heat capacity of nitrogen (J mol^{-1} K^{-1}).

Correlations and models approximate physical phenomena and often the fitted parameters—coefficients and exponents—have no physical meaning. Thus, the number of significant digits for these values should seldom exceed 3 for coefficients or 2 for exponents. Coefficients with seven significant figures give a false sense of certitude.

---

[1] http://webbook.nist.gov/cgi/cbook.cgi?ID=C74828&Units=SI&Type=JANAFG&Table=on# JANAFG.

2.3 STATISTICAL NOTIONS

By how much has the mean temperature of the world increased over the last 50 or 100 yr and how has it changed? What is the mean temperature of a particular city in the world and can its variation from year to year be representative of the variation of the world's temperature? Figure 2.1 is a histogram of 1950 points of the mean daily temperature of the month of May in Montreal over 64 yr. The mean of this population is simply the sum of all the recorded temperatures divided by the number of days sampled (the result is 13.2 °C):

$$\mu = \frac{1}{n} \sum_{i=1}^{n} x_i. \tag{2.1}$$

Representing this data with a single value is only the first step in characterizing the population. The graph shows that the maximum measured average daily temperature was 26 °C and the minimum temperature was around 0 °C. Moreover, there are few occurrences at these two extremes. What is the likelihood that the mean daily temperature next year will be greater than 20 °C or conversely, what is the likelihood that the temperature will average less than 10 °C? These questions relate to the variability of the population and it is best characterized by the standard deviation, which equals the square root of the variance (σ^2):

$$\sigma = \sqrt{\frac{1}{n} \sum_{i=1}^{n} (x_i - \mu)^2} = \sqrt{\frac{1}{n} \sum_{i=1}^{n} x_i^2 - \mu^2}. \tag{2.2}$$

Data points are grouped closely together around the mean when the standard deviation is low and they are spread out for a large standard deviation. For a continuous set of data, the standard deviation equals σ and the mean is μ, but for a subset of the population or when the population standard deviation is unknown,

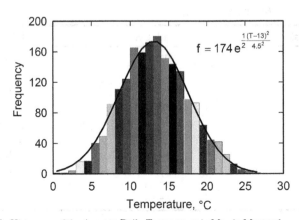

FIGURE 2.1 Histogram of the Average Daily Temperature in May in Montreal

the sample standard deviation and sample mean are used, s and $\bar{x}$, respectively. The sample standard deviation includes the Bessel correction $\sqrt{n/(n-1)}$. The sample standard deviation is higher than the standard deviation of the entire population but the difference between the two is less than 2% for a sample population of 30 or more:

$$s = \sqrt{\frac{1}{n-1}\sum_{i=1}^{n}(x_i - \bar{x})^2} = \sqrt{\frac{1}{n-1}\sum_{i=1}^{n}x_i^2 - \frac{n}{n-1}\bar{x}^2}. \tag{2.3}$$

Typically, a variable should be measured at least five times to have a good idea of the average and at least 10 times (if not 20) to have a good estimate of the standard deviation.

Example 2.1. One hundred milliliters of water was measured 10 times in a 100 ml volumetric flask, a 250 ml graduated cylinder, and a 150 ml graduated Erlenmeyer flask—Table E2.1. Each flask was washed, dried then weighed on a balance with a resolution of 0.01 g. The flask was filled with deionized/distilled water so that the bottom of the meniscus barely touched the top of the graduation mark. Water along the tube neck was then wiped with a paper towel to remove any water drops. Calculate the mean and standard deviation for the flasks and the graduated cylinder.

Solution 2.1. The density of water at 20 °C equals 0.998, so that 99.49 g of water—the mean mass of the 10 samples from the volumetric flask—is in fact equal to 99.69 ml. Using

$$\bar{x}_i = \frac{1}{n}\sum_{j=1}^{n}x_{i,j}$$

TABLE E2.1 Measurements (g)

Vol. Flask	Grad. Cyl.	Erlenmeyer
99.69	97.74	104.23
99.42	98.18	103.96
99.51	98.17	102.97
99.47	97.91	104.24
99.44	97.44	103.17
99.50	97.44	102.71
99.51	97.69	102.90
99.47	97.02	103.05
99.42	97.53	102.65
99.50	97.79	102.76

and

$$s_i = \sqrt{\frac{1}{n-1} \sum_{j=1}^{n} (x_{i,j} - \bar{x}_i)^2}.$$

results in $\bar{x}_1 = 99.69$ ml, $\bar{x}_2 = 97.89$ ml, $\bar{x}_3 = 103.47$ ml and $s_1 = 0.08$ ml, $s_2 = 0.35$ ml, $s_3 = 0.63$ ml.

The mean mass of the volumetric flask was closest to the target at 99.69 g with a sample standard deviation of 0.08 (including the Bessel correction).

2.3.1 Normal (Gaussian) Distribution

As shown in Figure 2.1, the daily average temperature in May appears to be uniformly distributed around a central point situated at about 13 °C, which happens to equal its mean temperature. This bell-shaped curve is common to all processes in which the variability (variance) in the data is random follows a normal, or Gaussian distribution. The continuous mathematical function that characterizes this is of an exponential form:

$$f(x) = \frac{1}{\sqrt{2\pi\sigma^2}} e^{-\frac{1}{2}\left(\frac{x-\mu}{\sigma}\right)^2}. \tag{2.4}$$

The mean value, μ, shifts the curve along the abscissa while the variance, σ^2, modifies the shape of the curve. In the case of a large variance—corresponding to a higher level of variation—the curve becomes more spread out and shorter, while a small variance indicates a lower level of variation and a correspondingly taller and narrower curve. These effects are demonstrated in Figure 2.2 in which histograms of the mean temperature in January and July are compared on the same graph with the same axis. The average daily temperature in January was -10 °C but varies from -30 °C to $+9$ °C. The standard deviation equals 7.7 °C. The standard deviation of the mean temperature in July was less than half that of January's at 3.1 °C with a much lower range of 13–30 °C. As a consequence, the data is more closely grouped together around the mean, which equals 21 °C, and the peak height is over double the mean (it is higher in proportion to the ratio of the standard deviation).

In order to compare data with different means and variance quantitatively, a parameter z is introduced to normalize the exponential factor:

$$z = \frac{x - \mu}{\sigma}. \tag{2.5}$$

With this transformation, the equation describing the Gaussian distribution becomes:

$$p(z) = \frac{1}{\sqrt{2\pi}} e^{-\frac{1}{2}z^2}. \tag{2.6}$$

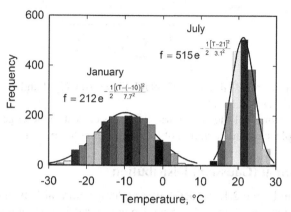

FIGURE 2.2 Mean Temperature in January and July

The area under the curve bounded by $\pm\infty$ equals 1 and this equation is referred to as the probability distribution. It has many powerful characteristics necessary to analyze data. The area of the curve bounded by one standard deviation ($\pm\sigma$) represents 68% of the total area while 95.4% of the area lies within $\pm2\sigma$. Another way of expressing the area with respect to probability is to say that 95.4% of the variance in the data is found within $\pm2\sigma$ of the mean—the measurement of a randomly distributed variable will be within $\pm2\sigma$ of the mean 95.4% of the time. Two other common reference points are the variance in the data bounded by $\pm2.57\sigma$ and $\pm3\sigma$, which equal 99.0% and 99.7%, respectively.

The probability, $p(z)$, that a random variable lies between an interval z and $z + \Delta z$ is given by:

$$P(z < z_m < z + \Delta z) = \int_z^{z+\Delta z} p(z)dz = \int_z^{z+\Delta z} \frac{1}{\sqrt{2\pi}} e^{-\frac{1}{2}z^2} dz. \quad (2.7)$$

Assuming there is global climate change but that it is imperceptible from one year to the next and that temperature is a randomly distributed variable, there is a 68% probability that the mean temperature July 1st next year lies between 18 °C and 24 °C; the probability of it lying between 15 °C and 25 °C is over 95%.

It is easier to derive the probabilities of events and measurements from a table rather than computing the integral of the probability distribution. Table 2.1 lists the value of the probability as a function of z to a precision of three digits after the decimal point. The value quoted in the table corresponds to only half the curve: the interval bounded from zero (the mean) to a positive or negative integer a, as shown in Figure 2.3. The value of the integer a is comprised of two parts. The first column corresponds to the value of a to two significant figures. The value of the probability to three significant figures corresponds to the intersection of the row with two significant figures and that of the corresponding digit in

TABLE 2.1 Probability as a Function of z

z	0	0.01	0.02	0.03	0.04	0.05	0.06	0.07	0.08	0.09
0	0	0.004	0.008	0.012	0.016	0.020	0.024	0.028	0.032	0.036
0.1	0.040	0.044	0.048	0.052	0.056	0.060	0.064	0.068	0.071	0.075
0.2	0.079	0.083	0.087	0.091	0.095	0.099	0.103	0.106	0.110	0.114
0.3	0.118	0.122	0.126	0.129	0.133	0.137	0.141	0.144	0.148	0.152
0.4	0.155	0.159	0.163	0.166	0.170	0.174	0.177	0.181	0.184	0.188
0.5	0.192	0.195	0.199	0.202	0.205	0.209	0.212	0.216	0.219	0.222
0.6	0.226	0.229	0.232	0.236	0.239	0.242	0.245	0.249	0.252	0.255
0.7	0.258	0.261	0.264	0.267	0.270	0.273	0.276	0.279	0.282	0.285
0.8	0.288	0.291	0.294	0.297	0.300	0.302	0.305	0.308	0.311	0.313
0.9	0.316	0.319	0.321	0.324	0.326	0.329	0.332	0.334	0.337	0.339
1.0	0.341	0.344	0.346	0.349	0.351	0.353	0.355	0.358	0.360	0.362
1.1	0.364	0.367	0.369	0.371	0.373	0.375	0.377	0.379	0.381	0.383
1.2	0.385	0.387	0.389	0.391	0.393	0.394	0.396	0.398	0.400	0.402
1.3	0.403	0.405	0.407	0.408	0.410	0.412	0.413	0.415	0.416	0.418
1.4	0.419	0.421	0.422	0.424	0.425	0.427	0.428	0.429	0.431	0.432
1.5	0.433	0.435	0.436	0.437	0.438	0.439	0.441	0.442	0.443	0.444
1.6	0.445	0.446	0.447	0.448	0.450	0.451	0.452	0.453	0.454	0.455
1.7	0.455	0.456	0.457	0.458	0.459	0.460	0.461	0.462	0.463	0.463
1.8	0.464	0.465	0.466	0.466	0.467	0.468	0.469	0.469	0.470	0.471
1.9	0.471	0.472	0.473	0.473	0.474	0.474	0.475	0.476	0.476	0.477
2	0.477	0.478	0.478	0.479	0.479	0.480	0.480	0.481	0.481	0.482
2.1	0.482	0.483	0.483	0.483	0.484	0.484	0.485	0.485	0.485	0.486
2.2	0.486	0.486	0.487	0.487	0.487	0.488	0.488	0.488	0.489	0.489
2.3	0.489	0.490	0.490	0.490	0.490	0.491	0.491	0.491	0.491	0.492
2.4	0.492	0.492	0.492	0.492	0.493	0.493	0.493	0.493	0.493	0.494
2.5	0.494	0.494	0.494	0.494	0.494	0.495	0.495	0.495	0.495	0.495
2.6	0.495	0.496	0.496	0.496	0.496	0.496	0.496	0.496	0.496	0.496
2.7	0.497	0.497	0.497	0.497	0.497	0.497	0.497	0.497	0.497	0.497
2.8	0.497	0.498	0.498	0.498	0.498	0.498	0.498	0.498	0.498	0.498
2.9	0.498	0.498	0.498	0.498	0.498	0.498	0.499	0.499	0.499	0.499

the top row for the third significant digit. For example, the probability of the interval bounded by the value of a between 0 and 1.98 equals 47.7%, which is the twentieth column after the top row and the ninth column after the leftmost column. The probability that the mean temperature on July 1st lies between 21 °C and 24 °C is 1.00σ, which is 34.1%.

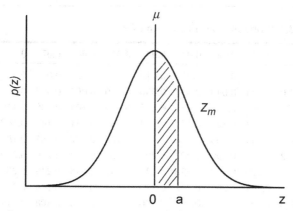

FIGURE 2.3 Gaussian Probability Distribution (One-Tailed)

Example 2.2. A gas chromatography column to evaluate the concentration of nepetalactone—a new mosquito repellent—has a 3.0 yr life expectancy with a standard deviation of 0.5 yr:

(a) What is the probability that the column will last 2.2 yr?
(b) What is the probability that it will last longer than 3.8 yr?

Solution 2.2a. $x = 2.2, \mu = 3, \sigma = 0.5, z = \frac{x-\mu}{\sigma} = \frac{2.2-3}{0.5} = -1.60, P(x < 2.2) = P(z < -1.60)$.

As shown in Figure E2.2Sa, the area of interest is to the left of the curve bounded by the value of $z = -1.60$. The probability that the column will last longer than 2.2 yr is to the right of -1.60 and the probability that it will last less than 2.2 yr is to the left. From Table 2.1, the value of $P(0 < z < -1.60)$ is 0.445, thus the

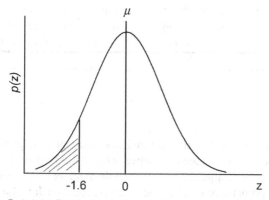

FIGURE E2.2Sa Probability Distribution for Columns Lasting Less than 2.2 yr

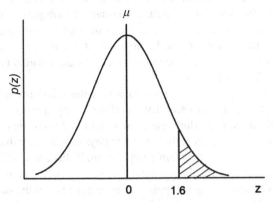

FIGURE E2.2Sb Probability Distribution for Columns that Last Beyond 3.8 yr

probability that it will last only 2.2 yr equals:

$$P(-\infty < z < -1.60) = P(-\infty < z < 0) - P(-1.60 < z < 0)$$
$$= 0.5 - 0.445 = 0.055 = 5.5\%.$$

Remember that the half curve from $-\infty$ to zero represents 50% of the variance. Another way of looking at it is to say that the probability that it lasts at least 2 yr is the sum of the probability from -1.60 to $+\infty$:

$$P(-1.60 < z < +\infty) = P(-\infty < z < +\infty) - [P(-1.60 < z < 0)$$
$$+ P(0 < z < +\infty)]$$
$$= 1 - [0.445 + 0.5] = 0.055 = 5.5\%.$$

Solution 2.2b. $x = 3.8$, $\mu = 3$, $\sigma = 0.5$, $z = \frac{x-\mu}{\sigma} = \frac{3.8-3}{0.5} = +1.60$, $P(x > 3.8) = P(z > +1.60)$.

Figure E2.2Sb demonstrates that the region of interest lies beyond the bound of 3.8 yr—from 3.8 to $+\infty$. The probability $P(-1.60 < z < 0)$ equals 0.455 and this is the same as it is for $P(0 < z < +1.60)$ but we are interested in the region of $+1.60$ to $+\infty$:

$$P(+1.60 < z < +\infty) = P(0 < z < +\infty) - P(0 < z < 1.60)$$
$$= 0.5 - 0.445 = 0.055 = 5.5\%.$$

2.3.2 Criterion of Chauvenet

The normal distribution characterizes data with random variability. However, during the course of running an experiment, recording the data, calculating a parameter, etc. one or more of the data points will appear appreciably different compared to the entire population or expectation. These data points are referred to as outliers. For very large data sets (perhaps 30 samples or more), a single outlier will have only a marginal impact on the sample standard deviation and even less on the sample mean. In the case of small data sets (less than 10 samples), outliers may affect the mean noticeably and the effect on the standard deviation could be substantial.

Before adopting statistical tests to assess the reliability of data, outliers should be first analyzed carefully to identify any anomaly in instrument fidelity, calibration, procedure, environmental conditions, recording, etc. The first objective is to reject an outlier based on physical evidence that the data point was unrepresentative of the sample population. If this exercise fails to identify probable cause (or if details on the experimental methods are unavailable), Chauvenet's criterion may be applied to assess the reliability of the data point (Holman 2001). Simply put, the criterion recommends rejecting a data point if the probability of obtaining the deviation is less than the reciprocal of two times the number of data points—$1/(2n)$. For an outlier smaller than the mean, the data point may be rejected when:

$$P(-\infty < z < -a) < \frac{1}{2n}$$

or

$$0.5 - P(-a < z < 0) < \frac{1}{2n}.$$

An outlier greater than the mean is rejected when:

$$1 - P(-\infty < z < a) < \frac{1}{2n}$$

or

$$0.5 - P(0 < z < a) < \frac{1}{2n}.$$

If the data point is rejected, a new mean and standard deviation should be calculated. Eliminating more than one outlier from the population is generally discouraged.

Example 2.3. The viscosity of transparent liquids can be measured with a falling ball viscometer. A glass tube is filled with the liquid of interest. A ball is allowed to drop through the liquid in the tube and the time it takes the ball to travel between two lines separated by an precisely defined distance is measured. The results (in s) were: 12.4, 11.6, 13.9, 11.8, 12.4, 10.0, 11.6, 12.8, 11.5, and 11.9.

(a) For the 10 measurements taken, apply Chauvenet's criterion and state whether or not any points may be rejected.
(b) Report the mean and sample deviation of the time of descent that approximates the data best.

(c) Would a stopwatch with a resolution of 0.01 s improve the precision?

Solution 2.3a. $x_o = 10.0$ s, $\bar{x} = 12.0$ s, $s = 1.0$ s, $z = \frac{x_o - \bar{x}}{s} = \frac{10.0 - 12.0}{1.0} = 2.0$, $P(-a < z < 0) = 0.477$ from Table 2.4, $0.5 - 0.477 = 0.023 < 0.05$, therefore, the data point should be rejected.

Solution 2.3b. $\bar{x} = 12.2$ s, $s = 0.77$ s $\cong 0.8$ s.

Solution 2.3c. The measuring system comprises the reaction time of the person recording the data and the chronometer. The reaction time of an individual is on the order of 0.1 s. So, increasing the chronometer resolution to 0.01 s does not improve the overall precision (of the measurement) and thus it should have no effect.

2.3.3 Uncertainty (Type B)

Measuring liquids in flasks demonstrates the concept of uncertainty: a 250 ml graduated cylinder has marks on the side at intervals of 2 ml. The bottom of the meniscus corresponding to the adjacent graduation mark when read at eye level represents the volume. The resolution of the cylinder is 2.

The uncertainty in any given measurement may be related to the resolution. Figure 2.4 demonstrates the interval that characterizes uncertainty. If Z is the measured value (the volume of liquid measured in a graduated cylinder), then the uncertainty in the measured volume lies between a lower limit Z_- and an upper limit Z_+.

The height of a meniscus in a graduated flask may be read with a greater precision than 2 ml and most probably better than 1 ml (the midpoint between two graduation marks). However, to calculate the interval requires as many as 10 measurements. In the absence of these values, the uncertainty may be assumed to be equal to the half point of the resolution of the instrument. Therefore, as a first approximation, the uncertainty of the 250 ml graduated cylinder with 2 ml graduation marks equals ± 1 ml. The uncertainty of a 250 ml graduated

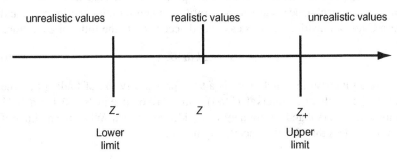

FIGURE 2.4 Uncertainty Interval

Erlenmeyer flask with marks at 25 ml intervals would be ± 13 ml, which should be truncated to ± 10 ml.

The simplistic approach to uncertainty is to assign a value based on a physical characteristic of the instrument or measuring device. However, to assign an accurate value requires a statistical approach—standard deviation or confidence intervals, which are described in the next sections.

2.3.4 Confidence Intervals and Uncertainty (Type A)

Thus far, we have related the measurement of uncertainty to the resolution—the scale divisions on a thermometer, the graduation marks on a beaker. It may also be defined by an engineer or a scientist with a significant amount of experience. In circumstances where measurements of the mean and standard deviation have been made, it is often simply expressed as the standard deviation $\sigma : \mu \pm \sigma$.

A more rigorous definition of uncertainty (Type A) relies on the statistical notion of confidence intervals and the Central Limit Theorem. The confidence interval is based on the calculation of the standard error of the mean, $s_{\bar{x}}$, which is derived from a random sample of the population. The entire population has a mean μ and a variance σ^2. A sample with a random distribution has a sample mean and a sample standard deviation of $\bar{x}$ and s, respectively. The Central Limit Theorem holds that the standard error of the mean equals the sample standard deviation divided by the square root of the number of samples:

$$s_n = \frac{s}{\sqrt{n}}. \qquad (2.8)$$

Confidence intervals are somewhat arbitrary but the generally accepted criterion requires a 95% probability of the true value falling within the confidence interval. For very large samples, $(n > 30)$, the confidence interval is calculated assuming the distribution is Gaussian. For sample sizes less than 30, the value of s^2 fluctuates substantially from sample to sample and thus the distribution is no longer a standard normal distribution. For this case, we represent the distribution with a statistic that is known as the Student's t-statistic.

The uncertainty, Δ, is defined as the product of the standard error of the mean and a confidence interval (arbitrarily defined by the experimenter) and for a Gaussian distribution it is calculated according to the following relation:

$$\Delta = \pm k(\alpha)\sigma. \qquad (2.9)$$

For an interval in which there is a 95% probability (α) of finding the true value, the confidence interval $(k(\alpha))$ equals 1.96. It equals 1.0 for a 68% confidence interval and 2.6 for a 99% confidence interval. When the number of samples is less than 30, the uncertainty becomes:

$$\Delta = \pm t(\alpha, n - 1)s_n = \pm t(\alpha, n - 1)\frac{s}{\sqrt{n}}. \qquad (2.10)$$

TABLE 2.2 Values of the Student's t-Statistic

α, %	t_{50}	t_{80}	t_{90}	t_{95}	t_{98}	t_{99}	$t_{99.9}$
$n-1$							
1	1.000	3.078	6.314	12.706	31.821	63.657	636.619
2	0.817	1.886	2.920	4.303	6.965	9.925	31.599
3	0.765	1.638	2.353	3.183	4.541	5.841	12.924
4	0.741	1.533	2.132	2.777	3.747	4.604	8.610
5	0.727	1.476	2.015	2.571	3.365	4.032	6.869
6	0.718	1.440	1.943	2.447	3.143	3.707	5.959
7	0.711	1.415	1.895	2.365	2.998	3.500	5.408
8	0.706	1.397	1.860	2.306	2.897	3.355	5.041
9	0.703	1.383	1.833	2.262	2.821	3.250	4.781
10	0.700	1.372	1.813	2.228	2.764	3.169	4.587
11	0.697	1.363	1.796	2.201	2.718	3.106	4.437
12	0.696	1.356	1.782	2.179	2.681	3.055	4.318
13	0.694	1.350	1.771	2.160	2.650	3.012	4.221
14	0.692	1.345	1.761	2.145	2.625	2.977	4.141
15	0.691	1.341	1.753	2.132	2.603	2.947	4.073
16	0.690	1.337	1.746	2.120	2.584	2.921	4.015
17	0.689	1.333	1.740	2.110	2.567	2.898	3.965
18	0.688	1.330	1.734	2.101	2.552	2.878	3.922
19	0.688	1.328	1.729	2.093	2.540	2.861	3.883
20	0.687	1.325	1.725	2.086	2.528	2.845	3.850
21	0.686	1.323	1.721	2.080	2.518	2.831	3.819
22	0.686	1.321	1.717	2.074	2.508	2.819	3.792
23	0.685	1.320	1.714	2.069	2.500	2.807	3.768
24	0.685	1.318	1.711	2.064	2.492	2.797	3.745
25	0.684	1.316	1.708	2.060	2.485	2.787	3.725
26	0.684	1.315	1.706	2.056	2.479	2.779	3.707
27	0.684	1.314	1.703	2.052	2.473	2.771	3.690
28	0.683	1.313	1.701	2.048	2.467	2.763	3.674
29	0.683	1.311	1.699	2.045	2.462	2.756	3.659
inf	0.675	1.282	1.645	1.960	2.326	2.576	3.291

Values of the Student's t-statistic are summarized in Table 2.2. For a sample size of six (five degrees of freedom, $n-1$) and a 95% confidence interval ($\alpha = 95\%$), the value of the Student's t is 2.571 ($\sim$2.6). It is approximately equal to 2 for a Gaussian distribution. Table 2.3 compares the Student's t with the Gaussian distribution for several common combinations of sample numbers

TABLE 2.3 Comparison of Student's t with Gaussian Distribution for Common Combinations

α	90.0	95.0	99.0
$t(\alpha,5)$	2.020	2.570	4.030
$t(\alpha,10)$	1.810	2.230	3.170
$t(\alpha,25)$	1.708	2.060	2.787
$k(\alpha)$	1.640	1.960	2.570

and confidence intervals. As the number of samples increases, the Student's t-statistic approaches the values of the Gaussian distribution.

Example 2.4. The average residence time distribution (RTD—also known as the contact time) of a fluid in a vessel is a parameter indicative of the inhomogeneities in the fluid flow and is particularly useful for troubleshooting. An inert fluid is fed to the vessel at a constant rate and is switched to another fluid while the effluent is monitored at a high frequency. Figure E2.4 shows the exit concentration of oxygen from a vessel. The feed gas was switched from argon to air at $t = 0$ and it took more than 5 s before the oxygen was first detected. The average residence time is equivalent to the first moment and is calculated according to:

$$\bar{t} = \frac{\sum_i C_i t_i}{\sum_i C_i}. \tag{2.11}$$

The gas phase RTD of a reactor was measured six times and the average residence times (in s) are: 34.4, 34.8, 33.7, 35.3, 34.0, and 34.1:

(a) Calculate the sample mean and standard deviation.

(b) Calculate the uncertainty of the mean residence time from the point of injection to the point of detection.

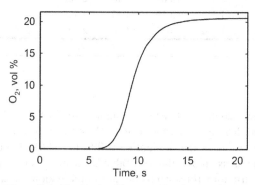

FIGURE E2.4 Residence Time Distribution (Heaviside Step Impulse Function)

Solution 2.4a. The sample mean and sample standard deviation are $\bar{x} = 34.4$ and $s = 0.585$.

Solution 2.4b. To calculate the uncertainty, first the standard error of the mean is calculated and the 95% confidence interval of the Student's t-statistic (with five degrees of freedom—$n - 1$) is read from the table: $\Delta = \pm t(\alpha, n - 1)s_{\bar{x}} = \pm t(\alpha, n - 1)\frac{s}{\sqrt{n}} = 2.57\frac{0.58}{\sqrt{6}} = 0.61$, $\bar{t} = (34.4 \pm 0.6)$ s.

In Example 2.1, the resolution of the graduated cylinder was 2 ml. The simplest assignment of the uncertainty interval would be ± 1 ml. A second way to assign the uncertainty interval would be to subtract the highest value of the measured volume (98.18 ml) from the lowest value (97.02 ml) and divide it by two. In this case, the uncertainty would equal 0.6 ml. The uncertainty calculated assuming a confidence interval of 95% (with 9 degrees of freedom) gives $\Delta = 0.25$ ml, which is the lowest value of the three methods. Often the uncertainty is confused with the standard deviation, σ. The standard deviation characterizes the population of a random variable. Note that the standard deviation of the 10 samples was 0.35, which is closer to Δ than the other two methods.

2.3.5 Uncertainty Propagation

Few measurements rely on a single factor; velocity, for example, is a product of time and distance—two factors. The volumetric flow rate through an orifice depends on the pressure drop, fluid density, and geometry, which adds up to three factors or more. Thermal conductivity, k, is the ratio of heat flux to temperature gradient—this measurement could have as many as six factors. All factors have an associated uncertainty that contributes to the overall uncertainty of the measurand.

To calculate the compounding factors on the measurand, f is expressed as:

$$f = f(x_1, x_2, x_3, \ldots, x_n). \tag{2.12}$$

The uncertainty Δ_f is simply the sum of the squares of the product of the uncertainty of the individual factor and the first partial derivative of f with respect to that factor:

$$\Delta_f^2 = \left(\frac{\partial f}{\partial x_1}\Delta_1\right)^2 + \left(\frac{\partial f}{\partial x_2}\Delta_2\right)^2 + \left(\frac{\partial f}{\partial x_3}\Delta_3\right)^2 + \cdots + \left(\frac{\partial f}{\partial x_n}\Delta_n\right)^2. \tag{2.13}$$

Rather than differentiating the functions to derive the uncertainty, there are two common cases with simple expressions for the uncertainty: the first is for functions that are products of the factors and the second is for arithmetic functions that are simply the addition and subtraction of the factors.

$$f = x_1^{a_1} x_2^{a_2} x_3^{a_3} x_4^{a_4} \cdots x_n^{a_n} \tag{2.14}$$

and

$$f = a_1 x_1 + a_2 x_2 + a_3 x_3 + a_4 x_4 + \cdots + a_n x_n = \sum_i a_i x_i. \qquad (2.15)$$

In the case of a function equal to the product of, let us, three factors, the first derivative of f with respect to each is:

$$\frac{\partial f}{\partial x_1} = a_1 x_1^{a_1-1} x_2^{a_2} x_3^{a_3}, \qquad (2.16)$$

$$\frac{\partial f}{\partial x_2} = a_2 x_1^{a_1} x_2^{a_2-1} x_3^{a_3}, \qquad (2.17)$$

$$\frac{\partial f}{\partial x_3} = a_3 x_1^{a_2} x_2^{a_2} x_3^{a_3-1}, \qquad (2.18)$$

and so:

$$\Delta_f^2 = (a_1 x_1^{a_1-1} x_2^{a_2} x_3^{a_3} \Delta_1)^2 + (a_2 x_1^{a_1} x_2^{a_2-1} x_3^{a_3} \Delta_2)^2 + (a_3 x_1^{a_2} x_2^{a_2} x_3^{a_3-1} \Delta_3)^2. \qquad (2.19)$$

By dividing both the right- and left-hand sides of this equation by the original function squared, we can simplify the expression to:

$$\frac{\Delta_f}{f} = \sqrt{\left(\frac{a_1}{x_1}\Delta_1\right)^2 + \left(\frac{a_2}{x_2}\Delta_2\right)^2 + \left(\frac{a_3}{x_3}\Delta_3\right)^2}. \qquad (2.20)$$

The general expression for the uncertainty of a function that is a product of n factors is:

$$\frac{\Delta_f}{f} = \sqrt{\sum_{i=1}^{n} \left(\frac{a_i}{x_i}\Delta_i\right)^2}. \qquad (2.21)$$

For an arithmetic expression involving addition and subtraction, the uncertainty becomes:

$$\frac{\Delta_f}{f} = \sqrt{\sum_{i=1}^{n} (a_i \Delta_i)^2}. \qquad (2.22)$$

The next example combines the concepts of uncertainty propagation as well as confidence intervals. The problem is related to measuring the viscosity of a transparent fluid in a falling ball viscometer, in which the time it takes a ball to cross two lines separated by a distance is measured. The ball falls through a tube filled with the fluid and is assumed to reach a steady velocity by the time it reaches the first line. The velocity is determined by the geometry and density of the ball as well as the density and, most importantly, the viscosity of the liquid:

$$\mu = \frac{(\rho_b - \rho_f) V g}{6\pi r L} t, \qquad (2.23)$$

where μ is the viscosity, V is the volume of the falling ball, r is its radius, ρ_b is its density, ρ_f is the density of the fluid, and L is the distance between the two marks.

Example 2.5. The viscosity of a new polymer was measured at 85 °C with a falling ball viscometer. Nine measurements were made and the time it took the ball to travel 300 mm was measured to be:

$$t_i(s) = 23.5,\ 21.9,\ 22.8,\ 20.8,\ 24.8,\ 23.3,\ 26.6,\ 23.7,\ 22.3.$$

The density of the polymer equaled 852 kg m^{-3} ± 80 kg m^{-3}. The ball was made of stainless steel ($\rho_b = 7850$ kg m^{-3}), with a diameter of 2.00 cm.

(a) Calculate the viscosity of the polymer.

(b) Calculate its uncertainty assuming a confidence interval of 95%.

Solution 2.5a. $\bar{x} = 23.3$ s, $s = 1.69$, $s_{\bar{x}} = 0.56$, $r = 0.01$ m, $V = 4.19 \times 10^{-5}$ m^3, $\rho_b = 7850$ kg m^{-3}, $\rho_f = 852 \pm 80$ kg m^{-3}, $L = 0.300$, $\mu = \frac{(\rho_b - \rho_f)Vg}{6\pi rL}t = \frac{(7850-852)4.19\times 10^{-5} 9.81}{6\pi 0.010.3} 23.3 = 1185$ Pa s $= 1.19$ cP.

Solution 2.5b. The calculation of the uncertainty involves both an arithmetic function and a function of the product of the variables. The equation below relates the uncertainty of the viscosity with the difference in density between the steel ball and the polymer fluid as well as the measurement of time:

$$\frac{\Delta\mu}{\mu} = \sqrt{\left(\frac{\Delta_{\Delta\rho}}{\Delta\rho}\right)^2 + \left(\frac{\Delta_t}{t}\right)^2}. \tag{2.24}$$

The uncertainty in the difference of the densities is calculated according to the arithmetic function:

$$\Delta_{\Delta\rho} = \sqrt{\sum_{i=1}^{n}(a_i \Delta_i)^2} = \sqrt{(a_{\rho_b}\Delta_{\rho b})^2 + (a_{\rho_f}\Delta_{\rho f})^2}$$
$$= \Delta_{\rho f}$$
$$= \pm 80 \text{ kg m}^{-3}.$$

Since the uncertainty around the density of the steel was not quoted, we assume it equal to zero ($\Delta_{\rho_b} = 0$) and the coefficient a_{ρ_f} is equal to one. So the uncertainty in the difference in density is simply equal to the uncertainty in the measure of the density of the polymer.

The uncertainty with respect to time is calculated using the Student's t-statistic, with 8 degrees of freedom and a 95% confidence interval:

$$\Delta_t = \pm t(\alpha, n-1)s_{\bar{x}} = \pm t(95\%, 8)\frac{s}{\sqrt{n}} = 2.306 \cdot \frac{1.69}{\sqrt{9}} = 1.3 \text{ s.}$$

The uncertainty in the measure of viscosity is:

$$\Delta\mu = \sqrt{\left(\frac{\Delta_{\Delta\rho}}{\Delta\rho}\right)^2 + \left(\frac{\Delta_t}{t}\right)^2}\,\mu = \sqrt{\left(\frac{80}{6998}\right)^2 + \left(\frac{0.56}{23.4}\right)^2}\,1.19\ \text{cP} = 0.03\ \text{cP}.$$

When the relative uncertainty of one of the factors is smaller than the others, it may be neglected without the significance because the uncertainty depends on the sum of the squared terms. For the case of the product of two factors, the second factor represents only 10% of the uncertainty if the ratio of the relative uncertainties of the factor is less than one third and therefore it can be neglected (for the case where $a_2\Delta_2 > a_1\Delta_1$):

$$\frac{a_2}{x_2}\Delta_2 > \frac{1}{3}\frac{a_1}{x_1}\Delta_1.$$

In the case of an addition or a subtraction:

$$a_2\Delta_2 \leqslant \frac{1}{3}a_1\Delta_1.$$

Example 2.6. Compare the relative uncertainty in the flow rate through an orifice meter for the case where the volumetric flow rate of a fluid is 2.69 m^3 s^{-1}, a density of (450 ± 5) kg m^{-3}, and a pressure drop of 5190 ± 160 Pa. *Ariane Bérard*

$$Q = \frac{C_0 X_A}{\sqrt{1-\beta^4}}\sqrt{\frac{2}{\rho}\Delta P}.$$

Solution 2.6. Since the volumetric flow rate is related to the product of density and pressure drop:

$$\frac{\Delta_f}{f} = \sqrt{\sum_{i=1}^{n}\left(\frac{a_i\Delta_i}{x_i}\right)^2},$$

$$\frac{\Delta_Q}{Q} = \sqrt{\left(\frac{a_\rho\Delta_\rho}{\rho}\right)^2 + \left(\frac{a_{\Delta P}\Delta_{\Delta P}}{\Delta P}\right)^2},$$

$$\Delta_Q = Q\sqrt{\left(-\frac{1}{2}\frac{\Delta_\rho}{\rho}\right)^2 + \left(\frac{1}{2}\frac{\Delta_{\Delta P}}{\Delta P}\right)^2},$$

Including both the pressure and density terms give:

$$\Delta_Q = 2.69\sqrt{\left(-\frac{1}{2}\frac{5}{450}\right)^2 + \left(\frac{1}{2}\frac{160}{5190}\right)^2} = 0.04\ \text{m}^3\ \text{s}^{-1}.$$

Comparing the relative uncertainty of pressure versus density, we see that the relative uncertainty for density is less than one third that of pressure:

$$\frac{\frac{1}{2} \cdot 160 \cdot 450}{-\frac{1}{2} \cdot 5 \cdot 5190} = 2.77 \leqslant 3.$$

Thus it can be ignored and the uncertainty in the volumetric flow rate remains the same (to one significant figure):

$$\Delta_Q = \frac{1}{2} \frac{\Delta_{\Delta P}}{\Delta P} \cdot Q = \frac{1}{2} \frac{160}{5190} \cdot 2.69 = 0.04 \text{ m}^3 \text{ s}^{-1}.$$

2.4 INSTRUMENTATION CONCEPTS

The three common measurement techniques are direct, comparative, and indirect. Rulers and callipers are examples of instruments that assess length directly. The oldest instrument to assess weight is the balance and this is a comparative method: an unknown weight is placed in a pan hanging from a horizontal beam on one side and known weights are placed in a pan on the other side to balance the unknown weight. Although these two techniques are the oldest and simplest, most physical properties are measured indirectly. For example, the volume of a liquid in a cylinder or beaker is assessed by graduation marks. Graduation marks are positioned along the side of the vessel to represent a volume. The same principle is applied for a spring scale, which measures the deflection—distance—due to an applied load.

Indirect measuring instruments are generally comprised of three parts: a detector, an amplifier, and an indicator. A thermometer is an indirect measuring device that depends on the dilation of a fluid to represent a change in temperature. The fluid detects a change in temperature and either contracts or dilates. The fluid is connected to a narrow tube that amplifies the change. Graduation marks are positioned along the tube to indicate the temperature.

An instrument is characterized by several factors including its accuracy, robustness, sensitivity, etc. The following glossary will be used throughout this book (BIPM JCGM 200:2008):

2.4.1 Interval

The set of real numbers, x, that lie between two endpoints a and b is designated by $[a; b]$. The number 323 is halfway between the interval $[273; 373]$.

2.4.2 Range

The range is the difference between the two endpoints of an interval. The range of the interval $[273; 373]$ is 100. This term is also called the span or the full scale (FS). It is recommended not to operate sensors at less than 10% of full scale; some instruments will also specify a maximum allowable measure greater than the full-scale measure. Pressure instruments often quote a value greater than full scale: operating at a higher value (even momentarily) risks damaging the measuring device.

2.4.3 Resolution, Sensitivity, Detection Limit, Threshold

The resolution of an instrument is the smallest increment that can be displayed or recorded. The detection limit or threshold is often equal to the resolution but the ability to record a consistent value at the limit is very poor. The resolution of a 250 ml graduated cylinder with graduation marks every 2 ml is 2 ml. Typically, the first graduation mark on these cylinders is at 10 ml, which makes this its threshold value or detection limit.

The sensitivity of electronic instruments is quoted with respect to an electrical signal—14 bit analog-to-digital converter (ADC), for example. The sensitivity is calculated based on the full-scale input voltage of the data logger, V_{dl}, the full-scale output voltage of the instrument, V_I, the bit resolution, n, and a conversion factor, E:

$$S = \frac{V_{dl}}{V_I 2^n} E, \tag{2.25}$$

To illustrate the calculation, consider a pressure transducer with a 5 V full-scale range rated at 100 bar connected to a data logger with 8 bit ADC resolution and a 1 V full-scale range. According to the equation, the resolution equals 0.08 bar or about 0.08%:

$$S = \frac{V_{dl}}{V_I 2^n} E = \frac{1}{5 \cdot 2^8} \cdot 100$$
$$= 0.08 \text{ bar.}$$

Sensitivity is an absolute quantity that represents the smallest change detectable by an instrument and is often measured in mV, mA or $\mu\Omega$. In this case, the resolution of the instrument will be the lowest increment displayed by the data logger. If the ADC is configured for bipolar operation ($+-10$ V, for example), the exponent in Equation 2.6 becomes ($n - 1$) rather than n, thus reducing the sensitivity versus a unipolar range by a factor of 2.

2.4.4 Precision

The concepts of precision and resolution (sensitivity) are often confused. While resolution represents the smallest measurable unit that can be read, precision is the smallest measurable unit that can be read repeatedly and reliably. Consider the difference between a digital chronometer with a resolution of 0.01 s and an analog stop watch with tick-marks at intervals of 0.1 s. The total measuring system not only includes the chronometer but also the person using it. Since the

FIGURE 2.5 Standard Bourdon Gauge Pointer and Dial Face

reaction time of a human is known to be no better than 0.1 s, the overall precision
of the measurement system—human and stop watch—is 0.1 s. The precision
equals the resolution in the case of the analog stop watch while precision is
lower than resolution in the case of the digital chronometer.

The precision can be greater than the resolution, as illustrated in Figure 2.5.
The Bourdon gauge reads pressure from 0 bar to 6 bar and has ticks at 1 bar
intervals. The tip of the needle is very narrow and it is reasonable to estimate
the needle position to a precision of at least 0.2 bar, if not 0.1 bar.

The precision of a measure within a laboratory is referred to as the
repeatability while the precision measured between laboratories is referred to as
the reproducibility. Ideally, the precision within the same laboratory should be
as good as the precision between laboratories. Even better would be a precision
equal to the resolution. In practice, however, the precision is generally much
less than the resolution but much better than the repeatability or reproducibility:

Reproducibility < Repeatability < Precision < Resolution.

Precision is expressed mathematically as a percentage. For an individual
measurement, the precision is:

$$c_i = 1 - \frac{x_i - \bar{x}}{\mu}. \tag{2.26}$$

The overall precision of an instrument may be expressed as the average of
the individual measurements:

$$C = \frac{1}{n} \sum_{i=1}^{n} c_i. \tag{2.27}$$

Round-robin testing is a common technique used to evaluate the reproducibility of a measurement, instrument, standard, technique, procedure, etc.

Analytical equipment such as gas chromatographs and especially particle size distribution instruments require significant attention to achieve a high level of reproducibility. For particle size measurements, some chemical plants historically based all measurements on a single instrument and then used these values to compare with others—variation between instruments and even technicians was very high for particle fractions below 44 μm.

2.4.5 Error

Lord Kelvin stated that "to measure is to know" (Unnamed 2011). In fact, it would be more precise to say that "to measure is to know ... better." Almost all measurements have errors with respect to their absolute value, be it temperature, pressure, flow rate, or any of several other standard measurements necessary to evaluate a phenomenon, characterize an instrument, or run a chemical plant. Error is defined as the absolute or relative difference between the measured value and the true value as measured by a recognized standard. Note that the term "recognized standard" is mentioned, and not "true value" since the true value can rarely, if ever, be known exactly.

For an individual measurement the deviation of the value with respect to the true value is expressed by:

$$d_i = x_i - \mu, \tag{2.28}$$

where x_i is an individual measurement and μ is the true value of the mean of an entire population. When comparing a sample of the population with the true value of the mean, error is expressed in the following way:

$$\delta = \bar{x} - \mu. \tag{2.29}$$

At this point it would be useful to compare the concepts of uncertainty and error. Error is the difference between the measured value and a standard—the "true value"—while uncertainty is an interval or the range of values within which the true value has a probability to be found. The size of this interval is related to the degree of probability defined by the experimenter.

There are three classes of errors: systematic errors are reproducible and the cause is related to some physical law and may be eliminated by appropriate corrective actions; random errors are unpredictable and irreproducible and can be characterized by the laws of statistics; finally, inadmissible errors occur as a result of mishandling, incorrect instrument operation, or even poor record keeping. Figure 2.6 shows a hierarchical chart that outlines the three types of errors (Chaouki et al., 2007). Most treatments of error consider systematic and random errors. Inadmissible errors might somehow be related to outliers, which are values that appear to be completely unrelated to a rational data set.

An example of a systematic error would be reading the volume of a graduated cylinder from the top of the meniscus instead of the bottom. Systematic errors may also be introduced through improper calibration or changes in the

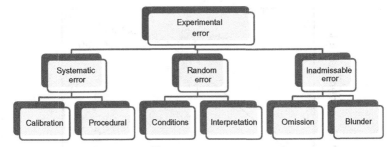

FIGURE 2.6 Types of Experimental Error

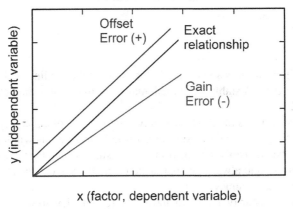

FIGURE 2.7 Offset and Gain Error

instrument due to use or environmental factors (scale build-up, rust, metal fatigue—localized damage as a result of cyclic loadings—and more).

The zero offset is a common error in measuring pressure with electronic transducers or even typical Bourdon gauges. When the gauge is operated close to the maximum, a random pressure surge can deform the transducer. The unit may still be operational (and perhaps give a linear signal) but when the pressure is released the gauge may appear to read a value (either positive or negative). Figure 2.7 demonstrates the zero offset error—at the origin, the reading is greater than zero (positive differential). The signal increases linearly with the measured variable and it is proportional to the true signal (Patience et al. 2011). Reading the top of a meniscus to represent volume would be an offset error. A scale error (gain error), like the zero offset error, may be caused by an overpressure (or underpressure), mechanical wear, etc. in the case of a pressure transducer. With electronic instruments, it corresponds to the difference between the output voltage (or current) at full scale and the calibrated output voltage. A negative gain is illustrated in Figure 2.7.

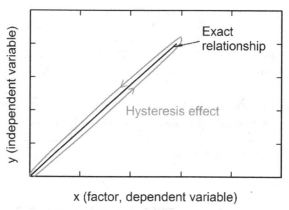

FIGURE 2.8 Instrument Hysteresis

Zero drift represents a variation in the value of the reading with no load as a function of time. A stable instrument maintains the system characteristics for long periods.

Hysteresis error is the difference between the value of a quantity when it is measured going up in scale versus going down in scale, as shown in Figure 2.8. To increase the precision of an instrument that suffers from hysteresis error, the measurement should always be approached from the same side. This of course may not increase accuracy.

Instrument linearity refers to the measurement deviation from an ideal straight-line performance. For instruments that have a nonlinear response, it must be calibrated along the whole range. The relationship between conductivity and acid concentration is nonlinear: the conductivity increases less than proportionally with acid concentration (see Figure 2.9).

Errors of omission result from an insufficient rigor in following strict laboratory protocols. For example, to determine the rate kinetics of a gas-phase reaction, the necessary measurements include composition at the entrance and exit, reactor volume, flow rate, pressure, and temperature. When the reactor is operated at "atmospheric" pressure—that is, the effluent exits to the atmosphere—the actual operating pressure is invariably different from 1.01325 bar. Not only does it depend on the elevation above sea level but also the meteorological conditions. Within hours, the barometric pressure may change by as much as 5%.

Consider the textile industry and assessing fabric weight (mass per unit area). It is a parameter used for quality purposes as well as to assess cost. It would seem to be a straightforward operation: cut a piece of fabric, calculate its surface area, and weigh it on a scale. However, procedures have been written by many (if not all) standards organizations for this apparently simple test. The procedure is covered in ISO 3801:1977, ASTM D3776:2007, and BS EN

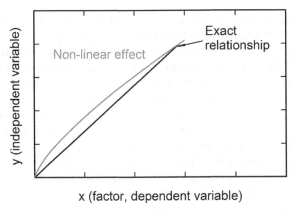

FIGURE 2.9 Nonlinear Response

12127:1998, to name a few. Measuring the fabric weight essentially considers six factors:

1. *Human factors:* The personnel making the measurement should be trained and qualified;
2. *Environmental conditions:* The measurement should be made in a controlled environment maintained at a temperature of $(21 \pm 2)\,°C$ and 65% relative humidity;
3. *Calibration:* Experimental instruments must be calibrated—balances, temperature, and relative humidity instruments—on a specified schedule and a record of the calibration documented;
4. *Instrumentation:* The measurement must be made with appropriate instruments—balances, scissors, and rulers;
5. *Sampling:* The sample of the fabric should come from the center of the roll;
6. *Conditioning:* The sample should be maintained at 21 °C and 65% relative humidity for a period of 24 h before taking the measurement.

The context of a chemical plant is quite different than that of an analytical or process laboratory: the focus in the laboratory might be to understand the minimum error allowable to interpret or extrapolate a phenomenon. The challenge in the chemical plant is to choose the appropriate measuring interval and the frequency at which it is measured and stored, how much data is needed, and what precision is necessary to take an action.

The most formidable decision that can be taken in a plant is to initiate an uncontrolled shutdown based on a perceived threat to personnel safety, to the environment, or to equipment integrity. Shutdowns represent a major economic penalty due to lost production but also due to the risk of contaminating equipment that might require a costly cleanup. Besides operator interventions, automatic triggers are implemented to halt production and bring plants into a stable and safe operating point: interlocks. A high-high (or low-low) interlock is the most severe trigger and must be carefully set to ensure personnel safety and at the same time set far enough from standard operating conditions to make sure it won't be "accidentally" tripped at a condition that is not particularly hazardous. Temperature and pressure are the two most common interlock triggers. Steel may be operated at very specific maximum pressures and temperatures, which are mandated by legislation. In the case of temperature, an interlock might be set at a value of 10–20 °C below this value. However, often production rates may be increased by increasing temperature and thus in times of high demand the operating envelope will be increased, thus bringing it closer to the interlock temperature. Clearly, in this case, the measurement precision must be high in order to operate with a minimum risk.

Instruments may record chemical plant data at a frequency of several hundred hertz but storing all these measurements would require enormous databases and is both impractical and unnecessary. High frequency data is recorded and stored temporarily before it is archived in a database management system at a lower frequency—every 15 s is typical. The frequency of the archived data is increased in the case where variables change suddenly above a predefined threshold value. This threshold value is higher than the resolution of the instrument and represents an arbitrary value chosen by the experienced instrument engineer. For example, temperature may be recorded at high frequency with a precision of 0.001 °C but generally a precision of ±0.2 °C is sufficient to monitor a process safely.

2.4.6 Accuracy

An instrument may be precise, but if the measurement deviates from the true value, it is inaccurate. Accuracy is a measure of an instrument's ability to reproduce the true value—it is the degree of closeness to it. It is also known as bias and may be expressed as a fraction or a percentage:

$$\beta = 1 - \frac{|\delta|}{\mu}. \tag{2.30}$$

In general, accuracy may be improved by increasing sample size—the number of measurements. The purpose of calibrating an instrument is to ensure that the measurement is accurate (Taraldsen 2006, ISO 1995). Calibration will not improve precision but will reduce the error or increase the exactness.

Figure 2.10 illustrates the difference between accurate precise measurements using the target analogy. In Figure 2.10a, the measured values are randomly off the target, so they are neither accurate nor precise. To improve the accuracy of the instrument, it must be calibrated and then the values will become more evenly distributed around the center, as shown in Figure 2.10b. Calibration will no longer improve the precision: this instrument is accurate but not precise. Increasing sample size will improve accuracy. In Figure 2.10c, the

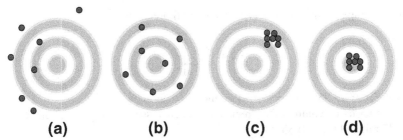

(a)   **(b)**   **(c)**   **(d)**

FIGURE 2.10 Accuracy Versus Precision: (a) Imprecise and Inaccurate Measurements; (b) Imprecise but Accurate Measurements; (c) Precise but Inaccurate Measurements; (d) Precise and Accurate Measurements

measured values are close to each other in one area of the target. In this case, the measurements are precise but they are inaccurate. Calibration would improve the accuracy of the instrument. Finally, in Figure 2.10d, the measurements are both precise and accurate: all measurements are close together and clustered around the target value.

Example 2.7. The accuracy written on the side of the Erlenmeyer flask of Example 2.1 was 5%, while it was 1.4 ml for the graduated cylinder and only 0.08 ml for the volumetric flask.

(a) What is the precision and accuracy of each?
(b) How do these values compare with the reported accuracy and the standard deviation?
(c) How would the accuracy and precision change if you were to read in between the graduation marks of the Erlenmeyer flask at a value of 110 ml?

Solution 2.7a. To calculate the accuracy, first the error of each sample must be calculated: $\Delta_1 = 100$ ml $- 99.69$ ml $= 0.31$ ml, $\Delta_2 = 100$ ml $- 97.89$ ml $= 2.11$ ml, $\Delta_3 = 100$ ml $- 103.47$ ml $= 3.47$ ml.

Together with the absolute error and the true value of the mean, the accuracy is determined from the following expression:

$$\beta_i = 1 - \frac{|\delta|}{\mu_i},$$

resulting in $\beta_1 = 99.69\%$, $\beta_2 = 97.9\%$, and $\beta_3 = 96.5\%$.

The overall precision of each measuring vessel is determined by calculating the difference between the values of the individual measurements and the sample mean:

$$c_i = 1 - \frac{|x_i - \bar{x}|}{\mu}.$$

The average of these values gives the overall precision:

$$C_i = \frac{1}{n} \sum_{j=1}^{n} c_{i,j}$$

with $C_1 = 99.95\%$, $C_2 = 99.73\%$, and $C_3 = 99.47\%$.

In terms of volume, the precision of each vessel is $C_1 = 0.05$ ml, $C_2 = 0.27$ ml, and $C_3 = 0.53$ ml.

Solution 2.7b. The precision of all measuring devices is close to an order of magnitude higher than the accuracy. The precision is lower than the standard deviation on average by 75%.

Solution 2.7c. Both the accuracy and the precision of the Erlenmeyer flask are better than expected because the volume was measured at a graduation mark. Moving away from the graduation mark would decrease both the accuracy and precision because of a lack of reference with which to assess the volume in a consistent manner.

2.4.7 Repeatability and Reproducibility

Both the concepts of repeatability and reproducibility are related to accuracy: repeatability is defined as "the closeness of agreement between the results of successive measurements of the same measurand carried out subject to all of the following conditions: the same measurement procedure, the same observer, the measuring instrument used under the same conditions, the same location, and repetition over a short period of time." The definition of reproducibility is: "the closeness of agreement between the results of measurements of the same measurand, where the measurements are carried out under changed conditions such as: different principles or methods of measurement, different observers, different measuring instruments, different locations, different conditions of use, or different periods of time."

The basics of inter-laboratory (between laboratories) and intra-laboratory (within a laboratory) studies have been summarized in the ISO-5725 standard (Feinberg 1995). Recognizing that measurements are often laborious to undertake, examples cited in the standard have often only three measurements per test and only four laboratories. A single reported measurand, y, is the sum of the true value, x, a component of the accuracy, B, and a component of random variability during any measurement, e (within the laboratory variance):

$$y = x + B + e. \tag{2.31}$$

Figure 2.11 illustrates the inter-laboratory variance, s_L^2, at the top and the intra-laboratory variance, s_r^2—the repeatability variance—, at the bottom, which is much narrower. Assuming equality of variance for each laboratory

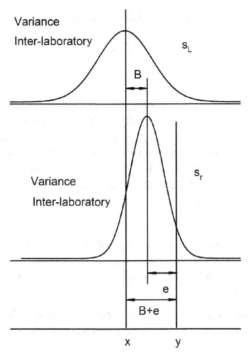

FIGURE 2.11 Intra- and Inter-Laboratory Variance

participating in an inter-laboratory study, the repeatability variance is equal to the mean of the repeatability variance of each laboratory:

$$s_r^2 = \frac{1}{n} \sum_{i=1}^{n} s_{r,i}^2. \tag{2.32}$$

The inter-laboratory variance is calculated based on the variance of the means of the individual laboratories, where $\bar{x}_i$ is the mean of each individual laboratory and $\bar{x}_L$ is the mean of the value of the means of the laboratories. However, the intra-laboratory variance includes part of the repeatability variance that must be subtracted out (the second term of the left-hand side):

$$s_L^2 = \frac{1}{n-1} \sum_{i=1}^{n} (\bar{x}_i - \bar{x}_L)^2 - \frac{s_r^2}{n}. \tag{2.33}$$

The reproducibility variance, s_R^2, equals the sum of the intra-laboratory variance and the inter-laboratory variance:

$$s_R^2 = s_L^2 + s_r^2, \tag{2.34}$$

from which the reproducibility standard deviation is calculated:

$$s_R = \sqrt{s_L^2 + s_r^2}. \tag{2.35}$$

Example 2.8. An international program has been undertaken to develop a mosquito repellent based on nepetalactone (NPL). A sample has been prepared to share between three laboratories to be analyzed by gas chromatography. The measurements (in mol%) of the first laboratory were 73.8, 71.4, 77.2, and 76.0; those of the second were 69.3, 72.1, 74.3, and 65.5; and those of the third were 74.9, 76.3, 73.7, and 73.0. Calculate the standard deviation of repeatability and reproducibility for the study.

Solution 2.8. The mean and variance for each laboratory are calculated: $\bar{x}_1 = 74.6$, $s_1^2 = 6.5$, $\bar{x}_2 = 70.3$, $s_2^2 = 14.4$, $\bar{x}_3 = 74.6$, $s_3^2 = 2.1$.
The repeatability variance equals the mean of the individual variances of each laboratory: $s_r^2 = \frac{1}{n} \sum_{i=1}^{n} s_{r,i}^2 = 7.7$, $s_r = 2.8$ mol%.
The inter-laboratory variance is derived from the variance of the mean values of each laboratory subtracted by the repeatability variance divided by the number of degrees of freedom:

$$s_L^2 = \frac{1}{n-1} \sum_{i=1}^{n} (\bar{x}_i - \bar{x}_L)^2 - \frac{s_r^2}{n} = 12.3 - \frac{7.7}{3} = 9.7$$

The reproducibility variance is the sum of the repeatability variance and the inter-laboratory variance: $s_R^2 = s_r^2 + s_L^2 = 7.7 + 8.7 = 17.4$, $s_R = 4.2$. Thus the standard deviation of repeatability, s_R, equals 1.6 mol% and that of reproducibility, S_R, is 4.2 mol%.

A more rigorous definition of repeatability has been described in the DIN 1319 standard. According to this definition, data is measured at the full-scale reading of the instrument and then reduced to 50% of full scale and measured again. This procedure is repeated for a total of 10 measurements each. The mean values at 100% FS and 50% FS are calculated:

$$\bar{X}_{50\%} = \frac{1}{n} \sum_{i=1}^{n} X_{i,50\%},$$

$$\bar{X}_{100\%} = \frac{1}{n} \sum_{i=1}^{n} X_{i,100\%}.$$

The relative standard deviations at 50% FS and 100% FS are:

$$S_{rel,50\%} = \frac{1}{\bar{\bar{X}}_{100\%} - \bar{\bar{X}}_{50\%}} \sqrt{\frac{1}{n-1} \sum_{i=1}^{n} (X_{i,50\%} - \bar{X}_{50\%})^2},$$

$$S_{rel,100\%} = \frac{1}{\bar{\bar{X}}_{100\%} - \bar{\bar{X}}_{50\%}} \sqrt{\frac{1}{n-1} \sum_{i=1}^{n} (X_{i,100\%} - \bar{X}_{100\%})^2}.$$

2.5 REPRESENTING DATA GRAPHICALLY

Graphs are an efficient means to identify trends, establish relationships, and evaluate the extent of errors or outliers. They are powerful tools to communicate ideas and concepts. The scales on the abscissa and ordinate are generally selected to cover the whole range of the data—from the origin to the maximum of the experimental data. By expanding the scales to fit the entire data range, the variation in the data becomes more evident.

Scatter plots are the most common type of graph used to show relationships between a dependent variable (responses) and independent variables (factors). Judiciously selecting and sizing the symbols in scatter plots allows one to communicate trends and the measured values' associated error. Frequently, to show the effect of more than one independent variable, different symbol types or colors can be used. Three-dimensional plots, surface plots, and contour plots are becoming more common to illustrate the effect of two or more factors on the dependent variable. Bar charts (and pie charts) are popular for presentations, and histograms are useful to compare the distribution of populations. Ternary plots are used in thermodynamics and to demonstrate explosion limits of different compositions of gases.

Plotting experimental data on ordinary rectangular coordinates is the first step to assess the quality of the data as well as to begin to understand the relationship between dependent variables and factors. Figure 2.12 illustrates the trend of the pressure drop as a function of volumetric flow rate of oxygen through multiple nozzles entering a reactor. It is clear that due to the curvature in the data a simple linear relationship is inapplicable. The shape of the curve is typical of a power law function that passes through the origin: the pressure drop equals zero when there is no flow.

$$Q \neq a\Delta P + b. \tag{2.36}$$

The next step is to plot the data on logarithmic scales for both the abscissa and ordinate as in Figure 2.13 (known as a log-log plot). A standard convention is that experimental data should be represented by symbols while correlations and models should be represented by lines. The plant data shows significant scatter at low volumetric flow rates but from approximately 1 m^3 h^{-1} and

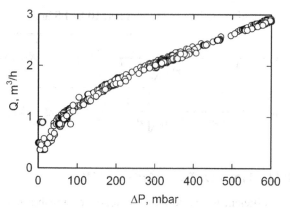

FIGURE 2.12 Linear Plot of Volumetric Flow Rate Versus Pressure Drop

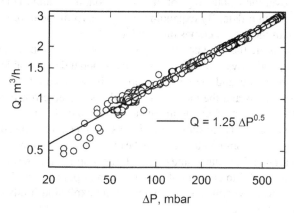

FIGURE 2.13 Log-Log Plot of Nozzle Volumetric Flow Rate Versus Pressure Drop

beyond, the data falls on a straight line. The slope of the line corresponds to the exponent. The concave down shape indicates that the exponent is less than 1 (and that the first derivative is negative) while a concave up shape would indicate an exponent greater than 1 (and a positive first derivative).

$$Q = a\Delta P^n. \tag{2.37}$$

The scatter in the data at the lower end may be a result of perturbations during start-up and shut-down, instrument error, or it may simply mean that the plant is out of the range of standard operation. Later, we will discuss under what conditions experimental measurements may be ignored.

Models should normally be included in a figure whenever possible. Often, lines may be introduced in a figure to demonstrate tendencies more clearly—trend lines—but they do not represent a physical phenomenon. In the case of

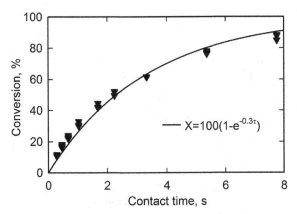

FIGURE 2.14 Butane Conversion Versus Contact Time

the pressure drop, when the exponent n equals 0.5, the volumetric flow rate increases with the square root of the pressure drop. This relationship agrees with orifice theory that relates the volumetric flow rate (Q), to pressure drop (ΔP), density (ρ), cross-sectional area (X_A), and the ratio of orifice diameter to pipe diameter (β):

$$Q = 0.61 \frac{X_A}{\sqrt{1 - \beta^4}} \sqrt{\frac{2\Delta P}{\rho}}. \tag{2.38}$$

Another exponential relationship common to reactor engineering is that between conversion and the ratio of the reactor (or catalyst) volume, V, and the volumetric flow rate, Q. This ratio is referred to as the contact time, τ. Maleic anhydride (MA) is an important chemical intermediate that is formed by the partial oxidation of butane over a vanadium pyrophosphate. Figure 2.14 plots n-butane conversion, X, against contact time, τ—collected in an ideal micro-reactor on ordinary rectangular co-ordinates. Since the shape of the data is concave down, we might assume that the relationship is a power law with a negative exponent. However, the data points do not fall on the trend line drawn in Figure 2.15.

In this case, we may be tempted to assume that the data has experimental error and simply calculate the slope of the line. Changing the scales of the ordinate and abscissa can squeeze the data together, enforcing a tendency to reject the possibility of a different functional relationship. However, in the case of an ideal reactor in which conversion is independent of the concentration of the reactants, the differential equation relating the conversion to the conditions is:

$$Q\frac{dX}{dV} = -k(1 - X), \tag{2.39}$$

where k is the first-order rate constant. The solution to this equation gives an exponential function. The value of the rate constant that best fits the data is

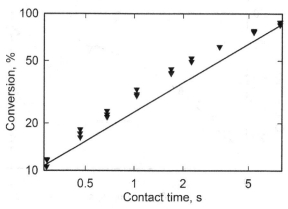

FIGURE 2.15 Log-Log Plot of n-Butane Conversion Versus Contact Time

TABLE 2.4 Graphical Transformations Commonly Used to Linearize Data

$y = a + bx^n$	In this case, the y intercept is first determined (graphically) and then a log-log plot of $y - a$ versus x will give a straight line with slope equal to n
$y = a + \frac{b}{x}$	The dependent variable, y, is simply plotted as a function of the inverse of the dependent variable. Rate constants of chemical reactions follow an Arrhenius-type expression, $k = A\exp\left(-\frac{E_a}{RT}\right)$. A plot of $\ln k$ versus $\frac{1}{T}$ gives a straight line with slope equal to $-\frac{E_a}{R}$
$y = ab^x$	This should be linearized by plotting $\ln y$ versus x

$0.3\ \mathrm{s}^{-1}$. The agreement between the experimental data and this model is good, as demonstrated in Figure 2.14.

$$X = 1 - e^{-k\tau}. \tag{2.40}$$

Other graphical transformations that can be used to linearize data are shown in Table 2.4.

Figure 2.16 shows the trend of selectivity versus conversion for the partial oxidation of n-butane to maleic anhydride in a fluidized bed reactor. Besides the by product acids (acrylic and acetic), CO and CO_2 are the principal products either formed directly from butane (parallel reaction) or via maleic

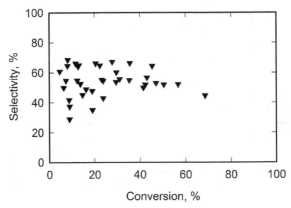

FIGURE 2.16 MA Selectivity Versus n-C_4H_{10} Conversion

anhydride (series reaction):

$$C_4H_{10} + \frac{7}{2}O_2 \rightarrow C_4H_2O_3 + 4H_2O,$$

$$C_4H_{10} + \frac{11}{2}O_2 \rightarrow 2CO_2 + 2CO + 5H_2O,$$

$$C_4H_2O_3 + 2O_2 \rightarrow 2CO_2 + 2CO + H_2O.$$

The range of conversion varies from about 5% to almost 80% and the selectivity is greater than 20% but less than 80%. The scale was chosen to represent the absolute limits of conversion and selectivity possible. They could also have been chosen to span only the range of the data, as shown in Figure 2.13 and Figure 2.14.

In Figure 2.17, the symbols of all collected data at 350 °C are circular and black. The symbols of the 380 °C data are red inverted triangles while the 400 °C data is represented by green squares. When figures are reproduced in black and white, the difference between white (or "empty") and black symbols is clearly obvious but even colors appear as different shades of gray. Using different colors and shapes ensures that the differences between separate sets of data are distinguishable. This graph shows that selectivity is highest at 350 °C and the lowest selectivity is at 410 °C—more information is communicated by adding color and shapes compared to Figure 2.16.

The third graph of the same data, Figure 2.18, is plotted to highlight the effect of changing feed concentrations (mol% or vol%) of butane and oxygen. The experiments were conducted with six combinations from low butane and oxygen (2% and 4%, respectively) to high butane and oxygen concentrations (9% and 10%, respectively). Again, independent symbols and colors were adopted for each condition. All symbols were outlined with a black edge for clarity. The trends in the data set become more obvious by plotting the data

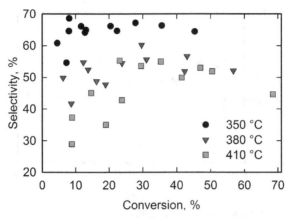

FIGURE 2.17 MA Selectivity Versus n-C$_4$H$_{10}$ Conversion and Temperature

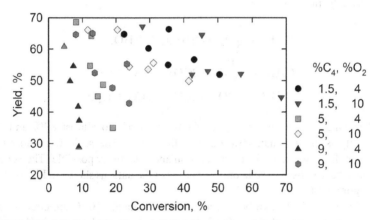

FIGURE 2.18 MA Selectivity Versus n-C$_4$H$_{10}$ Conversion and Composition

in this way: selectivity decreases as conversion increases at all conditions; the drop in selectivity with conversion is much higher at high butane concentration; at constant butane conditions, total conversion is higher with higher oxygen concentration. Finally, the temperature could be represented by color and the concentrations by symbols—three-dimensional phenomena collapsed into a two-dimensional plot.

Identifying cause and effect can also be attempted with graphs by plotting two separate dependent variables against the same independent variable. When the scales of the independent variables are similar, it suffices to use different symbols to distinguish the variables. When the scales are different, an additional axis is added either on the opposite side of the graph or offset from the first axis. Figure 2.19 illustrates the technique: the yearly average

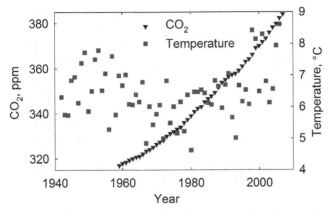

FIGURE 2.19 Global CO_2 Concentration and Local Temperature Versus Time

local temperature in Montreal, Canada (Environnement Canada 2011) is plotted coincidentally with the atmospheric CO_2 concentration (in ppm) as recorded at the Mauna Loa observatory in Hawaii (Keeling and Whorf 1999).

Global warming has attracted significant interest in scientific literature, for both government policy makers and the general public. It is commonly held that greenhouse gases (GHG), particularly CO_2, are a major contributor to a measurable increase in the average world temperature. A plot of average temperature and CO_2 concentration should demonstrate a similar rising trend. In North America, meteorological stations record local temperatures every hour. A yearly average consists of almost 9000 readings. The left ordinate represents the concentration of CO_2, the right ordinate depicts the average temperature while the abscissa is time. Whereas over the 50-yr period starting in 1959 the CO_2 concentration has steadily risen from below 320 ppm to over 380 ppm—a 21% increase—the average temperature varies between 4.6 °C in 1980 and 8.6 °C in 2006. The five highest temperatures were recorded between 1998 and 2006 but between 1959 and 2006, when the CO_2 concentration rose by 40 ppm, the temperature would appear to have been constant. In fact, from the mid-1950s up until the 1980s, the temperature arguably decreased.

2.5.1 Plotting Pitfalls

When plotting data, the extremes of the axes should be chosen so that the independent and dependent variables just fit within the graph. Choosing larger values for the axes' extremes will compress the data and give the appearance of linearity where none might exist. Another common error is to plot a variable that is itself a multiple of the factor being plotted. This will linearize the function and will again lead to a false impression with respect to the phenomenon.

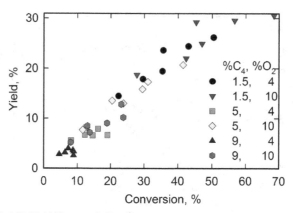

FIGURE 2.20 MA Yield Versus n-C_4H_{10} Conversion

In reaction engineering, yield, Y, is a product of conversion and selectivity. Thus, plotting yield versus conversion will have a tendency to give a straight line; such a plot's usefulness for identifying relationships is limited. However, in the case of multiple variables, value may still be derived from such a plot, as demonstrated in Figure 2.20: the yield increases linearly with conversion, which is expected since yield is a product of conversion. However, the graph also shows that yield is highest at lowest n-butane conversion and at the same n-butane inlet concentration; the conversion is generally higher with a higher oxygen concentration. The same conclusions may be deduced from Figure 2.18 but it is more evident in this plot.

2.5.2 3D and Contour Graphs

Commercial graphics software has made generating multi-dimensional graphs straightforward. They should be used with care and for illustrative purposes rather than as a means of communicating data quantitatively. Often it is difficult to retrieve exact values from a three-dimensional plot versus two-dimensional graphs. An alternative to three-dimensional plots are contour plots that are used in geography to illustrate the change in elevation over a surface. They can be used in engineering to show the spatial distribution of temperature in a reactor, concentration gradients, P-V-T, or even the temporal evolution of a heat exchanger.

As an example, consider Figure 2.21 in which a catalyst deactivated by carbon build-up on its surface is fed continually to a reactor. Air is fed to the reactor from the bottom through a distributor. As it rises through the bed of catalyst, the oxygen reacts with the surface carbon to form CO_2. The effluent gases exit at the top of the reactor through a cyclone whereas the solids exit the bottom through a standpipe. The radial concentrations of oxygen and carbon dioxide (reported as vol%) are plotted in three-dimensions as shown in

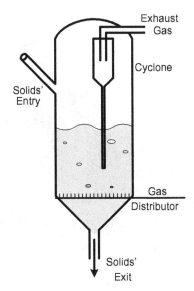

FIGURE 2.21 Fluidized Bed Regenerator

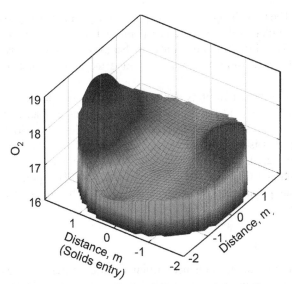

FIGURE 2.22 3D Plot of Oxygen Distribution Across a 4 m Diameter Regenerator (For color version of this figure, please refer the color plate section at the end of the book.)

Figures 2.22 and 2.23, respectively. The oxygen concentration is lowest in the center of the reactor parallel to the solids entry point and is highest on both sides perpendicular to the entry point. Figure 2.23 shows the variation of the

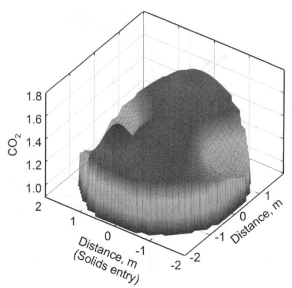

FIGURE 2.23 3D Plot of Carbon Dioxide Distribution Across a 4 m Diameter Regenerator (For color version of this figure, please refer the color plate section at the end of the book.)

vol% CO_2. The trend is opposite of that for the oxygen: the CO_2 concentration is highest parallel to the solids entry point and lowest perpendicular to the entry point. The plots clearly demonstrate that the catalyst is poorly distributed in the reactor and flows preferentially along the axis parallel to the solids entry.

The data in these two plots could have also been represented as contour plots. The colors would represent the concentrations and an additional key is required to represent the concentration as a function of color. An example of a contour plot is illustrated in Figure 2.24 for the STY (space-time-yield) data of Figure 2.20. The axes of the contour plot are temperature and butane concentration and the color key represents the STY. This plot clearly shows that maximum productivity is achieved at the highest butane concentration and temperature.

2.5.3 Bar Charts

Communicating data to audiences often requires different types of charts versus laboratory reports, journal articles, or theses. Bar charts, pie charts, and area charts are more qualitative and emphasize trends, differences, and relative quantities. They are particularly useful for representing market data— the market share of a company for a certain product, for comparing product distribution at different times, or the sales of a company by class of product— as shown in Figure 2.25: Arkema's three most important businesses represent 73% of sales.

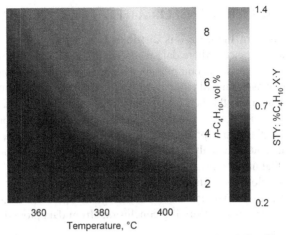

FIGURE 2.24 Contour Plot of STY Versus Temperature and vol% n-C_4H_{10} (For color version of this figure, please refer the color plate section at the end of the book.)

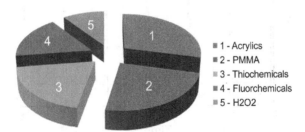

■ 1 - Acrylics
■ 2 - PMMA
■ 3 - Thiochemicals
■ 4 - Fluorchemicals
■ 5 - H2O2

FIGURE 2.25 Business Portfolio of Arkema (2010)

In education, the most common bar chart is a histogram. These charts can be used to group examination results into intervals also known as bins whose height corresponds to the number of people in the interval. It is also the standard to represent particle size distribution. The domain is divided into specified intervals and either the mass fraction or the number of particles is assigned to each interval. Figure 2.1 shows a histogram of the average daily temperature in Montreal in the month of May during a 64-yr period and includes 1950 data points. The chart shows that the temperature ranges from a high of 26 °C (occurred twice) to a low of 2 °C (again this happened only twice). The mode temperature (highest frequency) was 13 °C and it was recorded 180 times. During the 64-yr period, the average temperature was greater than 20 °C only 87 times—just over 6% of the time.

2.6 FAST FOURIER TRANSFORM (FFT)

Whereas the data in Figure 2.1 is distributed essentially evenly around the mean, there are instances where data repeats at certain intervals or at a certain frequency. For reciprocating equipment—compressors, pumps, blowers, etc.— the pressure downstream will typically vary with the period of rotation. Another example is heart rate: the heart beats at a regular frequency that may vary depending on the time of day—morning or evening—or the exertion at a given moment. When diagnosing problems in equipment or in a patient, it is useful to record the data and then analyze the characteristic frequencies to help identify any problem that might exist.

Figure 2.26 plots the voltage as measured by a cardiogram versus time for a healthy patient—the voltage increases at regular intervals. At a frequency of 1 Hz, the curve repeats. Deriving conclusions from data plotted in the time

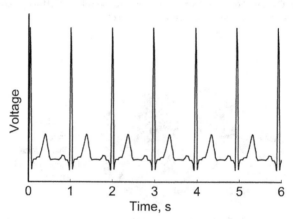

FIGURE 2.26 Cardiogram

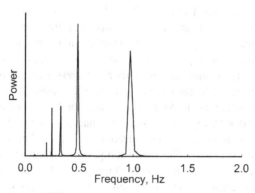

FIGURE 2.27 Cardiogram FFT

domain is difficult. Rather, the data may be compared in the frequency domain to identify anomalies. Figure 2.27 is a graph of the power, as calculated by FFT, versus the frequency. The peaks represent the dominant frequencies. By measuring the heart rate at different parts of the body and comparing their signals, cardiac diseases may be detected. Moreover, if the FFT has multiple peaks (beyond the standard for a healthy individual), that would also be indicative of heart disease.

2.7 EXERCISES

2.1 Calculate the numerical value of each operation, taking into account the number of significant figures.

(a) $(5.2 \times 10^{-4} \cdot 1.635 \times 10^6)/2.67$.

(b) $3.57 \cdot 4.286$.

(c) $1530 - 2.56$.

(d) $0.036 + 0.22$.

2.2 The mass of a single application of mosquito repellent is determined by spraying a tissue and subtracting the tare weight of the tissue from the total weight measured in a precision balance. The tissue is weighed immediately after spraying to minimize mass loss due to evaporation. The mass of a single pump action of a mosquito repellent was measured 10 times (in g): 0.1155, 0.1192, 0.1106, 0.1137, 0.1075, 0.1158, 0.1076, 0.0982, 0.1028, and 0.108.

(a) Calculate the average, variance, and standard deviation.

(b) If the true value of the mass of a single application (spraying once) is 0.1140 g, calculate the absolute and relative error of each measurement.

(c) What is the sensitivity of the instrument?

(d) Calculate the standard deviation of the error.

(e) Comment on the reproducibility and the accuracy of the measurements.

2.3 According to the Chauvenet criterion, could we reject one of the following experimental points: 32, 34, 25, 33, 37, and 10? Calculate μ, σ, and α. Calculate the value of P.

2.4 To increase the precision of the measurement of the mass of a single application of mosquito repellent, the spray is directed toward an aluminum boat, which is weighed after the 10th consecutive spray. The results after 10 runs (in g) are: 1.1516, 1.1385, 1.141, 1.1391, 1.1385, 1.148, 1.1271, 1.1354, 1.1439, and 1.1153.

(a) Calculate the average, variance, and standard deviation.

(b) Comment on this technique versus spraying a single time and measuring the weight.

(c) What are the sources of error of this technique versus spraying a tissue?

2.5 The cathode of new car batteries is manufactured with carbon-coated lithium iron phosphate (C-LiFePO$_4$). The target lifetime of the battery is 10 yr. An accelerated test was developed to assess manufacturing steps in which one month is equivalent to 1 yr. If the standard deviation of the catalyst life in the accelerated test is 1.4 month, what is the probability that the battery will last longer than 8 yr?

2.6 During the last several decades, global warming has become an important topic in the news media, the UN, and at international conferences. Indicators of climate change—global warming—such as rising sea levels, melting glaciers, increasing ocean and air temperatures, and increasing concentration of CO_2 in the atmosphere have been identified. The increase in the level of CO_2 in the atmosphere is a result of emissions from power generation, automobile exhausts, and the chemical industry as well as natural processes—volcanoes, decomposition of organic matter and the like. Greenhouse gases may absorb infrared radiation emitted from the Earth's surface (GreenFacts, 2010). A brainstorming session was organized and 18 environmentalist groups agreed to participate. Their task was to conceive independent experimental techniques to measure climate change. The number of techniques proposed by each group is the following: 8, 6, 9, 11, 11, 11, 9, 11, 9, 11, 10, 9, 8, 21, 15, 8, 9, and 16.

(a) Calculate the average number of techniques and the standard deviation.

(b) According to criterion of Chauvenet, can we reject the number of ideas proposed by one of the groups?

(c) What are the chances that there are less than five methods to measure the existence of climate change?

2.7 During the course of an experiment, five independent readings of a thermocouple were: 0.96 mV, 1.04 mV, 1.02 mV, 1.01 mV, and 0.97 mV. Calculate the confidence interval at 95% confidence level.

2.8 The maximum recorded temperature in °C on September 5th of each year from 1984 to 2009 in Montreal, Canada, is as follows: 17.5, 22.4, 20.4, 25.5, 18.6, 22.1, 21.7, 24.3, 23.3, 23.7, 19.5, 25.7, 29.8, 22.6, 22.7, 32.0, 16.7, 21.0, 22.8, 21.6, 22.7, 22.7, 21.2, 19.1, 33.3, and 21.5. *V. Guillemette*

(a) What was the average temperature during this time interval?

(b) What is the standard deviation?

(c) What is the likelihood that the temperature drops below 20 °C?

(d) According to the criterion of Chauvenet, can we reject some of the values for the temperature?

(e) Is there a statistical trend in the data (any evidence of climate change)?

2.9 A pharmaceutical company requires a ventilation system to purify the air of a biohazard containment laboratory. HEPA, ULPA, and SULPA technologies were considered but the ULPA filters were deemed sufficient for the particle size of the likely contaminants. The life span (in days) of the air filter given by the different suppliers is: 80, 110, 100, 150, 170, 60, 90, 90, 80, 110, 90, 110, 80, 90, 160, 110, 90, 210, 160, 120, 110, and 110. *A. R. Nassani*

(a) What are the average life, variance, and standard deviation of the filters?

(b) Based on the criterion of Chauvenet, are all of the life spans claimed by the suppliers credible?

2.10 Differential pressure in vessels and pipes containing gases is often measured with U-tube manometers. The difference in pressure from one point to another is proportional to the difference in height between the vertical legs of the manometer:

$$\Delta P = (\rho_{mano} - \rho_{fluid})g\Delta H,$$

where ρ_{mano} is the density of the fluid in the gauge and ρ_{fluid} is the density of the fluid in the vessel or pipe. Often, $\rho_{mano} \gg \rho_{fluid}$ and thus ρ_{fluid} may be neglected. Six measurements were recorded for air near STP with a U-tube: 0.154, 0.146, 0.149, 0.161, 0.152, and 0.144. The manometer fluid has a density of 830 kg m^{-3}. *N. Fadlallah*

(a) Calculate average pressure drop, variance, and standard deviation.

(b) Determine the confidence interval at a 99% confidence level.

(c) What is the uncertainty in the calculated pressure differential?

2.11 During a trek in the mountains, you feel lightheaded and presume it is due to lack of oxygen. Based on an altimeter reading, you are at an elevation of (2750 ± 50) m above sea level. Calculate the partial pressure of oxygen and the uncertainty. The relationship between pressure and elevation (Z in m) is given by:

$$P = P_o(1 - 2.255 \times 10^{-5}Z)^{5.255}.$$

2.12 Your laboratory is responsible for the development of a submersible to detect oil leaks at a maximum pressure of 250 MPa. You estimate the density of sea water as (1030 ± 15) kg m^{-3} and that your depth gauge is accurate to $\pm 1.00\%$ at a full-scale reading of 300 MPa.

(a) What is the maximum allowable depth at a 95% confidence level that you recommend for standard operation?

(b) What is the maximum allowable depth at a 99% confidence level?

2.13 Solvent tanks must be purged prior to maintenance operations to ensure that the level of oxygen is sufficient and that the vapor concentration is below the flash point and the limit authorized by OHSA (Occupational Safety and Health Association)—(500 ± 5) ppm. The volume of each tank is (8.0 ± 0.3) m^3, all liquid was drained beforehand, and the vapor is initially saturated. *G. Alcantara*

Determine the time and uncertainty needed to purge each tank to reach the maximum permitted by OSHA on the basis that the purge rate is (2.00 ± 0.05) m^3 m^{-1} at a pressure of 0.1 MPa and a temperature of 22 °C. Use the following expression:

$$y_{i,\text{final}} = y_{i,\text{initial}} \exp\left(-\frac{V_{\text{purge}}}{V_{\text{reservoir}}}t\right).$$

2.14 Five equations to correlate the molar heat capacity $(C_p$—J mol^{-1} K$^{-1})$ as a function of temperature are given in Section 2.2.

(a) Assuming that one of the equations is correct, calculate the error between this value and values predicted by the other correlations at 0 °C, 100 °C, 1000 °C, and 2000 °C.

(b) The fitted constants in the NIST equations carry seven significant figures. What is the error at 0 °C, 100 °C, 1000 °C, and 2000 °C with only six, five, and four significant figures?

2.15 A student must carry out a chemical engineering laboratory test featuring the following reaction: A + B → C. For this reaction to take place, he must boil the reagents. He collects 20 ml of A and B with a volumetric pipette of 20 ml (its uncertainty is 0.03 ml). After bringing the mixture to a boil, the student collects the product in a graduated cylinder. The reaction is complete (no reactants are left over). The experimental volume of C is 48.5 ml (the uncertainty of the specimen is 0.5 ml). The density of C is 0.825 kg l^{-1}, its molecular mass is 28.0 g mol^{-1}, and the theoretical molar yield is 1.50 mol. *S. Deacken*

(a) Calculate the theoretical volume of C.

(b) Calculate the relative error of the volume.

(c) Calculate the relative uncertainty of the experimental volume.

(d) Describe what kinds of errors could have been committed.

2.16 Calculate the molecular mass of air, assuming a relative humidity of 0%. The molar composition equals: 21.0% O_2, 78.1% N_2, and 0.9% Ar.

2.17 Nube wishes to purchase a laptop. His single criterion is the lifetime of its battery. He compares the battery lifetimes of 15 of his colleagues possessing a laptop and obtains the following results (in min): 182, 130, 167, 198, 145, 156, 165, 181, 176, 120, 90, 123, 179, 201, and 137. *M. Brière-Provencher*

    (a) What are the average, the standard deviation, and the variance of the batteries' lifetimes?

    (b) What is the probability that Nube's computer will operate for over 170 min if the samples respect a normal distribution?

    (c) Determine if data can be eliminated by applying Chauvenet's criterion.

2.18 As of 2005, the maximum permissible concentration of sulfur in gasoline is 30 ppm. Table Q2.18 summarizes the measured sulfur content for three grades of fuel in 2001—before the new regulations—and in 2005. *É. Michaud*

    (a) The units of parts per million (ppm) are 10^{-6} g m^{-3}. Compare the concentration of sulfur in gasoline permitted by law versus the measured values in 2001 and 2005. Express the differences in absolute and relative concentrations. Comment.

    (b) What is the 95% confidence interval for both ordinary and super fuels in 2001 and 2005?

TABLE Q2.18 Sulfur in Liquid Fuels (wt % Sulfur)

	2001		
Ordinary	0.028	0.038	0.029	0.025
Super	0.020	0.019	0.011	0.011
Aviation	<0.001	<0.001	<0.002	<0.001
	2005		
Ordinary	0.0018	0.0021	0.0026	0.0022
Super	0.0001	0.0002	0.0001	0.0002
Aviation	<0.0001	<0.0001	<0.0001	<0.0001

2.19 At high shear rates, the viscosity of a polymer melt behaves like a power law fluid characterized by the Carreau equation:

$$\eta = \eta_o(1 + \lambda^2\dot{\gamma})^{(\beta-1)/2},$$

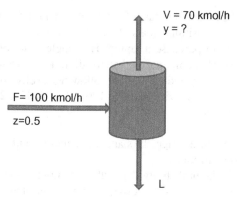

FIGURE Q2.20 Flash Separation

where η_o is the consistency index (25 ± 1) mPa s, λ is the characteristic time of fluid 4 $s \pm 10\%, \beta$ is the rheo-plasticizer index (0.72 ± 0.05), and $\dot{\gamma}$ is the shear rate (s^{-1}).

Calculate the polymer melt's viscosity and uncertainty at a shear rate of $0.112\ s^{-1}$. *E. Nguyen*

2.20 A mixture of benzene and toluene $(50 \pm 1\%)$ is fed to an evaporator at a rate of 100 kmol h^{-1}, as shown in Figure Q2.20. The vapor effluent stream is 70 kmol h^{-1}. The uncertainty of each flow rate is 1%. The vapor-liquid equilibrium constant equals 1.78. Determine the fraction of benzene in the vapor phase and the uncertainty. The fraction of benzene in the vapour phase maybe calculated from the formula: *L.-D. Lafond*

$$y = \frac{Fz}{V + L/K}.$$

2.21 A student who struggles to arrive on time for his first class in the morning decides to make a statistical study of the time it takes to travel to school on rollerblades. After 10 consecutive days of measurement, his results were: $58'39''$, $56'43''$, $59'32''$, $52'36''$, $54'42''$, $56'37''$, $57'25''$, $58'36''$, $54'39''$, and $52'27''$. How long before his class starts should he leave the house to arrive on time 90% of the time? *J.-S. Boisvert*

2.22 HadCRUT3[2] is a record of global temperatures that spans 160 yr and is compiled from both land measurements (Climate Research Unit— CRU—of the University of Anglia) and sea surface measurements (Hadley Center). Figure Q2.22 shows the temperature data together with the atmospheric CO_2 measurements collected at Mauna Loa.

---

[2] http://www.cru.uea.ac.uk/cru/data/temperature/hadcrut3gl.txt

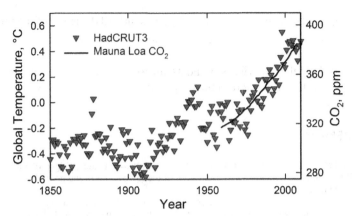

FIGURE Q2.22 HadCRUT3 Average Global Temperature

(a) Does the change in CO_2 concentration correlate with the rise in global temperature? Discuss.

(b) How is the temperature data different from that reported in Figure 2.19?

(c) What are the likely sources of error of the HadCRUT3 temperature data?

(d) The following equation characterizes the rise in CO_2 concentration in the atmosphere as reported by the Mauna Loa observatory:

$$\text{ppm}_{CO_2} = 316 + 0.38(t - 1958)^{4/3},$$

where t is the calendar time (yr). What was the rate at which CO_2 accumulated in the atmosphere in 1965 and 2005? What will the rate be in 2025? Estimate the concentration and temperature in 2020, 2050, and 2100.

(e) The mass of the atmosphere is 5×10^{18} kg and in 2011, 84 million barrels of oil were consumed per day. Assuming that the oil consumption represents 50% of the total carbon emissions, what should be the theoretical yearly rise in the CO_2 concentration? How does that compare with the value reported by Mauna Loa?

2.23 The following data was recorded to determine the viscosity of a fluid using a falling ball viscometer: 20.1, 19.5, 22.1, 18.7, 12.0, 21.1, 19.7, 23.4, 23.0, and 23.2. The equation used to determine viscosity is the following:

$$\mu = \frac{(\rho - \rho_o)Vgt}{6\pi rL},$$

where ρ is 7850 kg m^{-3}, ρ_o is (850 ± 4) kg m^{-3}, r is 1.5 cm, and L is 0.4 m. *C. Mayotte*

(a) Calculate the mean, variance, and standard deviation of the time.

(b) Can we reject a value of the time data according to Chauvenet's criterion?

(c) Calculate the viscosity and its uncertainty with a confidence interval of 95% without rejecting any values.

REFERENCES

ASTM D3776-09, 2007. Standard Test Methods for Mass Per Unit Area (Weight) of Fabric.

BIPM JCGM 200:2008, International vocabulary of metrology—basic and general concepts and associated terms (VIM).

BS EN 12127, 1998. Textiles. Fabrics. Determination of mass per unit area using small samples.

Bureau International des Poids et Mesures, 2006. The International System of Units (SI), eighth ed. (<http://www.bipm.org/utils/common/pdf/si_brochure_8_en.pdf>).

Chaouki, J., Klvana, D., Pepin, M.-F., Poliquin, P.O., 2007. Introduction à l'analyse des données expérimentales, fifth ed. Dept. Chemical Engineering, Ecole Polytechnique de Montreal.

Environnement Canada. (n.d.). Mean daily temperature of the month of May in Montreal over the last 60 years. Retrieved 2011, from Environnement Canada: <http://www.ec.gc.ca/default.asp?lang=fr>.

Feinberg, M., 1995. Basics of interlaboratory studies: the trends in the new ISO 5724 standard edition. Trends in Analytical Chemistry 14 (9), 450–457.

Himmelblau, D.M., 1962. Basic Principles and Calculations in Chemical Engineering, Prentice Hall.

Holman, J.P., 2001. Experimental Methods for Engineers, seventh ed. McGraw-Hill, Inc.

Howell, J.R., Buckius, R.O., 1992. Fundamentals of Engineering Thermodynamics, second ed. McGraw Hill.

ISO 1995, Guide to the Expression of Uncertainty in Measurement, International Organisation for Standardization, Geneva.

ISO 3801 1977. Textiles—woven fabrics—determination of mass per unit length and mass per unit area.

ISO 5725-2 1994. Accuracy (trueness and precision) of measurement methods and results— Part 2. Basic method for the determination of repeatability and reproducibility of a standard measurement method.

Jardine, L., 2011. The empire of man over things. Retrieved 2011, from Britain and the Rise of Science: <http://www.bbc.co.uk/history/british/empire_seapower/jardineih_01.shtml>.

Keeling, C.D., Whorf, T.P. 1999. Atmospheric CO_2 Concentrations (ppmv) Derived In Situ Air Samples Collected at Mauna Loa Observatory, Hawaii. Carbon Dioxide Information Analysis Center. <http://cdiac.esd.ornl.gov/ftp/ndp001/maunaloa>.co2.

Kelley, K.K., 1960. Contributions to the Data on Theoretical Metallurgy. XIII. High Temperature Heat-Content, Heat-Capacity, and Entropy Data for the Elements and Inorganic Compounds. US Dept. of the Interior, Bureau of Mines, Bulletin 584.

Kyle, B.J., 1984. Chemical and Process Thermodynamics, Prentice-Hall.

Patience, G.S., Hamdine, M., Senécal, K., Detuncq, B., 2011. Méthodes expérimentales et instrumentation en génie chimique, third ed. Dept. Chemical Engineering, Ecole Polytechnique de Montreal.

Reid R.C., Prausnitz, J.M., Sherwood, T.K., 1977. The Properties of Gases and Liquids, McGraw-Hill, p. 629.

Spencer, H.M., 1948. Empirical heat capacity equations of gases and graphite. Industrial and Engineering Chemistry 40, 2152–2154.

Taraldsen, G., 2006. Instrument resolution and measurement accuracy. Metrologia 43, 539–544.

Unnamed. (2008, December 14). *Quotations*. Retrieved 2011, from Lord Kelvin: <http://zapatopi.net/kelvin/quotes/>.

Van Wylen, G.J., Sonntag, R.E., 1978. Fundamentals of Classical Thermodynamics SI Version, second ed. Wiley.

Chapter 3

Experimental Planning

3.1 OVERVIEW

Tools and techniques to plan, execute, analyze, and apply results of experimental or developmental programs range from the most basic forms such as trial-and-error to detailed statistical strategies that begin with screening designs to identify important factors (variables) and then proceed to surface response models for optimization. Trial-and-error has been, and remains, one of the primary methods with which we learn, solve problems, and design experiments (Thomke et al., 1998). According to E. Thorndike, trial-and-error is the most basic form of learning. In some cases, it is as efficient or applicable as any other detailed statistical strategy. Finding the combination of a lock is an example where previous trials provide no insight in to subsequent trials (accept to not repeat). This is referred to as a flat topological information landscape; it is an extreme example. Most process topological landscapes include hills and valleys—these structural morphologies are used as a basis to choose future experiments: in the case of a chemical process, an example of a hill would be an increase in product yield with an increase in operating temperature; a valley would represent a decrease in product yield with increasing temperature. Trial-and-error combined with intuition (experience) is effective when beginning to understand a problem but in most cases, this strategy is inefficient versus an experimental design in which factors are modified in defined increments at specific levels.

Planning experiments statistically—design of experiments (DOE)—is not only driven by economic issues but also by the need to derive the correct (unique) solution. Running experiments is time consuming and/or expensive and thus it becomes imperative to minimize their number while maximizing the information generated. Various experimental methodologies have been

Experimental Methods and Instrumentation for Chemical Engineers. http://dx.doi.org/10.1016/B978-0-444-53804-8.00003-4

developed throughout history. Perhaps the first mention of a controlled experiment is that of J. Lind (1747), who identified the cause of scurvy (Dunn, 1997). He selected 12 "similar" individuals suffering from the disease and divided them into six pairs. He supplemented each pair's basic daily diet with different remedies previously proposed to treat scurvy:

1. Cider;
2. 25 drops of sulfuric acid (elixir vitriol) thrice daily on an empty stomach;
3. 125 ml of sea water;
4. A mixture of garlic, mustard, and horse radish;
5. 20 ml of vinegar three times a day;
6. Two oranges and one lemon.

Although all of the men improved somewhat, after 6 days one of the individuals treated with the sixth remedy became well enough to take care of the others.

Fisher was a pioneer in developing and using statistics to design and interpret experiments. He coined the term variance (Fisher, 1918) and wrote what became the standard reference entitled "Statistical Methods for Research Workers." He advocated the use of statistics to design experiments systematically.

3.2 DATA AND EXPERIMENTS

The choice between following a structured approach to collect and interpret data versus "intuition plus trial-and-error" depends on the type of data required, the availability of time, budget as well as the experience of the individuals conducting the tests. The pharmaceutical industry has well documented its product cycle together with the associated costs of developing new drugs. It involves three phases: discovery and optimization in which pre-clinical trials are conducted on promising new compounds; next, during clinical development the safety and efficacy of the drugs are established; finally, regulatory agencies review the clinical trials and grant or reject commercialization. In the chemical industry, similar approaches are adopted. The different categories of data and experiments include monitoring, qualifications, prove-outs, scouting, process development, and troubleshooting.

3.2.1 Monitoring

Environmental conditions—temperature, barometric pressure, accumulated precipitation, wind speed, direction, etc.—have been monitored and recorded for hundreds of years. The oldest continuous recording of monthly temperature is the Central England Temperature record (CET) that began in 1659 (Legg, 2012). This data is used to relate changes in climate to factors that include

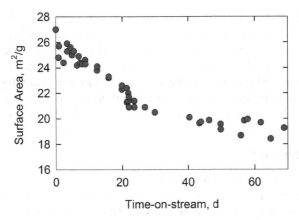

FIGURE 3.1 Monitoring of Catalyst Surface Area in a Pilot Plant

the increase in greenhouse gas (GHG) concentrations, sunspot activity, and other phenomena. Ice cores from the Antarctica ice sheet have been drilled to depths exceeding 3000 m to produce a historical record extending 890 000 yr (EPICA) (White, 2004). Temperature variations together with atmospheric concentrations of GHGs are derived from the core.

In modern chemical plants, thousands of measurements are recorded at frequencies that can exceed 1 Hz. Aside from plant operating conditions including pressures, temperatures, flow rates, and stream composition, other recorded variables include product purity, contamination levels (air, water, soil), and even safety compliance. All this information is stored in enormous databases. This historical record may be interrogated to monitor process performance, and control, for troubleshooting, to demonstrate environmental compliance and modeling. Often smoothing techniques are required to help identify trends in the data that may be masked by low signal-to-noise ratios.

Figure 3.1 is an example of the variation of the surface area of a catalyst withdrawn from a pilot plant over a 60-day period. The catalyst area started at 26 m^2 g^{-1} but declined steadily to 20 m^2 g^{-1} and was thereafter steady or at least declined less rapidly. Catalytic performance was independently shown to vary linearly with surface area and thus the question that needs to be answered is why does the surface area decline and what tests are required to identify the drop (troubleshooting).

3.2.2 Qualification

As in the pharmaceutical industry, when a new product, process, or ingredient is introduced, extensive testing is required to demonstrate the potential consequences. Testing costs may be elevated; for example, testing costs in the aerospace industry represent as much as 40% of the total cost of an aircraft. Qualifying leather for automotive interiors requires a battery of tests that include tensile strength and elongation, thickness, abrasion, resistance to water penetration, flame resistance, UV resistance, delamination, heat aging, and environmental cycling, to name a few. A key factor in qualifying materials is to determine testing frequency, which will affect cost directly. Statistical designs may be used in the selection of frequency and samples to test.

3.2.3 Prove-Out

While qualification tests may be conducted by a single customer or by the organization proposing the modification in question, prove-outs involve multiple customers simultaneously. It is equivalent to a limited market introduction. During this time, the product is tested not only for its operating performance by the customer but also for end-use applications. A significant amount of data is generated as a result of these trials and statistical tests are necessary to identify the adequacy of the product in the market.

3.2.4 Scouting/Process Development

Prove-outs and qualification testing are generally done as a result of new products developed from scouting tests. The pharmaceutical industry is continually seeking new compounds with which to treat diseases ranging from Amyotrophic Lateral Sclerosis (ALS) to tuberculosis. Besides developing processes for drugs, the chemical industry has been developing new products for consumers as well as adapting processes with organic renewable resources. Over many decades, the trend has been to replace saturated hydrocarbons with unsaturated hydrocarbons: ethane will replace ethylene that has largely displaced acetylene as a raw material. Economics is the main drive for this substitution and catalysis is one agent that has advanced this transition.

Scouting tests are conducted at the bench scale, in which production rates are on the order of mg h^{-1} to g h^{-1}. Depending on the complexity of the technology and market size, several scales of pilot facilities may be tested. Often scale factors of as much as 200:1 are adopted between scales. At each scale, costs increase while the uncertainty with respect to feasibility decreases. The object of piloting is to identify the factors that influence operability including productivity, selectivity, by-product formation, purity, energy consumption, and safety. As a result of the pilot study, the ideal scenario would be to develop a phenomenological model that perfectly characterizes all aspects of the process. However, the general philosophy is to generate a sufficient amount of data to understand and minimize the risk of commercialization. Risks may be analyzed at several levels—from a purely economic point of view (market penetration) to one that involves assessing the consequences of an earthquake or other event potentially causing damage to equipment and as a consequence to personnel, the public, and the environment. In the case of hydrocarbon partial oxidation (ethane to acetic acid, for example), the highest level of risk is that a flammable mixture forms and causes a detonation. A lower level of risk is that the hydrocarbon combusts to form carbon oxides. In the former, the personnel, environment, and public are at risk. In the latter, the risk is a lower product yield and thus lower revenues. Each of these possible scenarios should be analyzed with dedicated experimental programs.

Although structured experimental designs are preferred for scouting programs and process development, both trial-and-error and accidents have played a role in product development: the highly successful polymer Teflon® was discovered due to a laboratory accident.

3.2.5 Troubleshooting

Trial-and-error is the most common technique to troubleshoot the mundane problems we face every day from electronic equipment malfunctioning to perceived glitches in computer software. In the process industry, troubleshooting is the most stressful activity—when the plant is down or producing off-spec material, the pressure to bring it back on stream is enormous, as is the economic penalty for lost production. Experience drives the initial testing, which is based on trial-and-error or the process of elimination. The first tests are those that are easiest to conduct and have the least likelihood of causing further harm. When obvious solutions are rejected or the root cause is ambiguous even to the experienced engineer, the total process must be analyzed to identify possible cause-effect scenarios. Every instrument should be independently verified. Afterwards, the internal integrity of each piece of equipment should be examined. This task can be facilitated by using radioactive sources to scan columns. Otherwise, the process needs to be cooled down and inspected manually. Rarely are statistical methods applicable to troubleshooting plant operations; creativity is required to consider possible unobvious scenarios.

3.3 DATA ANALYSIS

As discussed in Chapter 2, after collecting data, it should be plotted to assess if there are any trends or relationships between response variables and factors that should be identified. Uncertainties should be calculated to verify that trends are due to physical phenomena and not random processes. Theories and relationships can then be derived between the responses—independent variables—and the factors—dependent variables—as shown in Figure 3.2.

Chauvenet's criterion should be applied to data significantly different than the mean but it should be used infrequently. Rejecting data should be limited to cases where evidence exists that it is incorrect: a malfunctioning instrument or one that was miscalibrated or suffered physical damage, etc.

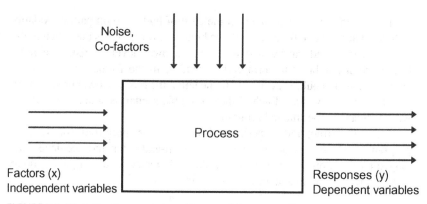

FIGURE 3.2 Black Box Representation of Inputs and Outputs to a Process

In process development, factors are purposely changed for a variety of reasons: reactor temperature may be increased to improve conversion; costly raw materials may be substituted by less expensive raw materials to reduce cost; mixing intensity of blenders may be changed to reduce cycle time. Under these circumstances, the change in the factor is obvious and the question is whether or not the desired effect is substantial—is the change in the response variable statistically different?

In this section, we first define hypothesis testing—a statement of whether or not a change in a factor resulted in a change in a response variable. This methodology is applied to many statistical tests: the Student's t-test compares the mean of two populations; analysis of variance (ANOVA) considers multiple populations; the Fisher test evaluates the difference between the variance of two populations; finally, the χ^2 (chi-square) test is used to determine whether or not the distributions of two populations are dissimilar. Whereas statistical tests are adapted for identifying differences in continuous data sets and populations, regression analysis is commonly used to analyse factors and response variables that vary continuously—the relationship between the pressure and volume of an ideal gas, between heat capacity and temperature, or between catalyst activity and time. As engineers, we seek to understand processes and thus attempt to develop mathematical and even physical models to characterize them. In this way, we are better able to identify factors that are most likely to optimize the response variables of interest. The level of complexity of the models should be sufficient to faithfully represent the relationship between the factors and response variables over the range of interest. Often establishing the relationships between factors and response variables is complicated due to interactions between the factors. For example, the productivity of a reactor depends on the composition of the reactants, temperature, flow rate, pressure, volume, residence time, etc. Increasing flow rate might increase pressure drop as well as modify the temperature profile—all three affect productivity. Simple models often help guide decisions but most often these models are inadequate to fully capture all trends in the data and thus it is necessary to use nonlinear regression analysis. Finally, in this section we introduce the concept of smoothing. When the signal-to-noise ratio is high, smoothing the data may be necessary to at least recognize trends.

3.3.1 Hypothesis Testing

A hypothesis test is an assertion (or assumption) concerning two populations that may or may not be true. The null hypothesis is a statement that asserts a relationship (together with a confidence interval) while the alternative hypothesis asserts the opposite relationship. As an example, compare the average heights between adults in North America and Europe. A statement of a null hypothesis would assert that the average height of adults in the two continents is the same. The alternative hypothesis is that the average heights are different. Another example of a null hypothesis is that the mean world temperature in 1950 is the same as the mean temperature in 2010. It is written as:

$$H_0 : \mu_1 = \mu_2, \tag{3.1}$$

where H_0 is the null hypothesis, μ_1 is the mean temperature in 1950, and μ_2 is the mean temperature in 2010.

The alternative hypothesis is that the mean temperatures are different and this is written as follows:

$$H_1 : \mu_1 \neq \mu_2. \tag{3.2}$$

A hypothesis statement should be accompanied by a confidence interval. A null hypothesis could be that the mean consumption of petroleum has remained unchanged over the last 5 yr at a 95% confidence level.

Rejecting the null hypothesis is equivalent to accepting the alternative hypothesis—that the two populations are unequal. Accepting the null hypothesis is equivalent to accepting the statement; accepting that the mean world temperature was not the same in 1950 and in 2010.

Note that "accepting" or "rejecting" a hypothesis does not necessarily imply truth, which leads to a discussion around two types of errors in statistics: a Type I error accepts the null hypothesis when in fact it is false; a Type II error is when the null hypothesis is rejected when in fact it is true.

British common law (the presumption of innocence) clearly demonstrates the consequences of Type I and Type II errors: "*ei incumbit probatio qui dicit, non qui negat*" is Latin meaning that the burden of proof rests with the person asserting and not the person denying. The null hypothesis in law can be stated as "innocent until proven guilty." Accepting the null hypothesis is equivalent to finding the accused person innocent. A Type I error is to accept the null hypothesis when it is false: the accused is in fact guilty but found innocent.

A Type II error is to reject the null hypothesis when it is in fact true: the accused is found guilty when the person is in fact innocent.

3.3.2 Statistical Tests

Ingredient substitution is an effective means to reduce the cost of manufacture. Suppliers compete by offering improved services, incentives, or cheaper and/or alternative ingredients. Substituting ingredients may reduce raw material costs or they may allow a process to operate at a higher rate. In the latter case, higher material costs are offset by increased productivity or performance for which the manufacturer may also command a higher price for the end product. Generally, in most manufacturing organizations, there is a tremendous resistance to change—their mandate is to produce and change implies uncertainty, which puts production at risk. Therefore, strict protocols are adopted to minimize risks and they often involve a non-objection from management as well as a detailed plan to execute the test schedule including a closure report containing statistical analysis demonstrating the benefit (or not).

In addition to ingredient substitutions, plant trials to improve productivity could include a change in catalyst formulation and modifying operating conditions or even vessel configuration. Aside from purposeful changes, monitoring plant operations is an important undertaking—has the product changed over the course of a specified time, are the physical properties the same, have the impurities increased (or decreased), and to what level?

These questions may be addressed with a Student's t-test. The independent one-sample t-test compares the sample population mean, $\bar{x}$, to a specified value, μ_0. An example of this would be to compare the concentration of dust in the air versus a target mandated by governmental environmental regulations in a spray drying facility to produce powders. A dependent t-test for paired samples compares data before and after a controlled test: modifying the concentration of carbon nanotubes on the mechanical properties of polymer composites is an example of a dependent t-test. A Student's t-test for unpaired samples compares data that was collected independently of each other. This test could be used to compare the performance of competing products or to compare the products made at two plant sites. The difference between the sample means of the two plants is expressed as $\bar{x}_1 - \bar{x}_2$. This difference is compared to a standard that is considered to be significant and is expressed as $\mu_1 - \mu_2$. A confidence interval of 95% is often chosen for comparing samples. The t-statistic, $t(\alpha, df)$, for unpaired samples that are normally distributed is:

$$t(\alpha, df) = \frac{(\bar{x}_1 - \bar{x}_2) - (\mu_1 - \mu_2)}{\sqrt{s_1^2/n_1 + s_2^2/n_2}}, \qquad (3.3)$$

where df represents the degrees of freedom, s_1 and s_2 are the sample standard deviations, and n_1 and n_2 are the number of measurements taken for each population.

The degrees of freedom are calculated according to the following relationship:

$$df = \frac{(s_1^2/n_1 + s_2^2/n_2)^2}{\frac{(s_1^2/n_1)^2}{n_1-1} + \frac{(s_2^2/n_2)^2}{n_2-1}}. \tag{3.4}$$

The following inequality is then used to determine whether or not the difference is significant:

$$\bar{x}_1 - \bar{x}_2 - t(\alpha,df)\sqrt{s_1^2/n_1 + s_2^2/n_2} < \mu_1 - \mu_2$$

$$< \bar{x}_1 - \bar{x}_2 + t(\alpha,df)\sqrt{s_1^2/n_1 + s_2^2/n_2}. \tag{3.5}$$

Alternatively, the t-statistic used to test whether the sample means are different may be calculated according to:

$$t(\alpha,df) = \frac{\bar{x}_1 - \bar{x}_2}{\sqrt{s_1^2/n_1 + s_2^2/n_2}}. \tag{3.6}$$

Table 2.3 (from Chapter 2) summarizes the t-statistic as a function of the "two-tailed" confidence interval. When a hypothesis is formulated, testing whether or not a mean is greater than or less than a specific value implies a one-tailed confidence interval. The t-statistic of a null hypothesis to test the equality of means at a 95% confidence interval with five degrees of freedom equals 2.571. To test if the same mean is greater than (or less than) the specified value at a 95% confidence interval is equivalent to the 90% confidence interval of a two-tailed test: the value equals 1.943.

Example 3.1. An elastic textile fiber is manufactured at two sites and their elongation at break—a characteristic parameter indicative of quality— is measured on a tensile measuring instrument (Instron, for example): a 10 cm sample is clamped at both ends and stretched and the distance traversed is recorded until the filament breaks. The two plants performed 10 tests to measure the percent elongation at break. The results of the first plant are: 725, 655, 715, 670, 660, 710, 690, 670, 680, and 690. The results of the second plant are: 600, 630, 650, 710, 620, 670, 650, 670, 720, and 610. Consider that a difference of 5% elongation is significant:

(a) Formulate a null hypothesis and an alternative hypothesis.

(b) Test the hypothesis.

(c) Are the fibers produced at one plant superior to the other?

Solution 3.1a. The null hypothesis is $H_0 : \mu_1 - \mu_2 \geq \bar{x}_1 - \bar{x}_2$. This hypothesis states that the sample mean between the two plants is less than 5% elongation at break. The alternative hypothesis is $H_1 : \mu_1 - \mu_2 < \bar{x}_1 - \bar{x}_2$. The alternative is that the difference between the mean of the data collected in the two plants is greater than 5% elongation at break.

Solution 3.1b. The sample mean and variances of the two populations are first calculated as well as the difference which is deemed to be significant, $\mu_1 - \mu_2 : \bar{x}_1 = 687\%$, $\bar{x}_2 = 653\%$, $\mu_1 - \mu_2 = 0.05(687 + 653)/2 = 34\%$, $s_1^2 = 23.9^2$, $s_2^2 = 40.3^2$. The next step is to calculate the degrees of freedom:

$$df = \frac{(s_1^2/n_1 + s_2^2/n_2)^2}{\frac{(s_1^2/n_1)^2}{n_1-1} + \frac{(s_2^2/n_2)^2}{n_2-1}}$$

$$= \frac{(23.9^2/10 + 40.3^2/10)^2}{\frac{(23.9^2/10)^2}{10-1} + \frac{(40.3^2/10)^2}{10-1}}$$

$$= 14.6 = 14.$$

Note that if the two population variances can be assumed to be equal, $\sigma_1^2 = \sigma_2^2$, the pooled estimate of the common variance becomes:

$$s_p^2 = \frac{(n_1 - 1)s_1^2 + (n_2 - 1)s_2^2}{n_1 + n_2 - 2}.$$

Further, if $n_1 = n_2$, the pooled sample variance becomes the average of the two sample variances and the degrees of freedom would be: $s_p^2 = 1/2(s_1^2 + s_2^2) = 33.1^2$, $df = n_1 + n_2 - 2 = 18$. In this example, we assume that the population variances are unequal and thus there are 14 degrees of freedom. The t-statistic for a two-tailed confidence interval of 95% with 14 degrees of freedom equals 2.145 and the inequality for the null hypothesis is expressed as:

$$t(\alpha, df) = t(95, 14) = 2.145,$$

$$\bar{x}_1 - \bar{x}_2 - t(\alpha, df)\sqrt{s_1^2/n_1 + s_2^2/n_2} < \mu_1 - \mu_2$$

$$< \bar{x}_1 - \bar{x}_2 + t(\alpha, df)\sqrt{s_1^2/n_1 + s_2^2/n_2},$$

$$\bar{x}_1 - \bar{x}_2 = 34,$$

$$\sqrt{s_1^2/n_1 + s_2^2/n_2} = \sqrt{23.9^2/10 + 40.3^2/10} = 14.8,$$

$$34 - 2.145 \cdot 14.8 < \mu_1 - \mu_2 < 34 + 2.145 \cdot 14.8,$$

$$2 < \mu_1 - \mu_2 < 66.$$

Since the difference between the means considered to be significant ($\mu_1 - \mu_2 = 34\%$) lies within the 95% confidence interval, we are therefore unable to reject the null hypothesis (which is a more correct way of saying that we accept the null hypothesis).

Solution 3.1c. Since the confidence interval is greater than zero on the low end (as well as on the upper end), one might be able to say that the elongation at the first plant is superior to that at the second plant.

3.3.3 Regression Analysis

While statistical tests are usually limited to comparing two populations for equality with respect to means, variances, distributions, etc., regression analysis is a methodology that characterizes the relationship between factors and response variables. Ideally, this analysis leads to an exact representation of the relationship but in practice the most we can hope for is to model the data over a narrow range. Extrapolating beyond the range of data is discouraged particularly when the exact relationship is unknown. Linear regression with a single factor implies that the response variable changes in proportion to the independent variable:

$$y = \beta_0 + \beta_1 x. \tag{3.7}$$

Examples of linear relationships in engineering include: the volume of an ideal gas in a closed vessel and the temperature (at constant pressure), $V = \beta_1 T$ (where $\beta_1 = nR/P$); the hydrostatic pressure head in the bottom of a tank and the height of the liquid, $\Delta P = \beta_1 h$ (where $\beta_1 = \rho g$); the heat flux through a solid slab of a fixed thickness and the temperature differential, $q/A = \beta_1 \Delta T$ (where $\beta_1 = k/\Delta x$).

At every single data point, the response variable is fitted to the experimental data by including the residual, e_i, which is the difference between the calculated value of y from the regression equation and the experimental value:

$$y_i = \beta_0 + \beta_1 x_i + e_i. \tag{3.8}$$

The regression coefficients are chosen to minimize the sum of the squares of the residuals. The minimization procedure is known as the method of least squares:

$$SS_e = \sum_{i=1}^{n} e_i^2 = \sum_{i=1}^{n} (y_i - \beta_0 - \beta_1 x_i)^2. \tag{3.9}$$

The values of the regression coefficients may be derived by differentiating the sum of the squares of the error (SS_e) with respect to β_0 and β_1:

$$\frac{\partial}{\partial \beta_0} SS_e = -2 \sum_{i=1}^{n} (y_i - \beta_0 - \beta_1 x_i) = 0, \tag{3.10}$$

$$\frac{\partial}{\partial \beta_1} SS_e = -2 \sum_{i=1}^{n} (y_i - \beta_0 - \beta_1 x_i)^2 x_i = 0. \tag{3.11}$$

Solving for the coefficients gives:

$$\beta_1 = \frac{n \sum_{i=1}^{n} x_i y_i - \sum_{i=1}^{n} x_i \sum_{i=1}^{n} y_i}{n \sum_{i=1}^{n} x_i^2 - \left(\sum_{i=1}^{n} x_i \right)^2}, \tag{3.12}$$

$$\beta_0 = \bar{y} - \beta_1 \bar{x}. \tag{3.13}$$

3.3.4 Coefficient of Determination

Deriving a regression equation is an important step in characterizing experimental data. The following step is to evaluate whether or not the equation represents the trends in the data. Graphs are useful to assess the goodness-of-fit qualitatively as well as to identify outliers. Calculating the coefficient of determination (also known as R^2) provides a more quantitative assessment of the model. It represents the proportion of variance in the dependent variable with respect to the independent variable:

$$R^2 = 1 - \frac{SS_e}{SS_{tot}}, \tag{3.14}$$

where SS_e is the residual sum of the squares of the error, shown above, and the total sum of the squares is simply the variance multiplied by the degrees of freedom:

$$SS_{tot} = \sum_{i=1}^{n} (y_i - \bar{y}^2)^2 = (n - 1)\sigma^2. \tag{3.15}$$

A perfect model—one in which the predicted values of the response variable equal the experimental values exactly—gives an R^2 equal to one. If the value of R^2 is 0.90, for example, then the regression model accounts for 90% of the variance in the data. The remaining 10% of the variance remains unexplained. This unexplained variance may represent scatter in the data or it could mean that the model poorly characterizes the relationship between factor and response variable.

Example 3.2. One liter of glycerol is produced for every 10 l of biodiesel. By 2019, the OECD estimates that the yearly production of glycerol will be 3600 kt, which far exceeds the current and projected demand. As a result of the worldwide glut, the price of glycerol will remain depressed and storage of unwanted glycerol will become problematic. Many research projects examine glycerol as a feedstock to produce fine chemicals; its dehydration to acrolein is an example. Glycerol is vaporized and then reacts with an iron phosphate catalyst:

$$C_3H_5(OH)_3 \rightarrow C_3H_4O + 2H_2O.$$

Carbon quickly builds up on the catalyst's surface and thus the reaction rate declines. To regenerate the catalyst, the glycerol feed is replaced with air to burn the coke and produce CO and CO_2.
Based on the data in Table E3.2:

(a) Derive a regression model for the relationship between acrolein production rate and coke deposition (as measured by the CO_x produced during the regeneration step).
(b) Calculate the coefficient of determination and the adjusted value, then comment.

TABLE E3.2 Acrolein Production Versus Carbon Deposition (mol h^{-1})

C_3H_4O production	0.11	0.45	0.94	0.95	1.21	1.43	1.55	1.91	2.28
CO_x deposition	0.0	0.10	0.13	0.13	0.23	0.19	0.27	0.42	0.42

FIGURE E3.2Sa Carbon Deposition Versus Acrolein Production

Solution 3.2a. The first step in examining this problem is to plot carbon deposition (the response variable) as a function of acrolein production rate, which is shown in Figure E3.2Sa.

The graph shows an approximately linear relationship between acrolein production and carbon deposition (as measured by the CO_x evolution during the regeneration step). Note that the data does not pass through the origin, although it should: evidently, when no acrolein is produced, no carbon is be deposited on the surface (unless all of the glycerol forms coke).

The second step is to calculate the sums of the factors and response variables and their product: $n \sum_{i=1}^{n} x_i y_i = 27.3$, $\sum_{i=1}^{n} x_i \sum_{i=1}^{n} y_i = 20.6$, $n \sum_{i=1}^{n} x_i^2 = 150.8$, $(\sum_{i=1}^{n} x_i)^2 = 117.3$.

These values are used to calculate the regression coefficient β_1. This represents the slope of the regression line which equals 0.20:

$$\beta_1 = \frac{n\sum_{i=1}^{n}x_i y_i - \sum_{i=1}^{n}x_i \sum_{i=1}^{n}y_i}{n\sum_{i=1}^{n}x_i^2 - (\sum_{i=1}^{n}x_i)^2} = \frac{27.3 - 20.6}{150.8 - 117.3} = 0.20.$$

Finally, the average of the dependent and independent variables is calculated to derive β_0: $\beta_0 = \bar{y} - \beta_1\bar{x} = 0.21 - 0.20 \cdot 1.2 = -0.03$.

Solution 3.2b. The next step is to add the regression line to the graph and then to calculate the coefficient of determination. As shown in Figure E3.2Sb, it fits

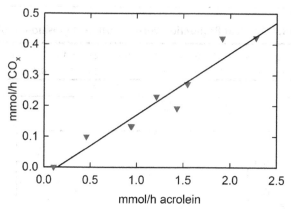

FIGURE E3.2Sb Carbon Deposition versus Acrolein Production with Regression Line

the data quite well. The sum of the squares of the error is 0.118 and the total sum of the squares is 0.161:

$$SS_{err} = \sum_{i=1}^{n}(y_i - b_0 - b_1 x_i)^2 = \sum_{i=1}^{n}(y_i - (-0.03) - 0.20 x_i)^2 = 0.012,$$

$$SS_{tot} = \sum_{i=1}^{n}(y_i - \bar{y})^2 = \sum_{i=1}^{n}(y_i - 0.21)^2 = 0.16.$$

The coefficient of determination, R^2, equals 0.93, which represents a good fit between experimental data and linear model. Only 7% of the variance is unexplained with this model:

$$R^2 = 1 - \frac{SS_{err}}{SS_{tot}} = 1 - \frac{0.012}{0.16} = 0.93.$$

In this example, the fit between the experimental data and a linear regression model was shown to be very good with a coefficient of determination equal to 0.93. Considering that the production of carbon oxides is expected to be equal to zero when the production of acrolein is zero, the regression analysis may be completed by assuming that the curve begins at the origin. In this case, the slope of the curve equals 0.18 and the coefficient of determination becomes 0.91.

3.3.5 Nonlinear Regression Analysis

In many situations, the relationship between response variables and factors is unclear. Linear regression analysis is a first step in identifying significant factors. However, many physical processes vary nonlinearly with factors and thus much of the variance in the data may not be accounted for with simple linear regression. To account for nonlinear behavior, response variables may be expressed as a power law function of the factors:

$$y = \beta_0 x_1^{\beta_1} x_2^{\beta_2} x_3^{\beta_3} \cdots x_n^{\beta_n}. \qquad (3.16)$$

Linear regression analysis may be applied to this relationship by taking the natural logarithm of both sides of the equation to give:

$$\ln y = \ln \beta_0 + \beta_1 \ln x_1 + \beta_2 \ln x_2 + \beta_3 \ln x_3 + \cdots + \beta_n \ln x_n. \qquad (3.17)$$

This is now a linear relationship in which the modified response variable is the natural log of the original, as are the factors. The exponents are simply the regression coefficients. Factors with regression coefficients close to zero may be ignored. Coefficients of less than zero indicate an inverse relationship with the response variable while coefficients greater than zero show a positive relationship between factor and response variable. Note that in most cases, regression coefficients should be expressed with at most two significant figures.

To illustrate a simple case of nonlinear regression analysis, consider the previous example but with additional experiments collected beyond the original data set, as shown in Figure 3.3.

The additional data indicates that the rate of carbon deposition appears to follow a parabolic relationship. The original regression equation derived in the previous example predicts that the carbon oxide production rate equals 0.54 mmol h^{-1} at a production rate of 3 mmol h^{-1} of acrolein as shown in Figure 3.4. The experimental data shows a production rate of double that predicted by this equation, thus indicating that the original linear regression is unsuitable. Regressing the new data set (with the additional two points) improves the model fit to the data; however, it overpredicts the evolution of CO_x at low acrolein production rates and underpredicts accumulated carbon on the surface at high acrolein production rates. By taking the natural logarithm of

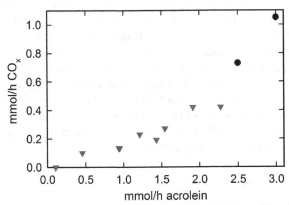

FIGURE 3.3 Carbon Deposition Versus Acrolein Production with Additional Data (Round Points)

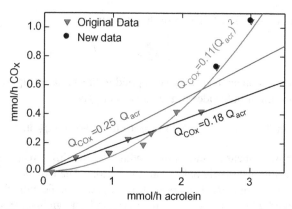

FIGURE 3.4 Nonlinear Regression Fit Relating Accumulated Carbon to Acrolein Production

the production rate of acrolein as well as that of the carbon monoxide and then applying the regression formula, the coefficient of regression approaches 0.96 and the best-fit parameters β_0 and β_1 are 0.11 and 2—a parabolic function fits the experimental data well. The best-fit parameter for the linear model, β_0, is 0.25 but the multiple R^2 is only 0.80.

3.3.6 Data Smoothing

Although many algorithms have been published to smooth data, generally it is discouraged because this operation causes a loss of information. Smoothing is most often used to identify trends and to locate approximate maxima and minima or points of inflection. It may also be used to interpolate. Determining if the fluctuations are real or is they are a low frequency component of the noise can also be important. These questions may be easily addressed by running more experiments—increasing the population of the data set but treating data numerically is a more efficient allocation of resources. Standard applications are related to enhancing the resolution of spectroscopic data or for the analysis of time-series radioactive data. Reducing the signal-to-noise ratio is one objective and this is particularly useful for radioactive data in which signal variation equals the square-root of the signal. Small changes in the signal are difficult to detect unless a large population is sampled. The moving average (also known as the rolling average or the running mean) is the simplest type of filter to smooth data. It consists of replacing all points with an average of neighboring data. A simple moving average (SMA) is the mean of the previous n data points. The equation for a five-point moving average is:

$$\mu_{SMA,5} = \frac{x_{i-2} + x_{i-1} + x_i + x_{i+1} + x_{i+2}}{5}. \tag{3.18}$$

The cumulative moving average (CMA) is calculated according to the following equation:

$$\mu_{CMA,i} = \frac{x_i + i x_{CMA,i-1}}{i+1}. \qquad (3.19)$$

In the expression for the simple moving average, the central point is calculated based on both previous and following points. The cumulative moving average relies solely on past points and thus any peaks or valleys will lag (the greater the number of points in the averaging, the longer the lag).

In a weighting moving average, coefficients are assigned to each value of i in the filtering formula. The most common weighted average polynomials are those proposed by Savitzky and Golay (1964) and are summarized in Table 3.1. Data with sharp changes are smoothed out with CMA and it becomes worse with an increase in the number of data points in the weighting factor. However, as shown in Figure 3.5, the Savitzky-Golay smoothing technique captures the trend in the heart over time extremely well: all features are captured including the initial peak, as well as the following two peaks. The CMA underestimates the first peak and the predicted peak height is shifted by 0.025 s. Furthermore, the second peak

TABLE 3.1 Savitzky-Golay Polynomial Coefficients

	i	-3	-2	-1	0	1	2	3
	35		-3	12	17	12	-3	
	21	-2	3	6	7	6	3	-2
First derivative	12			1	-8	0	8	-1
First derivative	252		15	-55	20	135	-30	55	-15

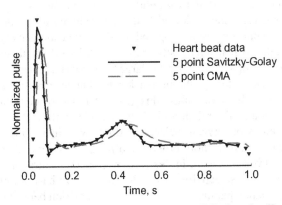

FIGURE 3.5 Comparing Savitzky-Golay Smoothing Versus CMA Over One Heart Beat

is shifted by the same amount and it is somewhat larger than the true data. The valley after the initial first peak is completely absent in the CMA approximation.

Other smoothing techniques include the low-pass filter, LOESS, gain filtering, kernel smoothing, and cubic spline interpolation. Another means to identify trends is to fit a regression equation to the data.

3.4 DESIGN OF EXPERIMENTS (DOE)

When rapid decisions are required to "fight fires" in a plant environment, planning and designing experiments may seem like a luxury. Experience, combined with trial-and-error, is the first level of "design." The next level may depend on the historical database to identify trends or correlations. The third level most often adopted is the "one-factor-at-a-time" experimental design. The highest level of design is when experiments are planned, executed, and analyzed with a systematic design of experiments: DOE. In general, DOE refers to the process of gathering information; however, it may also refer to other types of studies like opinion polls, statistical surveys, or natural experiments (epidemiology and economics are fields that rely on natural experiments).

Detailed experimental designs are undertaken at all levels of process development—discovery, scouting, product and process development and improvement, qualification, prove-outs, even start-up. With respect to start-up in an industrial environment, not only will DOE quantify the relationship between operating conditions and product specifications, it can also help develop standard operating procedures as well as optimize stand-by conditions (conditions at which the plant runs at reduced capacity due to poor market conditions or problems downstream). DOE is an efficient means to assess the effect and interaction of factors. The effect is defined as the variation of the response variable with respect to a change in a factor while interactions assess the relative change in the response variable with respect to two or more factors.

In pneumatic transport and gas-solids reactors operated at high gas velocity (known as Circulating Fluidized Beds), gas velocity and solids mass flux have a large effect on pressure drop and solids suspension density. As shown in Figure 3.6, the suspension density drops by 100 kg m^{-3} with an increase in gas velocity of 2.5 m s^{-1}. The effect is large but there is no interaction between mass flux and gas velocity: the proportionate change in suspension density is independent of the solids mass flux (for the range of conditions tested). Figure 3.7 illustrates the relationship between selectivity and temperature at two concentrations of oxygen. Selectivity drops with increasing temperature but the decrease is much more pronounced for the case with 5 vol% oxygen compared to 10 vol% oxygen. In this case, oxygen concentration and temperature strongly affect selectivity. When the two factors have an effect on the independent variable, parallel lines indicate no interaction between them while intersecting lines indicate the opposite.

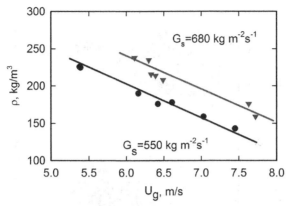

FIGURE 3.6 Solids Suspension Density Versus Gas Velocity at Conditions of Mass Flux: Illustration of Large Effect but No Interaction

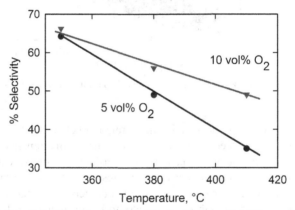

FIGURE 3.7 Selectivity Versus Temperature at Varying Degrees of Inlet Oxygen Concentration: Illustration of Large Effect with Interactions

Historical data, although pertinent to identify trends, at best only describes practice, not capability. Extrapolating performance or identifying optimal conditions requires a data set that covers a wide range of conditions. Historical data may also be compromised by correlated factors or by unrecorded control actions or process upsets that confound the data—making the determination of cause and effect ambiguous. Figure 3.8 illustrates the production of an organic compound (maleic anhydride) from the gas-phase partial catalytic oxidation of n-butane as a function of temperature and n-butane and oxygen vol% at the entrance. The data represents 523 days of operation and over 8000 hourly average data points.

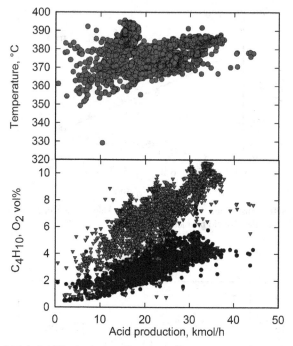

FIGURE 3.8 Maleic Acid Production as a Function of Temperature and Feed Concentrations

There is an "evident" positive linear correlation between both temperature and n-butane/oxygen feed concentrations and the production rate of maleic anhydride. The data contains what could be considered outliers, which are not necessarily due to "blunders" but may be a result of non-standard operating conditions (shut down, start-up, or stand-by) or defects in the analytical system or instrumentation. Although there appears to be an interaction between the n-butane and oxygen concentrations, the two are in fact directly related by stoichiometry:

$$n\text{-}C_4H_{10} + \frac{7}{2}O_2 \rightarrow C_4H_2O_3 + 4H_2O,$$

$$n\text{-}C_4H_{10} + \frac{9}{2}O_2 \rightarrow CO + 5H_2O,$$

$$n\text{-}C_4H_{10} + \frac{13}{2}O_2 \rightarrow CO_2 + 5H_2O.$$

So, if selectivity is relatively insensitive to feed composition, the slope of the oxygen curve will be higher in proportion to the ratio of oxygen to butane required to form maleic anhydride. At 100% selectivity, the ratio would be 3.5 and at 50% selectivity (assuming $CO:CO_2 = 1$) it is 4.5.

One-factor-at-a-time experimental designs are inefficient. Moreover, they overlook experimental error and interactions between variables. Although the understanding of one effect may become evident, typically the overall understanding is insufficient.

Good experimental strategy is based on several principles. Experience is important: the environment should be well understood—the number and nature of the factors and response variables. The objectives should be clearly stated so that all relevant responses and factors are considered. The design should be statistically balanced. To increase precision, experiments should be blocked—experimental units arranged into groups that are similar. For example, experimental results from two distinct catalysts should be collected and analyzed independently. For experiments that vary with time-on-stream, the time factor should also be blocked. Together with blocking, experiments should be randomized in order to minimize the possibility of confounding the data. The values of the factors should cover the broadest range possible. Just enough runs should be completed to achieve the desired resolution and to estimate the experimental error. Finally, repeat runs should be specified at intervals to account for the effect of time but also to estimate the experimental error.

Three classes of DOE include screening designs, interaction designs, and response surface designs. Screening designs are capable of identifying key factors for systems with more than six factors and include the Plackett-Burman and fractional factorial designs. Interaction designs are best for systems with between three to eight factors and include full and fractional factorial designs. Response surface designs are capable of generating models for prediction and optimization and are suited for situations with two to six factors. Box-Behnken and Central Composite Face-Centered Cube designs are examples.

Most corporations go beyond DOE and have developed strategies aimed at minimizing variability to eliminate (reduce) defects as well as to promote continuous improvement. These strategies are applied at all levels of business, from research to manufacturing to marketing. Quality control, TQM (total quality management), and Zero Defects were among earlier methodologies and in the late 1990s more than half of the Fortune 500 had adopted Six Sigma, which represents a level of performance in which there are only 3.4 defects per million observations.

3.4.1 Models

The relationships between response variables, y, and factors, x, are represented through mathematical expressions that take the form of correlations or regression equations. These linear or nonlinear mathematical expressions are often expressed as a polynomial, logarithmic, exponential, or trigonometric function. For example, the variation of the specific heats of gases, C_p, has been expressed as a third-order polynomial with respect to temperature:

$$C_p = \beta_0 + \beta_1 + \beta_2 T^2 + \beta_3 T^3. \tag{3.20}$$

Reaction rates are correlated with temperature exponentially following the Arrhenius equation:

$$k = k_0 \, e^{-E_a/R[1/T - 1/T_0]}. \tag{3.21}$$

The minimum number of experiments that should be conducted is at least equal to the number of parameters in the model. A linear model has two parameters, β_0 and β_1. (Strictly speaking, $y = \beta_1 x_1$ is a linear model and $y = \beta_0 + \beta_1 x_1$ is nonlinear but this nuance will be disregarded in our discussion.) A quadratic model has three parameters (β_0, β_1, and β_2) while there are four parameters in a cubic expression (see the relationship between specific heat and temperature above). The number of parameters for a simple polynomial is at most equal to $m + 1$, where m represents the order of the polynomial.

In the case of polynomials with multiple factors, $x_1, x_2, \ldots, x_n$, the minimum number of parameters required for a linear model excluding interactions is $n + 1$. When interactions are included, the total number of fitted parameter equals $2n$. For example, a model with three factors including interactions will have eight fitted parameters:

$$y = \beta_0 + \beta_1 x_1 + \beta_2 x_2 + \beta_3 x_3 + \beta_{12} x_1 x_2 + \beta_{13} x_1 x_3$$
$$+ \beta_{23} x_2 x_3 + \beta_{123} x_1 x_2 x_3. \tag{3.22}$$

For polynomials of order m and n factors, the maximum number of parameters would be $(m + 1)^n$. A third degree polynomial with four independent variables would have $(3 + 1)^4 = 256$ fitted parameters (and thus a minimum of 256 experiments would be required).

3.4.2 Experimental Designs

The potential inefficiency of the one-factor-at-a-time experimental plan is demonstrated in Figure 3.9. The factors are X_1 and X_2 and the topographical map of the response variable is depicted with ellipses that increase in shading depth with an increase in value. Test 1 is shown at the bottom left-hand corner. The response variable X_2 is increased with no change in the value of the response variable in Test 2. The factor is increased again in Test 3 and the response variable increases as well (since it enters the shaded region). In Test 4, the factor is increased with no apparent change in the response variable but a subsequent increase in Test 5 indicates a drop in the response variable. The next step in the test is to return to step 4 and decrease the value of the factor X_2. In this case the response variable decreases; therefore, in Tests 7–9, the response variable is increased and the "local" maximum for this design would then be identified as Test 7 or 8. Additional iterations with X_1 would then be required to locate the true maximum.

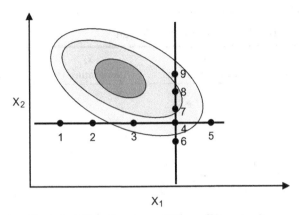

FIGURE 3.9 One-Factor-At-A-Time Experimental Plan

A statistical design should be capable of locating the absolute maximum with much fewer trials than the one-factor-at-a-time experimental design. An excellent example of DOE application is the Halon Replacement Program initiated by the USAF. The program objective was to replace the fire-extinguishing agent, Halon (a class of compounds consisting of various different alkanes with linked halogens that destroy the ozone), with one that is inert when released to the atmosphere. The Air Force identified many factors that lead to fires and the ability to extinguish them. They retained 14 for the experiments. In order to test each factor at two levels would require a total of $2 \times 10^{14} = 16\,384$ tests. By statistically selecting the experiments they constructed a program of 32 tests and were able to deliver an alternative to Halon on time and within budget.

3.4.3 Factorial Designs

A full factorial design is an experimental program that consists of two or more factors and each of these factors is tested at two or more levels. Most experiments are only conducted at two levels and they are designated as + and −, the former indicating the higher value of the factor and the latter representing the lower value. The designation for the case of three levels is +, 0, and −. Consider the extraction of the active phase of a spent precious metal catalyst in which the factors are solvent concentration and temperature. The full factorial design at two levels would consist of four experiments, 2^2 (an example of a square Hadamard matrix). Adding extraction time as a factor increases the number of experiments to 2^3, as shown in Table 3.2.

Full factorial designs may be contemplated for up to five factors at two levels but perhaps only three factors at three levels, which translate to 32 and 27 experiments, respectively (excluding repeat runs). For more than about

TABLE 3.2 Two-Level Three-Factor Design

Experiments	x_1 Concentration	x_2 Temperature (°C)	x_3 Time (h)
1	+	+	+
2	+	+	−
3	+	−	+
4	+	−	−
5	−	+	+
6	−	+	−
7	−	−	+
8	−	−	−

TABLE 3.3 Three-Level Two-Factor Design

Experiments	x_1 Concentration	x_2 Temperature (°C)
1	+	+
2	+	0
3	+	−
4	0	+
5	0	0
6	0	−
7	−	+
8	−	0
9	−	−

32 experiments, partial factorial designs should be considered. Whereas the number of experiments for two- and three-level full factorial designs is 2^k and 3^k respectively, the number of experiments required for partial factorial designs is 2^{k-p} and 3^{k-p} where $1/2^p$ represents fractional reduction: for $p = 1$, half the number of experiments are run; for $p = 2$, one-quarter the number of experiments in a full factorial design are undertaken.

A three-level experimental plan with two factors results in nine experiments (Table 3.3).

The theoretical basis for neglecting to conduct a full factorial design is the sparsity-of-effect principle, which states that systems are generally dominated by main effects and low-order interactions (Montgomery, 2008). Thus, a subset of the full factorial is usually sufficient to characterize the entire system. The resolution of a fractional design is the difference $k - p$ and is a property of a fractional design that indicates the ability to discriminate main effects $(x_1, x_2, x_3, \ldots, x_n)$ from low-order interactions $(x_i x_j)$. A resolution of two confounds main effects and is pointless. A resolution greater than five may differentiate higher-order effects but since most physical systems are best described by low-order effects, the effort to include them is unwarranted. A resolution of three $(k - p = 4 - 1$ or $k - p = 5 - 2$, for example) and four are most common. A resolution of three can estimate main effects but may be confounded if two-factor interactions are significant. For example, in the following equation, if the constant β_{12} is in the same order of magnitude or greater than β_1 / x_2 or β_2 / x_1 then a higher resolution is necessary:

$$y = \beta_1 x_1 + \beta_2 x_2 + \beta_{12} x_1 x_2. \tag{3.23}$$

A resolution of four is adequate to identify main effects and two-factor interactions that are unconfounded with the main effects.

The assignments of the levels for each experiment are similar to those of a full factorial design except for columns greater than $k - p$. These columns are assigned based on a generator function. This function is typically a product of the columns less than or equal to $k - p$. Two restrictions apply to the columns generated: their sequence must be different from any other column (or the matrix will be confounded) and it must not be equal to any term that is a product of two or more factors already defined in the equation.

The Plackett-Burman design is a special case of partial factorial design that applies to experiments in which the number of runs (n) is a multiple of four with up to $n - 1$ factors. The Plackett-Burman design, as with other partial factorial designs, is considered a screening design: it is adequate to identify the main effects but higher-order effects and interactions are difficult to assess (Jeff Wu and Hamada, 2009). In this design, as shown in Table 3.4, each row is a cyclic permutation of the previous row (except for repeat runs or supplement runs needed to complete the square matrix).

Example 3.3. For the following equation, create a fractional factorial design with a resolution of three and define three generator functions:

$$y = \beta_1 x_1 + \beta_2 x_2 + \beta_2 x_3 + \beta_4 x_4 + \beta_{12} x_1 x_2 + \beta_{23} x_2 x_3.$$

Solution 3.3. The design is shown below in Table E3.3. The generator function can be one of the following: the product $x_1 x_3$ or $x_1 x_2 x_3$, a constant value, or $+$ or $-$. Although $x_1 x_2$ and $x_2 x_3$ would normally be possible generator functions, they are excluded because their products are part of the equation. If β_0 were

TABLE 3.4 Plackett-Burman Design for Eight Experiments and Six Factors

Experiments	x_1	x_2	x_3	x_4	x_5	x_6
1	−	+	+	−	+	−
2	+	+	−	+	−	−
3	+	−	+	−	−	+
4	−	+	−	−	+	+
5	+	−	−	+	+	−
6	−	−	+	+	−	+
7	+	+	+	+	+	+
8	−	−	−	−	−	−

TABLE E3.3 Partial Factorial Design

Experiments	x_1	x_2	x_3	x_4 (1) $x_1 x_3$	x_5 (2) $x_1 x_2 x_3$	x_6 (3)
1	+	+	+	+	+	−
2	+	+	−	−	−	−
3	+	−	+	+	−	−
4	+	−	−	−	+	−
5	−	+	+	−	−	−
6	−	+	−	+	+	−
7	−	−	+	−	+	−
8	−	−	−	+	−	−

part of the function then a constant value for x_4 would also be excluded to avoid confounding.

3.4.4 Response Surface Designs

Response surface designs are the advanced stage of design. Screening designs—Plackett-Burman or fractional factorial, for example—identify the important factors; interaction designs—full factorial or fractional factorial with a resolution of five—substantiate the interactions between factors; finally, response surface designs allow for optimization, prediction, calibration, and even process control adjustments. Examples of response surface designs include Box-Behnken, Central Composite, and Face-Centered Cube. These topics are examined in more advanced books on statistics.

3.5 EXERCISES

3.1 (a) How many fitted parameters are there in the NIST equation for the specific heat of methane?

$$C_{p,CH_4} = -0.703029 + 108.4773T - 42.52157T^2 + 5.862788T^3 + 0.678656T^{-2},$$

$$C_{p,CH_4} = 85.81217 + 11.26467T - 2.114146T^2 + 0.138190T^3 - 26.42221T^{-2}.$$

(b) How many parameters are there for the correlation proposed by Van Wylen and Sonntag?

$$C_{p,CH_4} = -672.87 - 439.74T^{-0.25} - 24.875T^{0.75} - 323.88T^{-0.5}.$$

3.2 The local television network wished to examine the accuracy of gas pumps and collected 5 l samples from 10 different gas stations in the city. They brought gasoline samples to the university and were poured into a (2×2) l graduated cylinder and a 1 l graduated cylinder and the temperature was measured. After pouring the gasoline to the 2 l mark, the cylinder was emptied then dried with paper towel before measuring again. Avoiding spillage was difficult because of the cramped space in the fume hood. The measured volume of gasoline, V_m, and the temperature data are summarized in Table Q3.2. The reference temperature for the gasoline dispensed is 15 °C. To account for the thermal expansion of the gas, the following equation is used to correct the measured volume to the volume at the reference temperature, V_{ref}:

$$V_{ref} = V_m(T_m - 15) \cdot 0.00124.$$

The NIST requires an accuracy of 0.3 vol%:

(a) Formulate a hypothesis statement for this experiment.
(b) Are the gas stations compliant with the NIST regulation (i.e. does the television station have something newsworthy to report)?
(c) How could this experiment be improved?
(d) Recommend five areas to improve the accuracy of this experiment.

TABLE Q3.2 Volume of Gasoline Collected at 10 Stations

V_m (ml)	4978	5013	5000	5010	5010	5048	5000	5000	5020	4974
T_m (°C)	21.8	21.9	21.9	22.4	22.5	22.4	22	21.8	21.8	21.4

(e) As a gas station operator and knowing that the uncertainty of the pump is 0.05%, how would you set the pump to avoid prosecution but to maximize profit?

3.3 Engineering students preferred Basmati rice over another well-known brand of long grain rice (Brand UB) and the reason cited for the preference was because the former was longer. The class was divided into 17 groups and the grains were measured 10 times each. The measured length of the Basmati rice (in mm) was: 6.9, 7.0, 7.0, 6.9, 7.0, 6.8, 7.0, 6.5, 6.8, and 7.1. That of UB rice (also in mm) was: 6.5, 6.1, 6.6, 6.7, 6.2, 6.3, 6.9, 6.3, 6.4, and 6.3:

(a) Formulate a null hypothesis and an alternative hypothesis statement to compare the length of the grains of rice for each type.
(b) Test the hypothesis and conclude if it should be accepted in favor of the alternative.
(c) As much as 10% of the Brand UB is broken in half. Should these grains be included in the study? If they were rejected, would this change the conclusion?

3.4 (a) Propose a regression equation for a subset of the HadCRUT3 global temperature data given in Table Q3.4.
(b) Based on this equation, what will be the average global temperature in 2020, 2050, and 2010?
(c) What is the likelihood that the global temperatures rise as high as the predicted values from the regression equation? What are the possible consequences?

3.5 Let there be three factors whose variations are normalized from -1 to 1. The base case uses the combination $(0,0,0)$. You can use eight additional experiments to build a regression model. Three models are considered:
Bala Srinivasan

$$Y = \beta_0 + \beta_1 X_1 + \beta_2 X_2 + \beta_3 X_3 + \beta_{12} X_1 X_2,$$
$$Y = \beta_0 + \beta_1 X_1 + \beta_2 X_2 + \beta_3 X_3 + \beta_{11} X_1^2,$$
$$Y = \beta_0 + \beta_1 X_1 + \beta_2 X_2 + \beta_3 X_3 + \beta_{12} X_1 X_2 + \beta_{11} X_1^2.$$

(a) Propose eight new experiments using a full factorial design at two levels.
(b) Propose eight new experiments using a fractional factorial design at three levels (take $X_3 = X_1 X_2$).
(c) What are the combinations of (a) and (b) with each of the models for which there is a confounding factor when using only new experiments?
(d) What do you observe when using the base case in addition to new experiments in the third model?

TABLE Q3.4 HadCRUT3 Global Temperature Data from 1960

Year	ΔT (K)	Year	ΔT (K)	Year	ΔT (K)
1960	−0.1247	1977	0.0174	1994	0.1718
1961	−0.0236	1978	−0.0646	1995	0.2752
1962	−0.0219	1979	0.0492	1996	0.1371
1963	0.0015	1980	0.0762	1997	0.3519
1964	−0.2958	1981	0.1203	1998	0.5476
1965	−0.2161	1982	0.0100	1999	0.2971
1966	−0.1481	1983	0.1766	2000	0.2705
1967	−0.1495	1984	−0.0208	2001	0.4082
1968	−0.1590	1985	−0.0383	2002	0.4648
1969	−0.0109	1986	0.0297	2003	0.4753
1970	−0.0677	1987	0.1792	2004	0.4467
1971	−0.1902	1988	0.1799	2005	0.4822
1972	−0.0578	1989	0.1023	2006	0.4250
1973	0.0768	1990	0.2546	2007	0.4017
1974	−0.2148	1991	0.2126	2008	0.3252
1975	−0.1701	1992	0.0618	2009	0.4430
1976	−0.2550	1993	0.1058	2010	0.4758

3.6 Identify all combinations of four experiments to develop the following model:
$$Y = \beta_0 + \beta_1 X_1 + \beta_3 X_3 + \beta_{12} X_1 X_2.$$
See Table Q3.6.

3.7 The Young's modulus (E) of an immiscible polymer depends on the following: stretch during injection (ϵ), injection temperature (T), and the addition of a fraction of fiberglass (F): B. Blais

TABLE Q3.6 Example Model

Experiments	Plan		
1	−1	−1	−1
2	−1	−1	1
3	−1	1	−1
4	−1	1	1
5	1	−1	−1
6	1	−1	1
7	1	1	−1
8	1	1	1

(a) How many parameters are there for a linear model with no interactions?

(b) How many experiments are required to estimate the parameters? Outline a full factorial design.

(c) During the first set of experiments, it is apparent that there is an interaction between the stretch during injection and the injection temperature. Propose a new model to account for this interaction.

(d) How many experiments are required to determine the new parameter based on a partial factorial experimental plan?

3.8 The fiber strength of electrospinning a new polymer depends on flow rate, solution concentration, electric potential, and temperature:

(a) Propose an experimental plan of eight experiments for a linear model with interactions. Assume the generator function for the fourth factor is $X_4 = X_1 X_2 X_3$.

(b) Discuss the problems of confounding for each of the following models:

$$Y = \beta_0 + \beta_1 X_1 + \beta_2 X_2 + \beta_3 X_3 + \beta_4 X_4 + \beta_{123} X_1 X_2 X_3,$$
$$Y = \beta_0 + \beta_1 X_1 + \beta_2 X_2 + \beta_3 X_3 + \beta_4 X_4 + \beta_{124} X_1 X_2 X_4,$$
$$Y = \beta_0 + \beta_1 X_1 + \beta_2 X_2 + \beta_3 X_3 + \beta_4 X_4 + \beta_{12} X_1 X_2.$$

3.9 Typing speed varies based on the total number of words (L_t) and the complexity of the vocabulary (C_t) as shown in Table Q3.9 (TypeRacer®). Develop a linear model with interactions. Your equation must be of the type:

$$P = \beta_0 + \beta_1 L_t + \beta_2 C_t + \beta_{12} L_t C_t.$$

TABLE Q3.9 Typing Speed

Typing Speed (wpm)	Length of Text (words)	Word Complexity (a.u.)
20	40	3
40	30	2.5
60	30	2.1
80	50	1.8
100	50	1.7
120	30	1
140	30	0.5

3.10 The distillation of ethanol and water solutions depends on concentration and temperature (assuming near atmospheric pressure operation). Derive a cubic model with interactions up to third order. How many experiments are needed? Would a fractional factorial design be sufficient to identify all model parameters?

3.11 A paper bag manufacturer wishes to improve paper strength. The tear resistance is a function of the concentration of hardwood pulp from 5%

to 20%. An experimental study designed to establish the relationship between tear resistance and concentration includes four levels with six repeats at each level. The 24 specimens are tested randomly by a laboratory dynamometer and the results are given in Table Q3.11. *M.B. Moudio*

(a) At a 99% confidence interval is the resistance independent of concentration?
(b) If not, which concentration provides the greatest resistance?
(c) Is this a multi-factorial problem?

TABLE Q3.11 Hardwood Concentration

Hardwood Concentration (%)	1	2	3	4	5	6	Total	Average
5	7	8	13	11	9	10	60	10.00
10	12	17	15	18	19	15	94	15.67
15	14	18	19	17	16	18	102	17.00
20	19	25	22	23	18	20	127	21.17
							383	15.96

3.12 The Kraft process is the most common papermaking technique. Wood is chopped into pieces and cooked in caustic soda in a digester (reactor) operating at a constant temperature of 180 °C. The performance of the digester (Y) depends on the type of wood:

$$Y = f(A, B, n),$$

where A, B, and n are factors related to the wood. *T. Yasser*

(a) Submit eight experiments using full factorial design at two levels.
(b) Propose eight experiments using fractional factorial design at three levels. Take $n = AB$.
(c) We propose the following models:

$$Y = \beta_0 + \beta_1 A + \beta_2 B + \beta_3 n + \beta_{12} AB + \beta_{13} An + \beta_{23} Bn + \beta_{11} A^2 + \beta_{22} B^2 + \beta_{33} n^2,$$
$$Y = \beta_0 + \beta_1 A + \beta_2 B + \beta_3 n + \beta_{13} An + \beta_{23} Bn,$$
$$Y = \beta_0 + \beta_1 A + \beta_2 B + \beta_3 n + \beta_{23} Bn.$$

(i) Write the regression matrix Φ for each case.
(ii) What are the models for which there is "confounding"? Justify your answer using the regression matrix Φ.
(iii) If there is more than one model with no "confounding," which do you choose if we want to get better accuracy for the parameters of regression? Why?
(iv) Can we neglect the interaction between A and n (β_{13})? Why?

(d) We have 10 species of wood. For each case we have a value of A, B, and n (see Table Q3.12d). We propose the following model:

$$Y = \beta_0 + \beta_1 A + \beta_2 B + \beta_3 n + \beta_{12} AB + \beta_{13} An + \beta_{23} Bn + \beta_{11} A^2 + \beta_{22} B^2 + \beta_{33} n^2.$$

TABLE Q3.12d Model Parameters for Wood

Species of Wood	A	B	n
Western Hemlock (1)	96.8	4.84	0.41
Western Red Cedar (2)	84.1	4.68	0.35
Jack Pine (3)	94.3	5.88	0.35
Trembling Aspen (4)	85.5	1.37	0.76
Alpine Fir (5)	89.3	4.46	0.40
Balsam Fir (6)	87.5	4.32	0.40
Douglas Fir (7)	83.1	3.84	0.40
Hard Maple (8)	73.4	0.57	0.95
Red Alder (9)	66.1	0.41	0.95
Yellow Birch (10)	68.9	0.15	1.35

(i) Verify that the data allows us to find the regression coefficients. Justify your answer.

(ii) Find the values of regression coefficients.

(iii) Propose two alternatives to reduce the variance of regression parameters.

3.13 The relationship between the temperature and the pressure and the quantity of solvent to produce a pesticide follows a linear relationship with interactions of the form: $T.-V.$ Ton

$$Y = \beta_0 + \beta_1 x_1 + \beta_2 x_2 + \beta_{12} x_1 x_2.$$

(a) Build a factorial design at two levels: x_1 and x_2 represent the quantity of solvent.

(b) Determine the matrix Φ representing the problem.

(c) Write the system to be solved.

3.14 Students in a biology class are experimenting on growing medicinal plants. They note that, at maturity, the length, x_1, of the plant's leaves ranges from a minimum of 7 cm to a maximum of 15 cm. The height of the plant, x_2, ranges from a minimum of 110 cm to a maximum of 135 cm.
M.-M.L. Gilbert

(a) Develop a full factorial level three knowing that growth is governed by the following model:

$$y = \beta_0 + \beta_1 x_1 + \beta_2 x_2 + \beta_{12} x_1 x_2 + \beta_{11} x_1^2.$$

(b) Complete a partial factorial plan knowing that $x_3 = x_1^3 x_2$.

(c) For the model

$$y = \beta_0 + \beta_1 x_1 + \beta_2 x_1^2$$

determine $\Delta\beta_2$ knowing that $\Delta y = 0.03$ for all the experiments. Consider that the possible maximum is three and the minimum is one. Verify that the coefficient is negligible for a value of $\beta_2 = 0.15$.

3.15 In an assembly company, 120 workers perform a specific task (T). Management is interested in knowing whether the factor "operator" affects the variable "time taken to complete T." To this end, four operators are randomly selected to complete T several times—the time (in minutes) taken by the operators is summarized in Table Q3.15. *R. Miranda*

(a) Express the hypothesis test.

(b) Can you reject the null hypothesis at a confidence interval of 95%?

(c) Does the factor "operator" affect the variable of interest?

TABLE Q3.15 Time Required to Complete Task T

Operator 1		Operator 2		Operator 3		Operator 4	
48	37	24	18	37	43	19	13
31	29	16	6	40	40	26	21
31	24	22	24	51	35	31	26
36	38	10	30	49	33	13	24
39	41	25	24	36	39	12	12
		11	15	24	55	16	21
				35	40		

3.16 To determine the influence of the ingredients of a recipe on the stability of an emulsion of a cosmetic cream, an experimental plan is proposed with three factors (two levels): the nature of the cream (oil in water for a positive effect and water in oil for a negative one), an emulsifier (dilute or very dilute), and fatty acid concentration (high or low). The indices of the emulsion stabilities obtained are: 38, 37, 26, 24, 30, 28, 19, and 16, with an experimental error of ± 2. *Lamine*

(a) How many experiments would be required for a linear model with no interactions? With first-order interactions?

(b) Build an experimental plan for the model with first-order interactions.

3.17 The volume contraction, α, of aqueous glycerol solutions as a function of weight fraction of glycerol is given in Table Q3.17.

(a) Determine the constant a for following expression that best fits the data:
$$\rho = ax(1 - x^2),$$
where x represents the glycerol mass fraction.

(b) Calculate the coefficient of correlation, R^2.

TABLE Q3.17 Volume Contraction of Glycerol-Water Solutions at 20 °C

x	0	0.1	0.2	0.3	0.4	0.5	0.6	0.7	0.8	0.9	1
α (%)	0	0.225	0.485	0.705	0.948	1.08	1.101	1.013	0.814	0.503	0

Gerlach 1884.

3.18 The potential generated for a copper-constantan thermocouple versus temperature is given in Table Q3.18 (Holman, 2001):

(a) For each of the polynomial models given below, derive the coefficients:
$$E = \beta_0 + \beta_1 T,$$
$$E = \beta_0 + \beta_1 T + \beta_2 T,$$
$$E = \beta_0 + \beta_1 T + \beta_2 T + \beta_3 T.$$

(b) For each relationship, calculate the coefficient of correlation.

TABLE Q3.18 Copper-Constantan Thermocouple Potential Versus Temperature

°C	mV	°C	mV
−250	−6.18	100	4.279
−200	−5.603	150	6.704
−150	−4.648	200	9.288
−100	−3.379	250	12.013
−50	−1.819	300	14.862
0	0	350	17.819
50	2.036	400	20.872

3.19 Derive a relationship between the emf and temperature differential of a Type K thermocouple (see Table Q3.19).

TABLE Q3.19 emf and Temperature Differential of Type K Thermocouple

ΔT (°C)	emf (mV)	ΔT (°C)	emf (mV)
0	0	300	12.209
25	1	400	16.397
50	2.023	500	20.644
75	3.059	600	24.906
100	4.096	800	33.275
150	6.138	1000	41.276
200	8.139		

3.20 A brigantine commutes between Tortuga Island and Cuba. The crossing time depends on the load (in tonnes of rum), the phase of the moon, and the age of the crew (in years). The boatswain, using a full factorial experimental design, noted the times in hours (Table E3.20): *R. Taiebi*

(a) Propose a linear model and a linear model with interactions.

(b) Derive the constants for the linear model.

(c) Is the age of the crew a significant factor?

TABLE E3.20 Crossing Time Between Cuba and Tortuga

Moon	New	Full	New	Full	New	Full	New	Full
Loading	20	20	100	100	20	20	100	100
Age	30	30	30	30	50	50	50	50
Time (h)	24	16	23	31	18	14	17	22

REFERENCES

Dunn, P., 1997. James Lind (1716–94) of Edinburgh and the treatment of scurvy. Archive of Disease in Childhood Foetal Neonatal, United Kingdom: British Medical Journal Publishing Group 76(1), 64–65. doi: http://dx.doi.org/10.1136/fn.76.1.F64. PMC 1720613. PMID 9059193. Retrieved 17.01.2009.

Fisher, R.A., 1918. The correlation between relatives on the supposition of mendelian inheritance. Philosophical Transactions of the Royal Society of Edinburgh 52, 399–433.

Gerlach, G.T., 1884. Chemistry Industry 7, 277–287.

Jeff Wu C.F., Hamada, C.J., 2009. Experiments Planning, Analysis, and Optimization, 2nd ed. Wiley.

Legg, T., 2012, April 4. Hadley Center Central England Temperature Dataset. Retrieved from Met Office Hadley Centre observations datasets: <http://www.metoffice.gov.uk/hadobs/hadcet/>.

Montgomery, D.C., 2008. Design and Analysis of Experiments, 7th ed. Wiley.

Savitzky, A., Golay, M.J.E., 1964. Analytical Chemistry 36, 1627–1639.

Thomke, S., Von Hippel, E., Franke, R., 1998. Modes of experimentation: an innovation process—and competitive—variable. Research Policy 27, 315–332.

White, J.W., 2004, June 11. VOL 304. Retrieved 2011, from sciencemag: <http://www.climate.unibe.ch/stocker/papers/white04sci.pdf>.

Chapter 4

Pressure

4.1 OVERVIEW

In April 2010, gas and liquids from the Macondo oil well unexpectedly rose through the production casing (riser) to the surface and detonated on the Deep Horizon ultra-deepwater offshore drilling rig. Two days later, the rig sank and gas and oil continued to gush from the well on the sea floor 1600 m below the surface for several months causing the worst environmental disaster in American history. The blow out was caused by a combination of mechanical failure, poor judgment, and operational design, but correctly interpreting the pressure test could have averted this catastrophe. During the preparation for the temporary abandonment, BP replaced drilling mud (density of 1740 kg m^{-3}) with sea water (density of 1030 kg m^{-3}) (Graham et al, 2011). The hydrostatic head of the sea water ($\rho g h$) was presumably insufficient to avoid hydrocarbons from entering the well. Consequently, gas and liquids rose up through the well from several thousand meters below the surface of the floor bed. When it reached the platform, the methane gas entered the ventilation system and ignited, resulting in a fireball that was visible for over 35 mi. Oil and natural gas leaked through the riser for three months and the rate peaked at over 62 000 bbl d^{-1} and some estimate that it could have been more than 80 000 bbl d^{-1}.

Pressure measurement is a critical factor not only in the oil and gas industry but also for the operation of process equipment including reactors, distillation columns, pipelines, and for the measurement of flow rate. It is a scalar quantity and equals the perpendicular force exerted on a surface with the units of force per surface area:

$$P = \frac{F}{A} = \frac{N}{m^2} = Pa.$$
(4.1)

Experimental Methods and Instrumentation for Chemical Engineers. http://dx.doi.org/10.1016/B978-0-444-53804-8.00004-6

Pressure together with temperature, flow rate, and concentration are the essential macroscopic properties required to operate chemical processes. Various units are frequently used to report pressure and some common units and their conversions are reported in Table 4.2.

Atmospheric pressure is a result of the hydrostatic head of air $\rho g h$ and equals $101\,325$ Pa (at sea level and a clear day). However, the Aristotelian view was that air was weightless. It wasn't until the Renaissance that Toricelli challenged this postulate. To avoid being accused of witchcraft and sorcery, he worked secretly and invented the mercury ("quicksilver") barometer to conduct experiments in an instrument measuring only 80 cm in height compared to a water column of over 10 m that was commonly used at the time. Pascal was the first to demonstrate that pressure changed with elevation by measuring the height of mercury in a barometer at the base of the Puy de Dome and along its height to the summit at 1464 m.

Together with columns of liquids, other forms of pressure common to the process industry include the pressure of gases in vessels, vapor pressure, dynamic pressure, and pressure drop across pipes, flow meters and through porous materials. At constant pressure and a fixed mass, Boyle's law states that for an ideal gas, the volume and absolute pressure are inversely proportional:

$$PV = c \tag{4.2}$$

or

$$P_1 V_1 = P_2 V_2. \tag{4.3}$$

As the pressure of an ideal gas increases, its volume decreases proportionately. Charles' law describes the relationship between the volume and temperature of an ideal gas at constant pressure:

$$\frac{V_1}{T_1} = \frac{V_2}{T_2}. \tag{4.4}$$

The volume increases proportionally with temperature. Amontons' law relates pressure and temperature and can be derived from Boyle's and Charles' laws:

$$\frac{P_1}{T_1} = \frac{P_2}{T_2}. \tag{4.5}$$

The ideal gas law relates the product of pressure and volume to temperature and the number of moles:

$$PV = nRT, \tag{4.6}$$

where n is the number of moles and R is the ideal gas constant (which is equal to 8.314 J mol^{-1} K^{-1}).

This relation may be derived from the kinetic theory of gases. Pressure is a result of momentum exchange between molecules and a confining wall (in the example of a vessel). This exchange varies with the frequency of collisions together with the speed of the molecules and is given by

$$P = \frac{1}{3}\tilde{n}mv_{rms}^2,$$ (4.7)

where $\tilde{n}$ is the number of molecules per unit volume (N/V), m is the molecular weight (mass) of the gas molecule, and v_{rms} is the root mean square molecular velocity.

The root mean square velocity is related to absolute temperature and molecular mass by the following relation:

$$\frac{1}{2}mv_{rms}^2 = \frac{3}{2}kT,$$ (4.8)

where k is Boltzmann's constant $(1.3806 \times 10^{-23}$ J mol^{-1} K$^{-1})$. Thus:

$$P = \frac{1}{3}\tilde{n}mv_{rms}^2 \cdot 3kT = \frac{1}{V}NkT = \frac{1}{V}nRT.$$ (4.9)

The ideal gas constant, R, is the product of Boltzmann's constant and Avogadro's number—the number of molecules in one mole (6.022×10^{23}).

Example 4.1. What is the temperature and the root mean square velocity of one mole of helium in a one liter flask maintained at

(a) 1 atm.

(b) 10 atm.

(c) At what pressure will the temperature in the flask reach 0 °C.

Solution 4.1a. First the temperature is calculated based on the ideal gas law. The value of the temperature is then used in the equation relating kinetic energy and temperature:

$$PV = nRT,$$

so

$$T = \frac{PV}{nR}.$$

We know that $V = 1$ l, $n = 1$ mol, and $R = 8.314$ J mol^{-1} K^{-1}, so:

$$T = \frac{1 \text{ atm} \cdot 101\,325 \text{ Pa atm}^{-1} \cdot 1 \text{ l} \cdot 0.001 \text{ m}^3 \text{ l}^{-1}}{1 \text{ mol} \cdot 8.314 \text{ J mol}^{-1}\text{K}^{-1}} = 12.2 \text{ K}.$$

Finally:

$$\frac{1}{2}mv_{rms}^2 = \frac{3}{2}kT,$$

so

$$m = \frac{4 \text{ g mol}^{-1} 0.001 \text{ kg g}^{-1}}{6.023 \times 10^{23} \text{ molecule mol}^{-1}} = 6.641 \times 10^{-27} \text{ kg molecule}^{-1}$$

and

$$v_{rms} = \sqrt{\frac{3}{m} kT}$$

$$= \sqrt{3 \frac{1.3806 \times 10^{-23} \text{ J molecule}^{-1} \text{ K}^{-1} 12.2 \text{ K}}{6.641 \times 10^{-27} \text{ kg molecule}^{-1}}}$$

$$= \sqrt{76\,088 \text{ m}^2 \text{ s}^2} = 276 \text{ m s}^{-1}.$$

Solution 4.1b. We can apply Charles' law to calculate the temperature at the higher pressure:

$$\frac{P_1 V_1}{T_1} = \frac{P_2 V_2}{T_2},$$

$$T_2 = T_1 \frac{P_2 V_2}{P_1 V_1} = 12.2 \text{ K} \frac{10 \text{ atm}}{1 \text{ atm}} = 122 \text{ K},$$

$$v_{rms,122} = v_{rms,12.2} \sqrt{\frac{122}{12.2}} = 872 \text{ m s}^{-1}.$$

Solution 4.1c. In this example, pay attention to the units: there are 4 g of He in one liter, which translates to 4 kg of He in 1 m³. SI units require kg and m³ but all the values are given in cgs:

$$v_{rms,273} = v_{rms,12.2} \sqrt{\frac{273}{12.2}} = 1300 \text{ m s}^{-1},$$

$$P = \frac{1}{3} \tilde{n} m v_{rms,273}^2$$

$$= \frac{1}{3} 6.023 \times 10^{26} \text{molecule m}^{-3} \cdot 6.641 \times 10^{-27} \text{ kg molecule}^{-1}$$

$$\cdot (1300 \text{ m s}^{-1})^2$$

$$= 2253 \text{ kPa} = 22.2 \text{ atm}.$$

The ideal gas law characterizes the relationship between pressure, temperature, and volume for gases. Both the Clausius-Clapeyron and Antoine equations characterize the vapor-liquid equilibrium of pure components and mixtures. At atmospheric pressure and ambient temperature, water is a liquid but an equilibrium exists with its vapor phase concentration—its vapor pressure. The vapor pressure is a function of temperature. The formula for the Clausius-Clapeyron equation is:

$$\ln P^\circ = -\frac{\Delta H_v}{RT} + B_{CC}, \tag{4.10}$$

TABLE 4.1 Antoine Constants of Various Compounds (T in $°C$, P in kPa)

Compound	Formula	A	B	C	Temp.	T_b
n-Butane	n-C_4H_{10}	13.6608	2154.7	238.789	−73–19	−0.5
n-Pentane	n-C_5H_{12}	13.7667	2451.88	232.014	45–58	36
n-Hexane	n-C_6H_{14}	13.8193	2696.04	224.317	19–92	68.7
Cyclohexane	C_6H_{12}	13.6568	2723.44	220.618	9–105	80.7
n-Heptane	n-C_7H_{16}	13.8622	2910.26	216.432	4–123	98.4
n-Octane	n-C_8H_{18}	13.9346	3123.13	209.635	26–152	125.6
Benzene	C_6H_6	13.7819	2726.81	217.572	6–104	80
Toluene	$C_6H_5CH_3$	13.932	3056.96	217.625	13–136	110.6
Ethylbenzene	$C_6H_5C_2H_5$	13.9726	3259.93	212.3	33–163	136.2
Methanol	CH_3OH	16.5785	3638.27	239.5	11–83	64.7
Ethanol	C_2H_5OH	16.8958	3795.17	230.918	3–96	78.2
Isopropanol	i-C_3H_7OH	16.6796	3640.20	219.610	8–100	82.2
1-Butanol	1-C_4H_9OH	15.3144	3212.43	182.739	37–138	117.6
Phenol	C_6H_5OH	14.4387	3507.8	175.4	80–208	181.8
Glycol	$C_2H_4(OH)_2$	15.7567	4187.46	178.65	100–222	197.3
Acetone	C_3H_6O	14.3145	2756.22	228.06	26–77	56.2
Diethyl ether	$C_2H_5OC_2H_5$	14.0735	2511.29	231.2	43–55	34.4
MEK	$C_2H_4OC_2H_4$	14.1334	2838.24	218.69	8–103	79.6
Chloroform	$CHCl_3$	13.7324	2548.74	218.552	23–84	61.1
CCl4	CCl_4	14.0572	2914.23	232.148	14–101	76.6
Water	H_2O	16.3872	3885.7	230.17	0–200	100

where $P°$ is the vapor pressure, ΔH_v is the heat of vaporization ($J\ mol^{-1}$) and B_{CC} is a constant specific to each compound.

The Antoine equation is an empirical relationship between vapor pressure and temperature with three fitted parameters—A_A, B_A, C_A:

$$\ln P° = A_A - \frac{B_A}{T + C_A}. \qquad (4.11)$$

Note that this expression is most commonly expressed as a log (base ten) with the units of pressure in mmHg and temperature in $°C$. Table 4.1 summarizes the values of the Antoine constants for some common solvents—the units for $P°$ are kPa and temperature is in $°C$ (Poling et al., 2001).

Example 4.2.

(a) Calculate the vapor pressure of ethanol at $15\ °C$ and at $60\ °C$.

(b) How would your answer change if only three significant figures were carried for the Antoine constants?

Solution 4.2a. The Antoine constants for ethanol are $A_A = 16.8958$, $B_A = 3795.17$, $C_A = 230.918$:

$$\ln P^\circ = A_A - \frac{B_A}{T + C_A} = 16.8958 - \frac{3795.17}{15 + 230.918},$$
$$P^\circ_{15} = e^{1.46314} = 4.3 \text{ kPa},$$
$$P^\circ_{60} = e^{3.85030} = 47 \text{ kPa}.$$

Solution 4.2b. This is a clear example where too many significant figures are carried with respect to the constants in the correlation (with respect to the number of significant figures reported for temperature)

$$\ln P^\circ = A_A - \frac{B_A}{T + C_A} = 16.9 - \frac{3800}{15 + 231},$$
$$P^\circ_{15} = e^{1.453} = 4.3 \text{ kPa},$$
$$P^\circ_{60} = e^{3.841} = 47 \text{ kPa}.$$

4.2 UNITS OF PRESSURE

The SI unit for pressure is the Pa but for many calculations the kPa is adopted to reduce the number of digits in calculations. Until 1999, 1 atm had been accepted as the standard reference pressure in chemistry, in the oil and gas industry, for equipment specifications, etc. However, the IUPAC (International Union of Pure and Applied Chemistry) redefined standard pressure to equal 100 kPa and this is designated as 1 bar. In the Imperial system of units, the measure of pressure is psi and, as shown in Table 4.2, 14.696 psi equals 1 atm; 14.504 psi equals 1 bar. For assessing vacuum, standard units are Torr, mmHg, and μm. There are 760 mmHg in 1 atm and 1 Torr equals 1 mmHg; 1 μm is 0.001 Torr. While mercury is commonly used in barometers to measure atmospheric pressure, water is used in manometers; its lower density offers a greater precision (for the same pressure drop). In the Imperial system, the units of differential pressure are often quoted in are inH_2O (1 inH_2O = 25.4 mmH_2O).

4.3 TYPES OF PRESSURE

Pressure in the chemical industry covers a range of at least 12 orders of magnitude: natural gas pipelines operate at 200 bar; in three-phase gas-liquid-solid hydrogenation reactors, pressures can exceed 150 bar to achieve a high concentration of hydrogen in the liquid phase; many catalytic processes operate between 1 bar and 10 bar; vacuum distillation towers in oil refineries may be as large as 15 m in diameter and operate at 25 mmHg; freeze drying is common for preparing vaccines and antibiotics and to store blood and plasma

TABLE 4.2 Common Pressure Units and Their Conversions

	kPa	mmH$_2$O	mmHg	psi	inH$_2$O	inHg
kPa	–	101.975	7.5006	0.14504	4.0147	0.2953
mmH$_2$O	0.009806	–	0.073556	0.001422	0.03937	0.002896
mmHg	0.133322	13.595	–	0.019337	0.53524	0.03937
psi	6.894757	703.0875	51.7151	–	27.6787	2.03601
inH$_2$O	0.249082	25.4	1.86833	0.036129	–	0.07356
inHg	3.3864	345.313	25.4	0.49116	13.595	–

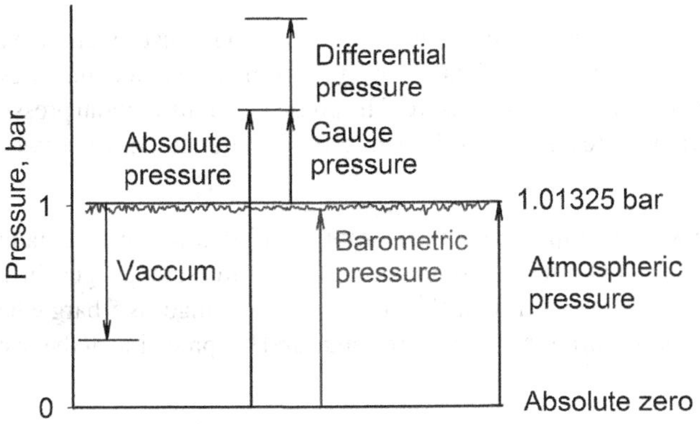

FIGURE 4.1 Types of Pressure

at pressures around 10^{-4} Torr; pressures below 10^{-6} Torr are required for analytical equipment such as electron microscopes and mass spectrometers. The lowest recorded pressure (up until 1991) was by a Japanese team who reached 7×10^{-13} mbar. Standard definitions of pressure are illustrated in Figure 4.1.

4.3.1 Atmospheric Pressure

Atmospheric pressure at sea level is 101 325 Pa and this is defined as 1 atm. The pressure reported by meteorological stations is usually the mean sea level pressure (MLSP): this is the pressure that would be calculated if the station were at sea level. Thus, the MLSP reported by the meteorological service at Denver, Colorado (at 1665 m) is much higher than the actual reading of a barometer located in the city.

4.3.2 Gauge Pressure

Pressure gauges often read zero at atmospheric conditions, whereas in reality they should read one atmosphere (or the barometric pressure); therefore, when a pressure indicator reaches a value of 2 atm, for example, the true pressure is in fact 3 atm. The value of 2 atm is referred to as the gauge pressure. To distinguish between gauge and absolute pressure, the letter "g" or "a" is added to the end of the unit: 4.5 barg = 5.5 bara = 65 psig = 80 psia. Remember that there is a relative difference between bar and atm of 1.3%, which is negligible when carrying two significant figures but must be accounted for when carrying more significant figures or calculating differential pressure or very high pressures. Thus, if the gauge is quoted in bar, 4.500 barg = 5.513 bara = 5.4410 atma.

4.3.3 Differential Pressure

Ideally differential pressure is measured with a single instrument: the upstream pressure in a flowing fluid is at the higher pressure and the pressure tap downstream is at the lower pressure. The precision of differential pressure gauge is significantly greater than subtracting the difference between two pressure gauges.

Example 4.3. Compare the precision of a differential pressure gauge versus that of subtracting the measurements of two pressure gauges in a system operating at 3 barg. The full-scale pressure on the gauges is 5 barg whereas it is 100 mbar for the differential pressure gauge and the precision of the instruments is ±0.01%.

Solution 4.3. Manufacturers may quote the precision either as an absolute value or as a percentage. When it is quoted as a percentage, the reference point is the full-scale measurement. In this case, the precision of the differential pressure gauge is 0.01 bar, whereas it is 0.05 bar for the two pressure gauges. The uncertainty of the differential pressure measured by the two gauges independently is calculated according to:

$$\Delta_f = \sqrt{\sum_{i=1}^{n} (a_i \Delta_i)^2}. \tag{4.12}$$

So,

$$\Delta_{\Delta P} = \sqrt{\Delta_{P_1}^2 + \Delta_{P_2}^2} = \Delta_{P_1} \sqrt{2} = 0.05 \text{ bar} \cdot \sqrt{2} = 0.07 \text{ bar}.$$

4.3.4 Vacuum Pressure

Vacuum is measured relative to atmospheric pressure. Outer space is close to absolute vacuum: the pressure is close to 0 bar but the vacuum is approximately

1.01325 bar. Various analytical instruments operate below 10^{-6} Torr and some processes in the pharmaceutical industry may operate at these low pressures. Chemical processes—distillation columns, for example—may operate at 25 Torr.

Example 4.4. A process is designed to operate at a vacuum of 0.60 bar. What is its operating pressure in Torr?

Solution 4.4. Taking atmospheric pressure equal to 1.01 bar, the pressure of the process equals:

$$P = (1.01 \text{ bar} - 0.60 \text{ bar}) \cdot \frac{1 \text{ atm}}{1.0133 \text{ bar}} \cdot \frac{760 \text{ Torr}}{1 \text{ atm}} = 308 \text{ Torr} \sim 310 \text{ Torr}.$$

The vacuum pressure is seldom used for instrumentation equipment—the absolute pressure is quoted.

4.3.5 Static vs. Dynamic Pressure

Pressure manifests itself under both static and dynamic conditions—the bottom of a lake or the sea is an example of static conditions while both wind blowing or liquid flowing through a pipe are examples of dynamic pressure. The pressure (Pa) at the bottom of a tank or reservoir is equal to the hydrostatic head of the liquid:

$$P = \rho g h, \tag{4.13}$$

where ρ is the fluid density (kg m^{-3}), g is the gravitational constant (m^2/s) and h is the height of liquid in the vessel (m).

Note that the pressure exerted is independent of the geometry: for each of the vessels shown in Figure 4.2 the pressure at the bottom of the column of liquid is equal. The general integral to calculate pressure is expressed as a function of vertical distance, z (m):

$$\int_1^2 dP = \int_1^2 \rho(z) g \, dz. \tag{4.14}$$

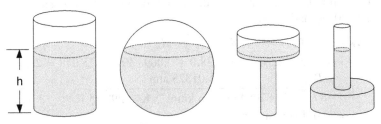

FIGURE 4.2 Equivalence of Static Head

In many cases, the fluid density is constant along the column length. Thus the integral becomes:

$$P_2 - P_1 = \rho g(z_2 - z_1),$$
$$\Delta P = \rho g \Delta z.$$

For the case where the vessel is open to atmosphere, $P_1 = 0$ atmg and thus the gauge pressure is simply equal to the hydrostatic head.

For multi-phase fluid flow (gas-solid, gas-liquid, gas-solid-liquid, etc.) through vertical pipes and vessels, the density often decreases with height. As a first approximation, the pressure drop is assumed to be due entirely to the hydrostatic head of the fluid and the solids (frictional and inertial contributions to the pressure drop are assumed to be negligible—which can introduce an error of up to 10%). The suspension density for a gas-solids system is:

$$\bar{\rho} = \rho_g \varepsilon + \rho_p(1 - \varepsilon),$$

where ρ_g is the gas density (kg m^{-3}), ρ_p is the particle density (kg m^{-3}), and ε is the void fraction.

In the case of air transporting a catalyst at near-atmospheric conditions, this equation may be simplified by neglecting the contribution of the gas density. The transport of petroleum from oil reservoirs is more complicated: (a) gas density in deep wells may be significant; (b) the gas density increases along the length of the pipe; and (c) because of the higher gas velocity the velocity of the liquid phase also increases and thus the void fraction also increases.

Dynamic pressure is measured normal to the direction of a flowing fluid and increases with both an increase in velocity and fluid density:

$$P = \frac{1}{2}\rho u_f^2. \tag{4.15}$$

Example 4.5. When you extend your arm outside the window of an automobile traveling at a high velocity, the air pushes the arm backwards—it experiences dynamic pressure. At what velocity would a car have to travel for a hand extended out of a window to feel the same dynamic pressure as what it would feel in water?

Solution 4.5. The first assumption is that the pressure is atmospheric and the temperature is 25 °C (and the relative humidity is zero). We calculate the density of the air from the composition of the major constituents—nitrogen (78.1%), oxygen (20.%), and argon (0.9%).

$$M_m \approx y_{O_2}M_{m,O_2} + y_{N_2}M_{m,N_2} + y_{Ar}M_{m,Ar}$$
$$= 0.210 \cdot 32.0 + 0.781 \cdot 28.0 + 0.009 \cdot 40.0 = 28.948 = 28.9,$$
$$\tilde{\rho} = \frac{P}{RT} = \frac{1.01325 \text{ atm}}{0.082059 \text{ m}^3 \text{ atm kmol}^{-1} \text{ K}^{-1} \text{ 298.15 K}}$$
$$= 0.041415 \text{ kmol m}^{-3}$$
$$\rho = M_m\tilde{\rho} = 28.9 \text{ kg kmol}^{-1}0.0414 \text{ kmol m}^{-3} = 1.20 \text{ kg m}^{-3}.$$

The density of water is 1000 kg m^{-3} and to achieve the same dynamic velocity in air as water, the air velocity will equal the product of the water velocity and the square root of the ratio of the densities:

$$P_{H_2O} = P_{air} = \frac{1}{2}\rho u_f^2,$$

$$u_{air} = u_{H_2O}\sqrt{\frac{\rho_{H_2O}}{\rho_{air}}} = u_{H_2O}\sqrt{\frac{1000}{1.20}} = 28.9\, u_{H_2O} \approx 30\, u_{H_2O}.$$

The sum of the static and dynamic pressures is termed the stagnation pressure or total pressure.

4.3.6 Barometric Pressure

Barometric pressure changes throughout the day and may even change several percent in an hour: skies are clear at high pressure whereas low pressures are accompanied by clouds and often precipitation. Figure 4.3 illustrates the change in barometric pressure in Montreal during an entire year. The maximum pressure was 103.1 kPa, while the minimum pressure was 96.5 kPa. The maximum change in pressure from one day to the next during the year was 4.5 kPa. The lowest barometric pressure ever recorded at sea level was during a typhoon in the Pacific Ocean at 87.0 kPa in 1979.

Together with barometric pressure, two other factors that are generally neglected in experimental data analysis are the effects of elevation and gravity. The gravitational "constant," g, varies with latitude. Its value is greatest at the poles at approximately 9.83 m^2 s^{-1}, as shown in Figure 4.4, and it is lower by 0.5% at the equator at 9.78 m^2 s^{-1}. The following equation adequately approximates the variation of gravity with latitude for most engineering calculations:

$$g = 9.780372(1 + 0.0053024\sin^2\theta - 0.0000058\sin^2(2\theta)). \qquad (4.16)$$

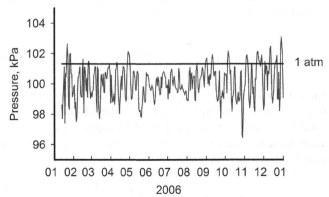

FIGURE 4.3 Variation of Atmospheric Pressure During 2006 in Montreal (Environment Canada, 2011)

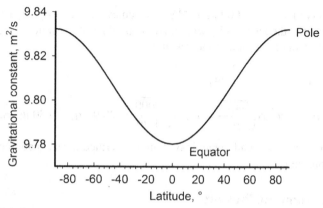

FIGURE 4.4 Relationship Between the Gravitational Constant, g, and Latitude

Elevation has a more appreciable impact on barometric pressure compared to the gravitational constant as shown in Table 4.3. Pascal proved that barometric pressure was caused by the weight of air and that as the elevation increases the pressure decreases.

$$\frac{dP}{dZ} = \rho g.$$

For an ideal gas, the molar density is a function of temperature, pressure, and gas composition:

$$\rho = \frac{M_m P}{RT} = \frac{P}{\bar{R}T}.$$

The molar mass, M_m, of air is 28.95 kg kmol^{-1} and thus the value of $\bar{R}$ is 287 J kg^{-1} K^{-1}. Substituting this relation into the differential equation relating pressure and elevation gives:

$$\frac{dP}{dZ} = \frac{P}{\bar{R}T} g.$$

This equation, when integrated, approximates the pressure poorly because it ignores the change in temperature with height: Figure 4.5 illustrates the variation of atmospheric temperature with elevation. The temperature decreases linearly with elevation in the troposphere and mesosphere but it increases with elevation in the stratosphere and thermosphere:

$$T = T_o + m_\theta Z.$$

TABLE 4.3 Atmospheric Pressure Variation with Elevation

z (m)	P (Pa)	z (m)	P (Pa)
0	101 325	12 192	18 822
457	95 950	15 240	11 665
1067	89 150	21 336	4488
1524	84 309	30 480	1114
1676	82 745	40 000	270
1981	79 685	42 672	201
2591	73 843	45 000	130
3048	69 693	50 000	70
6096	46 599	58 000	30
9144	30 148	63 000	13
10 668	23 988	77 000	1.3

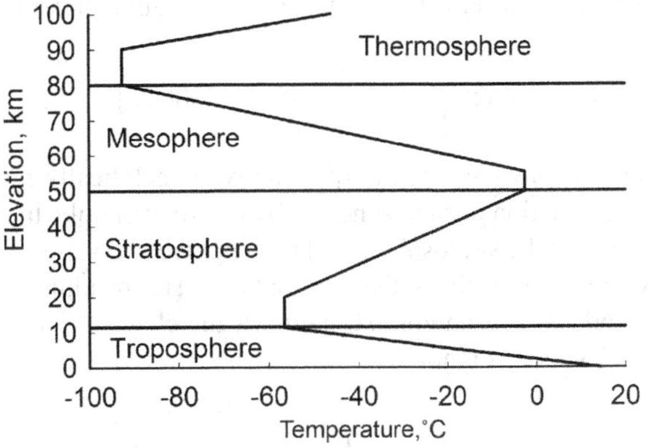

FIGURE 4.5 Atmospheric Temperature Variation with Elevation

To accurately calculate the pressure variation, the temperature variation with elevation must necessarily be included:

$$\frac{dP}{dZ} = \frac{P}{\bar{R}(T_o + m_\theta Z)} g.$$

Isolating pressure and elevation gives:

$$\int_{P_o}^{P} \frac{dP}{P} = \int_{0}^{Z} \frac{g}{\bar{R}(T_o + m_\theta Z)} dZ.$$

The integral becomes:

$$\ln \frac{P}{P_o} = -\frac{g}{m_\theta R} \ln(T_o + m_\theta Z) - \ln T_o.$$

This expression may be rearranged to give the following:

$$P = P_o \left(1 + \frac{m_\theta Z}{T_o}\right)^{-m_\theta \bar{R}/g}.$$

In the troposphere, the temperature drops by about 6.5 °C for every 1000 m ($m_\theta = -0.0065$ K m^{-1}) while it increases by about 1.8 °C for every 1000 m in the stratosphere ($m_\theta = 0.0018$ K m^{-1}).

So, from 0 km to 11.5 km, the equation relating pressure to elevation is:

$$P = P_{\text{baro}}(1 - 0.0000226Z)^{5.25}. \tag{4.17}$$

The value of P_{baro} equals the barometric pressure (which equals 1 atm on sunny days!). From 20 km to 50 km, the equation relating pressure to elevation is:

$$P = 0.0628\left(1 + \frac{0.0018}{216}(Z - 20000)\right)^{-19.0}. \tag{4.18}$$

The pressure at any elevation may be derived analytically by considering the temperature variation in each zone. At 70 km, for example, the temperature is constant in part of the stratosphere and mesosphere and varies with elevation in both these zones as well as the troposphere. The resulting expression is cumbersome and remains an approximation if the change in the gravitational constant with elevation is ignored:

$$g_Z = g\left(\frac{r_e}{r_e + Z}\right)^2, \tag{4.19}$$

where r_e is the radius of the earth (6.371×10^6 m). Rather than calculating the exact pressure variation analytically, an approximate model may be derived through a regression analysis—curve fitting. Parameters in models represent physical phenomena, whereas in curve fitting models they generally have no physical significance. The advantage of curve fitting is that simple expressions may be derived that adequately approximate the data.

Figure 4.6 demonstrates the pressure as a function of elevation and in Figure 4.7 the data is replotted in terms of the log of pressure. (Note that independent variables are usually reserved for the vertical—y-axis. In this case, the dependent variable, P, is shown on the x-axis and elevation is on the y-axis. This designation was adopted to represent the physical phenomena—elevation is in the vertical direction.) While it would be difficult to read the data at

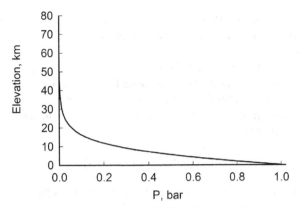

FIGURE 4.6 Atmospheric Pressure Variation with Elevation

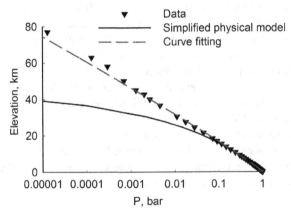

FIGURE 4.7 Atmospheric Pressure (log) Variation with Elevation

elevations above 25 km, the data is more obvious in Figure 4.7—the log of pressure appears to drop linearly with elevation.

The fit between the experimental data and the expression for the elevation in the troposphere is incredibly good with an $R^2 = 0.9999996$. (The agreement between data and model is so good that it is unlikely that the data is experimental but rather derived from the model.) The CRC reports an expression for the elevation as a function of pressure:

$$P = 100 \left(\frac{44331.514 - Z}{11880.516} \right)^{1/0.1902632}. \tag{4.20}$$

This expression has several fundamental deficiencies: in the first place, there are too many significant figures. Considering that atmospheric pressure may vary by as much as 4% in a single day, it is unreasonable to carry eight significant figures for any fitted parameter (ever). Also, there are too many fitted parameters: the coefficients 100 and 11 880.516 are 100% correlated. In fact, this expression is almost exactly equivalent to the model derived for the troposphere!

Regardless of the formulation, Figure 4.7 shows that the expression deviates significantly from the data at elevations above 12 000 m, which is expected since the model was derived for the troposphere. Recognizing that log P decreases linearly with elevations up to 70 km, we propose a two-parameter model (where the units of Z are km):

$$P = P_o \exp\left(-0.114Z^{1.07}\right). \tag{4.21}$$

At sea level ($Z = 0$), the pressure equals 1.01325 kPa and it increases exponentially with Z to the power 1.07. Based on Figure 4.7, we could conclude that this relationship fits the data very well up to 80 km. Figure 4.8 shows a deviation of no more than 10 mbar up until 20 km and then the absolute deviation approaches a very small number. The deviation between the predicted and experimental pressure of the physical model reaches 2 mbar up until the stratopause and then reaches a maximum of 12 mbar before dropping again. The percent deviation in the troposphere is superior for the physical model but the curve fit characterizes the entire data set very well. The error is as high as 50% in the mesosphere but the absolute difference between data and curve fit is only 0.006 mbar at 77 km.

As a practical example of the significance of elevation in chemical processes, consider the steam extraction of essential oils. Low temperatures are preferred to extract oil from plants because of their susceptibility to degradation. Thus, an alternative to vacuum extraction would be to operate at higher elevation.

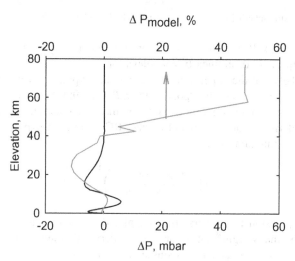

FIGURE 4.8 Model Comparisons versus Elevation (mbar and %)

Example 4.6. Using the Clausius-Clapeyron equation, calculate at what elevation water boils at 90 °C. The heat of vaporization of water is 40.66 kJ mol^{-1}.

Solution 4.6. Since the boiling point of water at atmospheric pressure is known, the Clausius-Clapeyron equation can be rearranged to replace the constant B_{CC}:

$$\ln \frac{P_2^\circ}{P_1^\circ} = -\frac{\Delta H_v}{R}\left(\frac{1}{T_2} - \frac{1}{T_1}\right).$$

Therefore, the pressure at which water boils at 90 °C becomes:

$$P_2^\circ = P_1^\circ \exp\left(-\frac{\Delta H_v}{R}\left(\frac{1}{T_2} - \frac{1}{T_1}\right)\right)$$

$$= 1 \cdot \exp\left(-\frac{40\,660 \text{ J mol}^{-1}}{8.314 \text{ J mol}^{-1} \text{ K}^{-1}}\left(\frac{1}{363 \text{ K}} - \frac{1}{373 \text{ K}}\right)\right)$$

$$= 0.70 \text{ atm.}$$

Expressing the elevation as a function of pressure gives:

$$Z = \frac{1}{0.0000226}\left(\left(\frac{P_1}{P_2}\right)^{0.19} - 1\right) = \frac{1}{0.0000226}\left(\left(\frac{1}{0.7}\right)^{0.19} - 1\right) = 3100 \text{ m.}$$

4.4 PRESSURE MEASUREMENT INSTRUMENTATION

As with most physical properties such as pressure, temperature, humidity, viscosity, etc. analytical instruments measure something other than the property of interest. A signal is recorded—current, voltage, change in height, or deflection—and is then compared to a standard. A basic limitation of all measurements is the necessity of a standard or reference point. In the case of pressure, a good reference point would be absolute vacuum, which is, in practice, impossible to achieve and thus a non-zero quantity is adopted as a standard. The standard for most industrial applications is atmospheric pressure, which is itself a poor standard as we have already seen that atmospheric pressure changes by as much as 6% throughout the year.

Several terms for pressure measuring devices are used interchangeably including transmitters, transducers, gauges, sensors, and manometers. More precisely, a gauge is a self-contained device that converts a force from the process to a mechanical motion of needle or other type of pointer. A manometer is a term reserved for an instrument that measures the hydrostatic head of a liquid and generally operates near atmospheric pressure. A transducer or transmitter combines the sensor with a power supply and a converter—generally mechanical-to-electrical or mechanical-to-pneumatic. The sensor refers to the element that is sensitive to changes in pressure and may either be a mechanical deviation or a change in electrical current.

In general, pressure measurement devices may be classified as either electrical or mechanical. The first sensors were mechanical and relied on flexible elements. Because of the limitations of mechanical motion devices, wire strain gauges were then adapted to measure pressure. Capacitance transducers were developed for vacuum applications. A potentiometric device is attached to a bellows element and converts a mechanical deflection into an electrical signal using a Wheatstone bridge circuit. Resonant wire transducers measure a wide range of pressures from vacuum at 10 Torr up to 42 MPa. They consist of a wire attached to a sensing diaphragm at one end and fixed at the other. The wire vibrates at its resonant frequency due to an oscillator circuit. Changes in pressure result in a change in the resonant frequency of the wire.

Factors influencing the choice of an instrument depend primarily on the pressure of the application and whether the gauge is required for control or simply as an indicator. For control, electronic transmitters are preferred whereas mechanical devices are suitable as indicators. Mechanical indicators are easy to read and interpret but may be difficult to install in tight areas or may require extensive tubing. Electronic transmitters may be easier to install in tight areas or remote connections but require a power supply and wire, as well as additional displays. The dynamic response of electrical instruments is superior to mechanical indicators since the data can be recorded at high frequency.

4.4.1 Barometer

The barometer is perhaps the simplest pressure gauge and it is still commonly found in laboratories to assess barometric pressure. It is one of the few instruments that measures absolute pressure. Before Toricelli, water columns of 10.3 m were necessary to study pressure. Toricelli invented the mercury barometer with a density 13.6 times that of water thus the column height is never much more than 0.9 m. The barometer is composed of a tube, a pool of mercury as well as a scale to measure distance and a vernier used to assess the exact position on the scale to a precision of 0.1 mm.

The tube filled with mercury is immersed in the reservoir of mercury. The height of mercury in the column increases as the pressure increases (the atmospheric pressure exerts itself on the pool of mercury forcing the liquid level to rise). At 1 atm of pressure, the height of the Hg column should equal 760 mm at sea level. At 1 bar pressure, the height of the column is 750.06 mm.

4.4.2 U-Tube Manometer

In undergraduate laboratories, the most common differential pressure measurement device is the U-tube manometer. As shown in Figure 4.9, the

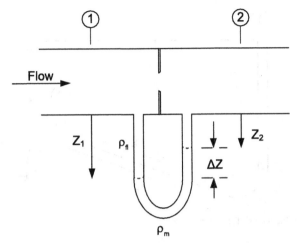

FIGURE 4.9 U-Tube Manometer

principal component is a tube in the shape of a **U** half filled with a fluid with a density, ρ_m. In the figure, a restriction in the pipe causes a pressure drop such that the pressure at point 1 is higher than at point 2. The difference in height, Δz, reflects the magnitude of the pressure differential. When the pressure at point 1 equals that at point 2 (when there is no flow), the height of the fluid in each leg of the tube is equal. As the flow increases, the height z_1 will begin to drop and z_2 will rise correspondingly (as long as the diameters on both sides of the tube are equal). The change in height, Δz, is calculated based on a force balance: the pressure at the top of the meniscus in leg 1 is equal to the sum of the pressure and the hydrostatic head of the fluid. In leg 2, the pressure at the same height equals the sum of the hydrostatic head of the fluid (z_2), the head of the fluid in the manometer (Δz), and the pressure at point 2:

$$P_1 + \rho_{fl}gz_1 = P_2 + \rho_{fl}gz_2 + \rho_m g \Delta z, \qquad (4.22)$$

where ρ_{fl} is the fluid density in the pipe (kg m^{-3}) and ρ_m is the fluid density in the manometer (kg m^{-3}).

The pressure differential is written as:

$$P_1 - P_2 = \rho_{fl}gz_2 - \rho_{fl}gz_1 + \rho_m g \Delta z,$$
$$\Delta P = (\rho_m - \rho_{fl})g \Delta z.$$

Under most conditions when measuring the pressure drop of a gas, its density may be ignored, as $\rho_m \gg \rho_{fl}$. When the second leg is open to atmosphere, the manometer reading represents the gauge pressure—the absolute pressure will equal the pressure differential plus the barometric pressure.

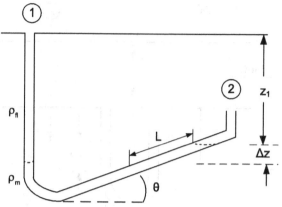

FIGURE 4.10 Inclined U-Tube Manometer

Common fluids used in the manometer are mercury, water, and oil. Maximum pressure differentials achievable are higher with mercury but the precision is much less than one-tenth compared to water. Certain types of oils have a density of around 800 kg m^{-3} and thus the measured ΔP has a 25% higher sensitivity versus water. In order to increase the sensitivity, the second leg may be inclined, as shown in Figure 4.10.

Example 4.7. The low pressure leg of a U-tube manometer is exposed to atmospheric pressure and is inclined at an angle of 30° from the horizontal. The tube diameter is 0.005 m and water is used as the measuring fluid. The resolution of the ruler to measure the displacement along the inclined leg equals 1 mm and the meniscus is 6 mm in the inclined leg.

(a) What is the resolution of the pressure differential in mbar?
(b) What are the differential pressure and its uncertainty, if the length of fluid in the inclined leg equals 140. mm (in mbar and %, respectively)?
(c) The percentage accuracy of the manometer may be improved by (True or False):

  (i) Reducing the angle.
  (ii) Using a tube with a larger diameter.
  (iii) Using oil with a specific gravity of 0.8.
  (iv) Increasing the precision of the ruler.

Solution 4.7a. The difficulty in determining the resolution resides in the measure of the height in the vertical leg and length in the inclined leg. The meniscus is reported to be equal to 6 mm but may not necessarily correspond to the uncertainty in the height. The bottom of the meniscus is the reference point: in the vertical leg, the height might be read to ± 0.5 mm (since the ruler has a resolution of 1 mm) and most likely ± 1 mm in the length direction in the inclined leg. Therefore, the uncertainty of the height of the inclined leg will be:

$$\Delta z_2 = 1 \text{ mm} \cdot \sin{(30°)} = 0.5 \text{ mm}.$$

The uncertainty in the length direction corresponds to a lower uncertainty in the vertical direction of the inclined leg.

$$\Delta z_1 = 0.5 \text{ mm}.$$

According to the theorem of error propagation:

$$\Delta_{\Delta z}^2 = \sum (a_i \Delta_{x_i})^2,$$

$$\Delta_{\Delta z} = \sqrt{\Delta_{\Delta z_1}^2 + \Delta_{\Delta z_2}^2} = \sqrt{0.5 \text{ mm}^2 + 0.5 \text{ mm}^2} = 0.5 \text{ mm} \cdot \sqrt{2} = 0.7 \text{ mm},$$

$$\Delta_{\Delta P} = \rho g \Delta_{\Delta z}$$
$$= 1000 \text{ kg m}^{-3} \cdot 9.81 \text{ m s}^{-1} \cdot 0.0007 \text{ m}$$
$$= 7 \text{ Pa} \cdot \frac{1 \text{ bar}}{100\,000 \text{ Pa}} \cdot 1000 \text{ mbar bar}^{-1}$$
$$= 0.07 \text{ mbar}.$$

Solution 4.7b. As a first approximation, gas density is neglected, so the pressure drop is calculated according to:

$$\Delta P = \rho g \Delta z.$$

The liquid displacement along the inclined tube equals 140 mm above the meniscus in the vertical leg. Its vertical height is:

$$\Delta z = 140 \text{ mm} \cdot \sin{(30°)} = 70 \text{ mm} = 0.070 \text{ m},$$

$$\Delta P = 1000 \cdot 9.81 \cdot 0.070 \cdot \frac{1000}{100\,000} = 6.9 \text{ mbar},$$

$$\Delta P = 6.9 \text{ mbar} \pm \frac{0.07}{6.9} \cdot 100 = 6.9 \text{ mbar} \pm 1\%.$$

Solution 4.7c.

   (i) True: Reducing the angle will increase the % accuracy (assuming that the ability to read the meniscus remains unchanged).
   (ii) False: Using a tube with a larger diameter should have little impact on the ability to read the meniscus and thus the precision should remain the same.
   (iii) True: Using oil with a specific gravity of 0.8 will increase the precision by 25% since the height change will increase 25% more for the same pressure differential.
   (iv) False: It would be difficult to be able to read the ruler better than 1 mm.

4.4.3 Bourdon Gauge

The Bourdon gauge is the most widely used pressure indicator in laboratories, the process industry, power generation, gas transmission, etc. Whereas U-tube manometers cover only a narrow range of pressures (from several mm of water to 1 bar), the range of the Bourdon gauge is very wide: from 10 mmHg (absolute pressure) up to 10^5 bar. As shown in Figure 4.11, the pressure is indicated by a needle that rotates around a pivot at the center. The precision of these gauges is typically on the order of $\pm 1\%$ but laboratory test pressure gauges have an accuracy of $\pm 0.1\%$FS. In many gauges, the precision is limited due to the number of graduation marks around the circumference of the outer edge. When the indicator is open to atmosphere, the needle points to zero—all positive values above zero are referred to as gauge pressure. Vacuum is read as positive values below zero gauge. For example, a needle indicating 0.3 bar below the zero represents 0.3 bar vacuum or 0.7 bara (absolute pressure).

The Bourdon gauge was developed in 1849 by the French engineer Eugene Bourdon. The gauge is based on the principle that a tube will expand when a pressure is applied to it. The tube is closed at one end and open to the process fluid at the other end. The tube is in the shape of an arc and the curvature of the arc increases when pressure is applied to it and it decreases when the pressure drops. The closed end of the tube is connected to a pivot and a spring that translates the movement of the arc into a rotation of a needle. The tube can be made of many different metals—bronze, steel, brass, etc.—depending on the application.

4.4.4 Diaphragm and Bellows

Two other types of aneroid gauges (aneroid meaning without liquid) include the diaphragm and bellows. The bellows gauge resembles the Bourdon gauge in that it relies on pressure to change the volume element. Figure 4.12 illustrates a bellows gauge expanding and contracting in a linear direction instead of in an arc like the Bourdon tube. The movement of the bellows may either be displayed

FIGURE 4.11 Bourdon Gauge

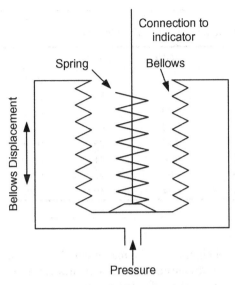

FIGURE 4.12 Bellows Gauge

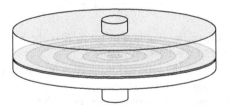

FIGURE 4.13 Diaphragm

mechanically with a needle or converted to an electrical signal. Because of the relatively large displacement, it is unsuitable for transient measurements.

The diaphragm gauge, on the other hand, has a low mass and a small displacement, which makes it ideal for transient applications. It is essentially a thin disk separating two regions as shown in Figure 4.13. When a pressure is applied to one side, the disk deforms elastically. The displacement is proportional to the applied pressure differential and can measure differentials as low as 1 Pa (approximately 0.01 Torr). The range and sensitivity may be increased by corrugating the disk but in this case a mechanical device is generally used as the sensing element. Otherwise, the deformation may be sensed with a strain gauge, which is a device bonded to the surface of the diaphragm. The deformation of the diaphragm causes a change in the electrical resistance of the strain gauge, which is measured with a Wheatstone bridge.

The deformation of the diaphragm depends on the physical geometry of the gauge as well as on the properties of the material of the disk. An important design

TABLE 4.4 Physical Properties of Materials

Material	E, GPa	G, GPa	ν	ρ, kg m^{-3}
Carbon steel	200	79	0.29	7760
Stainless steel	190	73	0.31	7760
Nickel steel	200	76	0.29	7760
Monel	179	66	0.32	8870
Inconel	214	79	0.29	8590
Cast iron	100	41	0.21	7200
Aluminum alloys	70	27	0.33	2720
Beryllium copper	124	48	0.29	8310

criterion of the diaphragm is that the maximum allowable displacement, y_{max} (mm), should be no more than one-third the thickness, t (mm). The displacement is calculated according to the following equation:

$$y = \frac{3\Delta P}{16Et^3} r^4 (1 - v^2),\qquad(4.23)$$

where r is the radius of the diaphragm (mm, E is the modulus of elasticity (Young's modulus) and v is Poisson's ratio.

Table 4.4 summarizes the physical properties of common metals used for diaphragms. Young's modulus is a measure of the stiffness of elastic materials and is defined as the ratio of stress (uniaxial) versus the strain (uniaxial), i.e. the ratio of the change in length of an elastic material as a function of a tensile or compressive load. The Poisson's ratio is a measure of a material's tendency to contract under a tensile load, i.e. the contraction in the direction normal to the direction in which the material is stretched. The symbol for the shear modulus of elasticity is G.

Example 4.8. Calculate the thickness of a diaphragm to measure a differential pressure of 10.0 kPa in air at 1.00 atm and 20.0 °C. The membrane is made of stainless steel and has a diameter of 1.0 in.

Solution 4.8. The properties of stainless steel are given in Table 4.4:

$$E = 200 \text{ GN m}^{-2} = 200 \times 10^9 \text{ N m}^{-2},$$
$$v = 0.3,$$
$$\Delta P = 10\,000 \text{ Pa},$$
$$D = 1 \text{ in} \cdot 25.4 \text{ mm in}^{-1} = 25.4 \text{ mm},$$
$$r = D/2 = 12.7 \text{ mm}.$$

The maximum deflection of the diaphragm should be no more than one-third the thickness:

$$y_{max} = \frac{3\Delta P}{16Et^3}r^4(1 - v^2) < \frac{1}{3}t.$$

Substituting y_{max} for $t/3$ and rearranging the expression in terms of thickness give:

$$t^4 = \frac{9\Delta P}{16E}r^4(1 - v^2).$$

So, the thickness to the fourth power is related to the radius to the fourth power:

$$t^4 = \frac{9\Delta P}{16E}r^4(1 - v^2) = \frac{9 \cdot 10\,000\ \text{Pa}}{16 \cdot 200 \times 10^9\ \text{Pa}}(12.7\ \text{mm})^4 \cdot (1 - 0.31^2)$$
$$= 6.61 \times 10^{-4},$$
$$t = 0.160\ \text{mm}.$$

4.4.5 Vacuum

The first gauge to measure very high vacuum (low absolute pressure) was invented by McLeod in 1874. The gauge is capable of assessing pressures as low as 10^{-6} bar or 10^{-4} Torr. They are still common to laboratories and serve well as calibration tools but they have been largely replaced by electronic vacuum gauges for analytical instruments.

Throughout antiquity, the notion of a vacuum has been abhorred. Greek philosophers posed the question "How can nothing be something?" The Catholic Church considered it heretical and against nature in medieval times since the absence of anything translated to the absence of God. In a practical sense, vacuum in the form of cavitation had restricted the ability to pump greater than ten meters of water—one atmosphere—up until medieval times. Cavitation is the rapid formation and implosion of voids—localized cavities. However, the Romans in Pompeii had a dual-action suction pump which allowed them to pump greater than ten meters of water. Suction pumps were also described by Al-Jazari in the thirteenth century. The Mayor of Magdeburg (Germany) Otto van Guericke invented the first vacuum pump. He demonstrated the concept by evacuating the air from two hemispheres: a team of eight horses was attached to each hemisphere, and they pulled until they could no more. From 1850 until 1970 the technology to create a vacuum improved continually—every ten years, the maximum vacuum produced dropped by an order of magnitude from 10^{-1} mbar to 10^{-13} mbar during this period.

The suction pressure of vacuum cleaners is about 80 kPa and the pressure in an incandescent light bulb varies from 1 Pa to 10 Pa. The vacuum in intergalactic space is estimated to be equal to 10^{-22} mbar, which is equivalent to 4 molecule m^{-3}. Table 4.5 classifies the different categories of vacuum together with the types of instruments generally used for each type. A single instrument is incapable of covering the full range of 10 mbar vacuum to 10^{-11} mbar. Several types of gauges are required to cover the range continuously.

TABLE 4.5 Vacuum Classification

Classification	Range	Measurement Technology
Low vacuum	$P > 3$ kPa	U-tube manometer, capsule gauge
Medium vacuum	10 mPa $< P <$ 3 kPa	McLeod, thermal, capacitive gauges
High vacuum	100 nPa $< P <$ 100 mPa	Ion gauge
Ultra-high vacuum	0.1 nPa $< P <$ 100 nPa	

FIGURE 4.14 BOC Edwards Capsule Pressure Gauges (Harris, 2005)

4.4.6 Capsule Pressure Gauge

Capsule pressure gauges are similar to Bourdon gauges but they are based on two diaphragms attached and sealed around the circumference whereas the Bourdon gauge relies on a sealed narrow tube. To measure very low pressures, several capsules may be combined. BOC Edwards Capsule pressure gauges are shown in Figure 4.14 with varying ranges—from vacuum pressures of 25 mbar to as high as 1040 mbar—or 1 mbar absolute pressure. Typically, these gauges have an accuracy of at least $\pm 2\%$FSD (full scale design) if not $\pm 1\%$FSD. They are for measuring the gas phase only. Note that as vacuum increases, the needle rotates clockwise.

A capacitance manometer refers to the device that senses the deflection of the diaphragm. Capacitance manometers are typically an order of magnitude more accurate than standard capsule dial gauges.

4.4.7 McLeod Gauge

The McLeod gauge consists of a finely calibrated capillary tube attached to a large spherical bulb whose volume, V_b, is known precisely. The base of the bulb is connected to both a reservoir of mercury as well as a reference

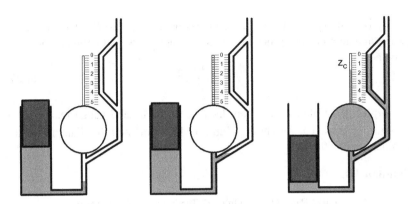

FIGURE 4.15 McLeod Gauge: (a) Open to Process Gases; (b) Trapped Process Gas P_1; (c) Compressed Pressure P_c

column that leads to the process. Initially, as shown in Figure 4.15, the bulb and capillary communicate with the reference column and process so that the pressure throughout equals P_1. Mercury is then pumped into the system until it reaches the cut-off point which effectively seals that bulb and capillary from the process; the pressure throughout remains at P_1. More mercury is pumped into the system such that the level increases in the bulb and the reference column. It is pumped into the system until the mercury reaches the zero point on the reference column. The height of the gap in the capillary, z_c, represents the mercury level difference on the reference column and thus the hydrostatic head in the capillary, P_c.

The volume of gas in the capillary, when the mercury reaches the zero point on the reference column, is equal to the product of the cross-sectional area and the gap distance, $X_{A,c}z_c$.

According to Boyle's law, as the volume decreases, the pressure increases proportionately. Therefore, we can equate the pressure and volume in the capillary when the mercury reaches the cut-off point to that in the capillary:

$$P_1(V_b + V_c) = P_c X_{A,c} z_c, \tag{4.24}$$

where V_c is the capillary volume and P_c is the capillary pressure.

Therefore, the process pressure is expressed as a function of the capillary pressure and volumes:

$$P_1 = P_c \frac{X_A z_c}{V_b + V_c}. \tag{4.25}$$

The gap distance in the capillary, z_c (the hydrostatic head of mercury), equals the difference in the capillary pressure and the process pressure (all units in mmHg):

$$z_c = P_c - P_1. \tag{4.26}$$

By design, $V_b \gg V_c$ and as a consequence, $P_c \gg P_1$. Therefore, the expression relating the process pressure to geometry and capillary pressure becomes:

$$P_1 = \frac{X_A}{V_b} z_c^2. \tag{4.27}$$

Example 4.9. Calculate the gap distance of a McLeod gauge measuring a pressure of 0.400 Pa. The bulb volume is 15.0 cm^3 and the diameter of the capillary is 0.300 mm.

Solution 4.9.

$$P_1 = 0.400\,\text{Pa} \cdot 760\,\text{mmHg}/101\,325\,\text{Pa} = 0.00300\,\text{mmHg},$$

$$z_c = ?$$

$$P_1 = \frac{X_a h^2}{V_B},$$

$$h = \sqrt{\frac{P_1 V_B}{X_a}} = \sqrt{\frac{0.00300\,\text{mmHg} \cdot 15\,000\,\text{mm}^3}{\frac{\pi}{4}0.300\,\text{mm})^2}} = 25.2\,\text{mm}.$$

4.4.8 Pirani Gauge

The Pirani gauge is based on heat transfer and thermal conductivity and has a wide operating range—from atmospheric to 10^{-4} mbar. A filament (platinum or tungsten) is placed in the measuring chamber and is connected to an electrical circuit (Wheatstone bridge). The circuit can be operated such that the wire is maintained at a constant temperature or so that it heats up or cools down with the operating pressure. When the pressure is high, there are more frequent collisions between molecules and the wire that cause the temperature to drop. Conversely, when the pressure is low, the wire temperature will be high. The resistance of the wire is proportional to its temperature. The thermocouple gauge is based on the same principle as the Pirani gauge except that the temperature of the filament is monitored by a thermocouple. The useful range of the thermocouple gauge is somewhat lower at between 1 mbar and 10^{-2} mbar.

4.5 PROCESS EQUIPMENT AND SAFETY

Pressure is measured at the exit of most vessels, at the discharge of pumps, inlet and outlet of compressors, and flow meters, along columns—distillation, extraction, reactors—sample lines and analytical equipment. Operating pressure has a significant impact on vessel design, metallurgy, control and operation. At moderate pressures (between 1 barg and 20 barg), the overall investment costs may be relatively insensitive to pressure. For gas systems, higher pressures result in smaller vessels, pipes, control valves, flow

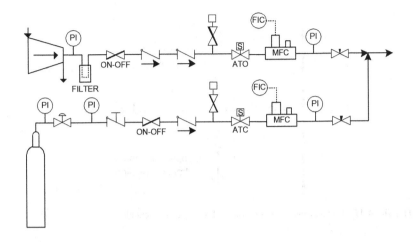

FIGURE 4.16 Elements of a P&ID to Feed Air from a Compressor and N_2 from a Cylinder

meters, etc. However, beyond 2 bara, compressors are required and cost an order of magnitude more than blowers and fans. Safety precautions and control systems must be more robust at higher pressures. Moreover, higher-grade steel may be required to conform to vessel code standards. With respect to carbon steel, 304SS and 316SS are approximately three times more expensive, while Monel and Titanium are between 5 and 10 times the cost. Besides the type of metal, investment cost depends on vessel weight, fabrication cost, complexity as well as operating pressure. The vessel wall thickness, t, from which the weight is derived, is calculated according to the following equation:

$$t = \frac{Pr}{2SE + 0.4P},$$ (4.28)

where r is the radius (mm), S is the maximum allowable stress (psi) (e.g. $S = 18\,800$ psi for SA-516 grade 70), and E is the weld efficiency (0.95 for radioagraphed first-class vessels and 0.6 for nonradiographed third-class vessels).

Figure 4.16 is a process flow diagram of the instrumentation for nitrogen and air leading to a reactor. Air is compressed then filtered and the discharge pressure is measured. Following the discharge of the compressor is an on-off valve and two check valves in series to prevent backflow of gas to the compressor. (Reliability of one-way valves is poor and so two have been placed in the line. Dust, oil, or other debris will block the seating of the valve and for this reason redundancy is recommended.) After the one-way valves, there is a pressure indicator and a ball valve to maintain the pressure at a certain level at the discharge of the compressor. An air-to-on (ATO) solenoid valve and a mass flow controller come after the needle valve: if electricity is lost, the solenoid valve automatically closes. Flow meters are calibrated at specified pressures and thus a needle valve is placed downstream of the meter plus a pressure gauge.

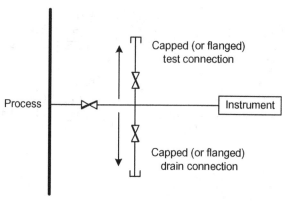

FIGURE 4.17 Typical Line Configuration for Process Instrumentation

The configuration shown in Figure 4.17 is recommended to connect an instrument to a process. The instrument line leads to a four-way connector; one line goes to the process, another to the vent (or drain), and a third to a test line. Each of the three lines has a valve to isolate the line from the instrument. During normal operation, the valves to the vent and test line (also capped) are closed and the line leading to the process is open. To calibrate the instrument (or replace/service it), the process line is first closed and the vent line is opened to evacuate the fluid and to drop the operating temperature. If the instrument is to be calibrated, the cap is removed, the valve is opened, and the line to the vent is then closed. To bring the line back on service, the test-line valve is closed and capped. The vent valve is opened, which drops the line pressure to atmospheric and then it is closed. Finally, the valve to the process line is opened. In some cases, the lines may be purged in order to clear the lines of any potential contaminates (like oxygen for example).

In many cases, opening and closing the vent lines must be done carefully to reduce sudden changes in pressure that could destroy the instrument. To bring the instrument back on service, for example, a secondary line may be attached to the vent line to pressurize the instrument slowly. The line may be overpressurized so that when the valve to the process line is opened, the purge fluid flows into the process instead of process fluid entering the instrument line.

4.5.1 Pressure Regulator

Instrumentation in the nitrogen line is similar to that of the air line. A gauge with a range of up to 4000 psig measures the cylinder pressure. It is an integral part of the pressure regulator that drops the line pressure closer to the desired operating pressure to maintain the line at safe and operable value. Often it will be set at 100 psig to 200 psig, which is measured by the second gauge on the regulator. Pressure regulators are specialized valves that interrupt flow when the line upstream drops below the set pressure: it will shut off when the cylinder pressure drains to the set point pressure. Two-stage regulators are common for gas cylinders, as is the case for nitrogen.

Check valves are infrequent on inert gas lines since there is no inherent safety risk if a hydrocarbon (or oxygen) were to flow back to the regulator. Backflow of hydrocarbon to an oxygen cylinder—or vice versa—presents a serious hazard.

As opposed to the air line, the nitrogen line has an air-to-close solenoid valve (ATC). In the case of an emergency, it is preferable to sweep the lines with an inert gas to reduce the possibility of forming a flammable mixture.

The air line and nitrogen line meet at a four-way valve. The valve has two entrances and two exits. One exit goes to the reactor and the other goes to a vent. To measure the hydrodynamics of the reactor—the gas flow pattern—the valve position is such that nitrogen sweeps the reactor. When the valve is switched, the air goes to the reactor and the nitrogen exits the vent. The oxygen in the air is analyzed at a high frequency and the signal is used to interpret the flow pattern or any flow abnormalities such as bypassing. Note that when the reactor and vent are open to atmosphere, the operating pressure equals the barometric pressure and this varies as much as 6% throughout the year with daily fluctuations of as much as 3%.

4.5.2 Back Pressure Regulator

When the valve switches from nitrogen to air, the flow of gas will change abruptly if the vent line and reactor line are inadequately equilibrated; the pressure at both exits of the four-way valve must be equalized. This may be accomplished by a back pressure regulator. Whereas a pressure regulator reduces the supply pressure at the inlet to a lower pressure at the outlet, a back pressure regulator throttles the flow downstream to maintain the inlet pressure. A needle valve or a back pressure regulator is required at the vent line to match the pressure across the reactor, process lines, and analytical equipment.

4.5.3 Relief Valves

Back pressure regulators provide steady-state control; relief valves provide on-off protection from overpressure or vacuum conditions. When the set pressure threshold is exceeded, the valve opens either to atmosphere or to an auxiliary line (where the fluid may go to a flare or storage tank or even to recycling). Once the vessel pressure drops to a predetermined pressure, the valve reseats. The difference in pressure between when the valve relieves and when the valve reseats is referred to as the blowdown; it is typically 2–20% lower than the relief pressure.

FIGURE 4.18 Vessel Failure Due to Vacuum

4.5.4 Rupture Disk

Most pressure vessels require rupture disks for which designs are specified according to ASME or other international standards codes. They may protect from either overpressure or vacuum conditions. It is composed of a membrane that will instantaneously fail (within milliseconds) at a specified differential pressure. Figure 4.18 shows a horizontal vessel that failed because it was poorly vented or due to a failure in the rupture disk.

Example 4.10. Methanol from a 10 m^3 reservoir at 20 °C is pumped to a rail car for shipment. The total volume of methanol in the reservoir is 9 m^3 and the vapor space is blanketed with nitrogen at a pressure of 1.2 bara:

(a) Calculate the vacuum pressure in the vessel when 1 m^3 of methanol remains in the reservoir (assuming the vent line was not opened, the rupture disk failed, and atmospheric pressure is 1 bar).

Solution 4.10. As a first approximation, assume that the methanol volatility is negligible, then:

$$P_1 V_1 = P_2 V_2,$$
$$P_2 = P_1 \frac{V_1}{V_2} = 1.2 \text{ bara} \frac{1 \text{ m}^3}{8 \text{ m}^3} = 0.15 \text{ bar}.$$

Vacuum pressure is the difference between atmospheric pressure and absolute pressure:

$$P_{vac} = 1 \text{ bar} - 0.15 \text{ bar} = 0.85 \text{ bar}.$$

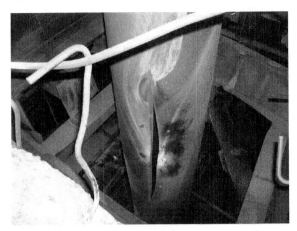

FIGURE 4.19 Burst Pipe

4.5.5 Pressure Test

The design, fabrication, operation, and inspection of pressure vessels and lines are strictly regulated by engineering agencies such as the ASME (American Society of Mechanical Engineers). The legal codes are very specific with respect to the materials of construction, wall thickness, design allowances, and operating conditions. Several tests may be used to verify the integrity of the vessel, welds, fittings, and ancillary equipment. They may be carried out at pressures lower than the specified operating pressure or at substantially higher pressures. Buried gas and oil pipelines are tested at 125% of the maximum operating pressure (MAOP). Some codes require vessels to be tested at 150% MAOP.

Many countries enact legislation regarding the frequency of testing the integrity of pressure vessels. Gas cylinders (high pressure) are generally tested every two years while fire extinguishers might be tested on a frequency of every five to ten years.

Generally, water or another incompressible fluid, like oil, is preferred over gases for pressure testing (Saville et al., 1998). If the vessel fails at high pressure, a sound wave (loud bang) may develop. When using a compressible gas, a shock wave (a pressure wave propagating at supersonic speeds) can be generated. An additional hazard in both cases is the possibility of forming high-speed missiles. Figure 4.19 is a picture of a ruptured pipe: the pipe had been mislabelled and was installed on a steam line at a higher pressure than was specified for the type of metal.

4.5.6 Leak Test

Leak testing is not only mandatory for testing pressure vessels, it is recommended for all process lines and analytical equipment. Besides the

FIGURE 4.20 Leak Test of a Glass Joint

obvious safety hazards of leaking gases or liquids, very small leaks on the lines leading to pressure gauges may have a profound effect on the recorded signal. In the laboratory, the most common technique to detect leaks is to bathe the outer surface of the connector with soapy water. Figure 4.20 shows a very large bubble forming on the ball joint connecting a quartz reactor to the feed line. Large bubbles are indicative of large gas velocities whereas small bubbles represent slow leak rates. Large bubbles are obvious immediately whereas small bubbles might take some time before they appear. Bubbles can also foam. At times, soapy water is insufficient to detect leaks and in these cases a very sensitive detector may be used. These leak detectors are described in the section concerning gas chromatography and are explained in Chapter 8.

4.6 EXERCISES

4.1 Derive an expression for the gas constant, R, as a function of Boltzmann's constant based on the kinetic theory of gases.

4.2 Calculate the temperature, V_{rms}, and concentration of 1 mole of O_2 and 1 mole of H_2 in a one L flask at 1 atm and 10 atm pressure.

4.3 What is the partial pressure of oxygen on the summit of K2 (elevation 8611 m)?

4.4 A reservoir 1 m in diameter open to atmosphere contains 10 m^3 of water at 12 °C. The barometric pressure is 770 mmHg. Calculate the absolute pressure at the bottom of the reservoir.

4.5 A U-tube manometer is filled with mercury at a specific gravity of 13.6 to measure the pressure of air exiting a pipe. The low pressure end of the pipe is open to atmosphere at a pressure of 99.2 kPa. The differential height of the manometer is 20.5 cm at 25 °C. What is the pressure in the pipe?

4.6 A mass spectrometer (MS) is an analytical instrument that determines the composition and concentration of compounds—mixtures of gases, liquids, or solids. A sample is ionized to form charged molecules or fragments and the mass-to-charge ratio (m/z) is then analyzed. Most mass spectrometers operate at 10^{-6} Pa to minimize collisions with other molecules. Dual pumps are required. The first brings the pressure down to 0.1 Pa while the second brings it down to the operating pressure:

(a) Calculate the number of molecules in an MS vacuum chamber operating at 10^{-6} Pa ($V = 2.0\,1, T = 25\ °C$).

(b) The New York City meteorological service reports a barometric pressure of 30.30 inHg. The pressure gauge after the first pump indicates a vacuum of 760.1 Torr. Is the pump operating correctly?

(c) If the instrument were located in Albany at an altitude of 99 m, would your conclusion in (b) change? (Note that the meteorological service corrects the barometric pressure to sea level.)

4.7 The level of oil in a reservoir is measured with a U-tube manometer (see Figure Q4.7. The pressure at the top of the liquid is 2.5 psig. The barometric pressure at sea level is 780 mmHg. The reservoir is located at Machu Pichu at an elevation of 2450 m. The manometer contains mercury ($\rho_{Hg} = 13\,600$ kg m^{-3}) and the density of the oil is 865 kg m^{-3}:

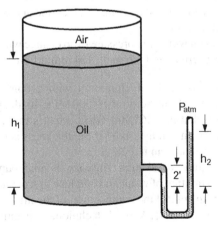

FIGURE Q4.7 Machu Pichu Oil Reservoir

(a) What is the atmospheric pressure at Machu Pichu in mbar?

(b) Calculate the height of the oil in the reservoir (h_1) when the height in the manometer (h_2) is equal to 1.45 m.

(c) What is the uncertainty in the measure of h_1 if the uncertainty in the measure of h_2 is 1.0 cm?

(d) You decide to replace the U-tube manometer with a diaphragm. What would be the ideal diameter for the reservoir if the maximum height, h_1, is 15 m? The diaphragm is made of steel ($E = 190$ GPa, $\mu = 0.29$, and thickness of 0.07 mm).

4.8 A U-tube manometer measures the pressure differential of air in a pipe operating at 20 MPa and 22 °C. The fluid in the manometer is a 50/50 (volumetric) mixture of water and glycerine. *C. Abou-Arrage*

(a) Calculate the weight fraction and density of the manometer fluid knowing that the mixture volume contracts thereby increasing the density according to the following relationship:

$$\rho_{mix} = \rho_{th}\gamma,$$

where:

$$\rho_{th} = \frac{\rho_{gl}}{x - (1-x)\rho_{gl}/\rho_{H_2O}},$$
$$\gamma = 1 + 0.0283x(1 - x^2),$$

and ρ_{gl} equals 1.262, ρ_{H_2O} is 1000, and x is the mass fraction of glycerine.

(b) What is the pressure differential for a differential height of 124 mm (in ψ) with a 50/50 mixture of glycerine and water as the fluid in the manometer.

(c) How much would the pressure drop change with pure water? With pure glycerine (and the same 124 mm differential height)?

4.9 A transducer is made of aluminum with a Poisson's ratio of 0.35 and a modulus of elasticity of 69 000 MPa. Its diameter is 1 cm. For a diaphragm that is 0.265 mm thick, what is the maximum pressure differential that it can measure? If the thickness is reduced by a factor of 5, what pressure can it measure?

4.10 The temperature of skating rink ice is maintained at −7 °C by circulating a mixture of water and ethylene glycol underneath the ice. It is important to control the composition of the mixture and samples are taken from time to time. A simple technique to quantify the composition is to measure the density, from which the composition may be inferred. Table Q4.10 summarizes voltage measurements for different liquid levels in a U-tube manometer of the sample. The calibration curve relating the pressure and voltage is given by:

$$P = 0.924V + 0.800,$$

where P is the pressure (kPa) and V is the voltage (mV). Assume that the densities of water and ethylene glycol are 1000 kg m^{-3} and 1132 kg m^{-3}, respectively. *F.-X. Chiron*

TABLE Q4.10 Pressure Voltage Readings as a Function of Liquid Level

Level (mm)	Pressure Transducer Voltage (mV)
21	−0.636
129	0.541
354	3.030
583	5.519

(a) Based on the measurements, calculate the weight percent of ethylene glycol.

(b) The target composition is 48 wt.% ethylene glycol. To reach this value, should water or ethylene glycol be added?

(c) What is the mass and volume of liquid required if the total volume in the reservoir equals 900l?

4.11 The summit of the mountain shown in Figure Q4.11 reaches an altitude of 3648 m (Point *A*). The lake is 20 m deep and is situated at an altitude of 1097 m (Point *B*). Point *D* is situated at a depth of 150 m below the surface of the water. Note that $\rho_{\text{air@20 °C}} = 1.20$ kg m^{-3}, $\rho_{\text{ocean}} = 1030$ kg m^{-3}, and $\rho_{\text{lake}} = 1000$ kg m^{-3}. *C. Mathieu*

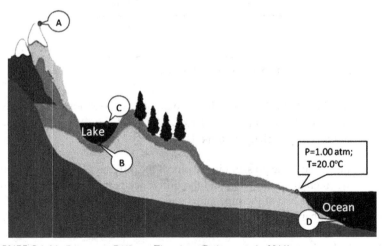

FIGURE Q4.11 Pressure at Different Elevations (Patience et al., 2011)

(a) Determine the absolute pressure at Points A, B, C, and D.

(b) If the uncertainty in the depth of the ocean is ± 10 m and that of the density of water is ± 10 kg m^{-3}, what is the uncertainty in the absolute pressure at Point D?

4.12 An inclined U-tube manometer is filled with water as the measuring fluid. The vertical leg is connected to the upstream of a venture meter and the downstream leg is at an angle of 30° with respect to the horizontal. The ruler has marks at intervals of 1 mm:

(a) Calculate the uncertainty in the pressure in mmH$_2$O if the uncertainty in the angle is $\pm 3°$.

(b) Calculate the pressure differential and uncertainty if the displacement of the fluid in the inclined leg reaches a distance of 1.110 m.

4.13 As shown in Figure Q4.13, a U-tube manometer measures the pressure drop of water in a pipe. If the height difference in the two legs of the manometer filled with mercury is 15.2 cm, what is the pressure difference between points A and B?

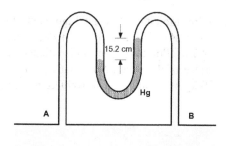

FIGURE Q4.13 U-Tube Manometer for Gas Phase Streams

4.14 A U-tube manometer is built with one end open to atmosphere as shown in Figure Q4.14. Oil, with a density of 0.750 kg m^{-3}, flows in the pipe and mercury is charged to the U-tube:

(a) What is the pressure in the pipe if the difference in elevation of the mercury is 340 mm and the difference between the height of the lower end of the pipe and the top of the mercury is 1.200 mm?

(b) What is the pressure if the mercury level increases by 10.0 cm?

4.15 In the eighteenth century, the Prince of Conde undertook renovations for the castle of Chantilly and built canals and fountains. The centerpiece of

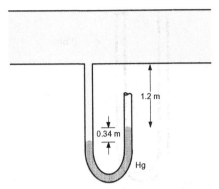

FIGURE Q4.14 U-Tube Manometer for Liquid (Low Pressure) Streams

the work was a fountain with a volumetric flow rate of 1 m³ h⁻¹ capable of projecting water to a height of 10 m. Water for the fountain was available from the Oise River to the west of the castle (at an elevation of 40 m above sea level) and the Nonette River to the east 105 m above sea level). The castle is at an elevation of 70 m. *B. Carrier*

(a) Calculate the absolute pressure of the water at the base of the fountain.

(b) Which of the two rivers should serve as the source for the fountain?

(c) To avoid the use of pumps, the Prince's architect, Le Nôtre, built a reservoir upstream at an elevation sufficient to project the water and to compensate for the pressure loss due to friction. If the friction loss equals 50% of the total head required, at what elevation should the reservoir be built?

(d) If the architect built the reservoir at a distance 40% further away at an elevation of 90 m higher, and the volumetric flow rate increased by 20%, calculate the height of the fountain. Note that $h_f = \frac{u^2}{2g}\left(\frac{L}{D}\cdot \text{const.}\right)$.

4.16 You wish to check your electronic altimeter with two portable U-tube manometers closed at one end (Figure Q4.16). The vertical height of the manometer is 0.50 m and you fill one with mercury and the other with water at sea level. Note that $\rho_{Hg} = 13\,600$ kg m⁻³, $\rho_{H_2O} = 1000$ kg m⁻³, $g = 9.81$ m s⁻². *J. Fortin*

(a) What is the uncertainty in the height and pressure measurement?

(b) What is the maximum altitude that can be read with the mercury U-tube?

(c) What is the maximum altitude that can be read with the water U-tube?

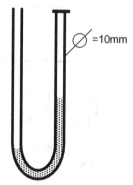

FIGURE Q4.16   Portable U-Tube Manometer Closed at One End

4.17 A spring type diaphragm pressure transducer is made of carbon steel to measure a pressure differential of 1000 psi. The diameter of the diaphragm is 0.5 in. Calculate the thickness of the diaphragm, in mm, so that the maximum deflection is less than one third its thickness.

4.18 The reservoir of a McLeod has a volume of 15 cm^3 and its capillary diameter is 0.3 mm. Calculate the height of mercury in the capillary at a pressure of 4 Pa.

4.19 A manometer contains an unknown liquid. To identify the liquid density, air pressure is applied at one end of the manometer and it is measured with a transducer. Table Q4.19 gives the reading of the transducer together with the corresponding height differential of the manometer. The relationship between the transducer reading and pressure (in psi) is given by:

$$P = 87.0V - 307.$$

(a) Calculate the pressure differential corresponding to each measurement (in kPa).

(b) Identify
the unknown liquid. Note that $\rho_{water} = 1000$ g l^{-1}, $\rho_{acetone} = 790$ g l^{-1}, $\rho_{glycerine} = 1260$ g l^{-1}, $\rho_{oliveoil} = 920$ g l^{-1}, and $\rho_{ether} = 710$ g l^{-1}.

TABLE Q4.19 Manometer Calibration Measurements

E, mV	4.00	5.00	6.00
ΔP, cm	16.0	49.7	83.4

4.20 A high pressure manometer is made to measure a pressure differential of air of up to 14.0 MPa 20.0 °C. Supposing the oil in the tube has a specific

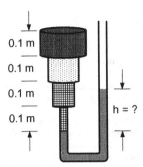

0.1 m

0.1 m

0.1 m

0.1 m

h = ?

FIGURE Q4.21 Mercury Manometer Connected to Multiple Cylinder Tank

gravity of 0.830, calculate the differential pressure corresponding to a differential height of 135 mm. If the fluid pressure was lowered to 3 MPa, how much would the differential pressure change with the same differential height?

4.21 A mercury manometer is connected to the bottom of a tank built with a series of cylinders. Each cylinder contains a different substance: oil (SAE 30), water or glycerin. The cylinder diameters are respectively 0.40 m, 0.30 m, and 0.20 m. Calculate the height h, shown in Figure Q4.21. *C.-G. Castagnola*

4.22 What is the pressure at an altitude of 5000 m on Venus?

4.23 A 20.0 ± 0.1 mm diameter beryllium copper diaphragm measures the differential pressure between a column of water and atmospheric pressure on Table Mountain (Cape Town, SA) at an elevation of 1086 m. The column of water causes the diaphragm to deflect by 0.030 ± 0.001 mm and its thickness is $.210 \pm 0.003$ mm. (a) What is the maximum permissible deflection of the diaphragm? (b) Calculate the pressure drop due to the column of water. (c) How high is the column of water? (d) What would the deflection of the diaphragm be if the column of water was brought down to Cape Town (at sea level)? (e) What is the uncertainty in the pressure drop. *V. Labonté*

4.24 Knowing that the gravitational constant may be approximated by the following relationship (in the range of 0–100 km, with Z in m):

$$g(z) = g - 0.00000303Z.$$

(a) Derive an exact expression for pressure as a function of elevation in the troposphere.

(b) What is the error in the predicted pressure at an elevation of 1000 m when the variation of the gravitational "constant" with elevation is neglected?

REFERENCES

Environnement Canada. (n.d.). Pressure measurement. Retrieved 2011, from Environnement Canada: <http://www.ec.gc.ca/default.asp?lang=fr>.

Graham, B., Reilly, W.K., Beinecke, F., Boesch, D.F., Garcia, T.D., Murray, C.A., Ulmer, F., "Deep Water: The Gulf Oil Disaster and the Future of Offshore Drilling", Report to the President, National Commission on the BP Deepwater Horizon Oil Spill and Offshore Drilling, January, 2011.

Harris, N., 2005. Modern Vacuum Practice, third ed. McGraw-Hill.

Haynes, W.M. (Ed.), 1996. CRC Handbook of Chemistry and Physics, 92nd ed. CRC Press.

Patience, G.S., Hamdine, M., Senécal, K., Detuncq, B., 2011. Méthodes expérimentales et instrumentation en génie chimique, third ed. Dept. Chemical Engineering, Ecole Polytechnique de Montréal.

Poling, B.E., Prausnitz, J.M., O'Connell, J.P. 2001. The Properties of Gases and Liquids, 5th ed., App. A, McGraw-Hill, New York.

Saville, G., Richardson, S.M., Skillerne, B.J., 1998. Pressure Test Safety, HSE Books.

Chapter 5

Temperature

5.1 OVERVIEW

The five senses that everyone recognizes are sight (ophthalmoception), hearing (audioception), taste (gustaoception), smell (oflacception), and touch (tactioception). Other senses that are less frequently cited include balance (equilibrioception), acceleration (kinesthesioception), kinesthetic sense (proprioception), pain (nociception) and, of course, temperature—thermoception. The receptors in our body sense heat and cold on our skin, but they are also sensitive to infrared radiation. Although the body senses temperature and changes in temperature, it is unreliable for quantitative measurements: many factors are required to constitute a reliable measure including reproducibility and a scale—both are lacking with respect to the body. The skin senses relative temperature—hotter and colder—but the range is narrow and unreliable. For example, if you place your left hand in ice water and your right hand in hot water, then put both hands in lukewarm water, the sensation will be different for each hand.

Although temperature measurement and control is critical in the development of many processes throughout history, a universally accepted scale was only developed in the last couple of centuries. Besides agriculture, temperature control and pyrotechnology were the most important developments responsible for the advancement of civilization. Pyrotechnology was used as an engineering tool in the development of prehistoric societies perhaps as far back as 164 000 yr ago (ka) in South Africa (Brown et al., 2009). Fire was used for warmth, light, and cooking up to 790 ka, as shown in Table 5.1. Evidence of the use of fire dates back more than twice that at over 1500 ka. The technological jump to using fire to alter and improve raw materials is poorly documented or understood. However, heat treatment of silcrete (a type of rock) to form

Experimental Methods and Instrumentation for Chemical Engineers. http://dx.doi.org/10.1016/B978-0-444-53804-8.00005-8

tools predominates in multiple sites in South Africa at about 71 ka. The first evidence of heated silcrete to make tools dates back as far as 160 ka, which is approximately the same time that the fire-hardening of wood (Thieme, 1997) to make tools and weapons seems to have emerged. Other processes have been developed in which temperature control has been critical, including mundane applications like drying wood for firewood or drying malted barley for brewing hops to highly sophisticated and demanding applications including firing clay to make pottery—earthenware, stone wares, and porcelain—or smelting copper and iron ores (Nicholson and Shaw, 2000).

The early applications could be easily achieved with the heat from fire in which temperatures are generally less than 600 °C such as fire setting for mining and quarrying (Weisberger and Willies, 2001). The production of earthen pottery requires temperatures higher than 800 °C and high-quality pottery requires temperatures as high as 1200 °C: to achieve and control these temperatures in 8000 BC necessitated an understanding of heat transfer, air flow, insulation, time, and temperature! Smelting (copper and iron), glass, cement, and porcelain are processes that were developed subsequently to the major innovations required to make pottery (Hummel, 2005).

How did the ancients achieve and control high temperatures for extended periods of time in order to make pottery? Modern kitchen ovens typically operate at less than 250 °C whereas gas barbecues can reach perhaps as high as 400 °C. The early use of fire to boil water, cook food, or harden wood would have required temperatures below 200 °C. Firing clay in order to make pottery requires much higher temperatures. To vitrify even low-fire clays requires temperatures in the vicinity of 1000 °C, while the coal bed of a campfire is typically around 600 °C it can get as high as 750 °C when it glows red hot (Berna et al., 2007).

Pottery requires both a high degree of precision around the temperature ramp as well as extremely high temperatures. In the first stage, "bone dry" clay is heated to drive off adsorbed water trapped in the interstices. If this is done too rapidly, steam is generated, which expands and will crack and destroy the pottery. In the second stage (from 300 °C to 800 °C), carbon and sulfur combust. In the third stage, kaolin clays dehydrate, beginning at 550 °C (and continuing up to 900 °C).

$$Al_2O_3 \cdot 2\alpha\text{-}SiO_2 \cdot 2H_2O \xrightarrow{573\,°C} Al_2O_3 \cdot 2\beta\text{-}SiO_2 + 2H_2O.$$

At 573 °C, the quartz inversion temperature where the α-SiO_2 crystal phase converts reversibly to the β-SiO_2 phase, the volume changes by 2%. This volume change can introduce cracks if the temperature on one side of the pottery is much different from the other side. At 900 °C, the clay sinters (the particles begin to fuse together)—fifth stage. In the sixth stage, alumina silicate forms during the process of vitrification.

TABLE 5.1 Pyrotechnology and Temperature in History (Prior to 1000 AD)

Year	Pyrotechnology	Temp. (°C)
790 ka	Demonstrable evidence for the use of fire	200–300
400 ka	Fire-hardening wood (for tools and weapons)	<200
70 ka	Widespread use of stone tools made from fire—South Africa	~ 450
70 ka	Smoking animal skins to produce leather	<100
50 ka	Ground, fired iron oxide producing various pigments (maybe as far as 400 ka)	300–350
23 ka	Baked clay figurine—Venus of Vestonice	300–400
14 ka	Pottery vessels	500–800
8800 BC	Line-plaster from limestone (quicklime—CaO)	870
8000 BC	Ceramic pottery	800–1200
7000 BC	Quarrying using fire-setting	600
5500 BC	Copper smelting	1084
4500 BC	Fire-setting malachite/azurite (copper ores) at Rudna Glava	480
3000 BC	Glazed earthenware to reduce porosity	
	Bread making (adding cold water, hot water, baking, hot stones, fire oven)	
	Brewing beer	
	Steam distillation (essential oil extraction—Frankincense)	100
2500 BC	Fire-setting by Chefryn to quarry stone for pyramids	480
2400 BC	Glass-making by Akkadians	
2000 BC	Fire-setting by Egyptians in gold mining	480
1400 BC	Stoneware	1200–1300
800 BC	Chinese glazed pottery	
300 BC	Distillation of sea water by Greek sailors	100
200 BC	Cement	1450
50 BC	Glassblowing—Phoenicia	
618 AD	Porcelain	1400

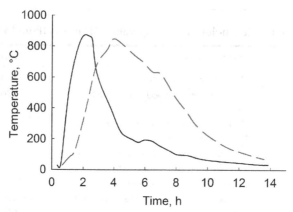

FIGURE 5.1 Pit Kiln Time-Temperature Profile

Open fires are insufficient to produce a ceramic material. Before brick kilns were developed (mud brick houses were common across the Indus valley before 3200 BC), the high temperatures were produced in pit-kilns: a fire was built in a hole in the ground, clay was placed in the pit and then covered with more combustible material, rocks, broken pottery, and dirt to reduce heat losses. Figure 5.1 demonstrates the time-temperature history of a pit-kiln built to produce lime plaster—CaO (Goren and Goring-Morris, 2008; Goren-Inbar et al., 2004) from limestone ($CaCO_3$). The experiment was conducted by digging a 2.5 m diameter hole 75 mm deep, then filling it with 500 kg of limestone and 1 t of fuel—logs, branches, and cow and horse dung. The logs were first placed radially, leaving a 0.5 m diameter circle in the center. Dung and branches were placed between the logs to fill the spaces and limestone pebbles were placed on top of the fuel. Four successive layers were made after which large boulders were placed around the outer perimeter and more boulders were added to form a dome. The dome in the middle was then filled with loosely packed fuel. As shown in the figure, a maximum temperature of 870 °C was achievable.

5.2 TEMPERATURE SCALES

Achieving high temperature for an extended period of time for earthenware, to smelt copper or iron and to manufacture cement is a tremendously energy-intensive operation. To produce 250 kg of lime clay (Figure 5.1), 1 t of fuel was necessary. Nowadays cement production, which requires calcination temperatures of 1450 °C, contributes between 2% and 5% of global CO_2 emissions. It is hard to imagine that ancient civilizations had limited or no means of assessing temperature considering the substantial economic impact

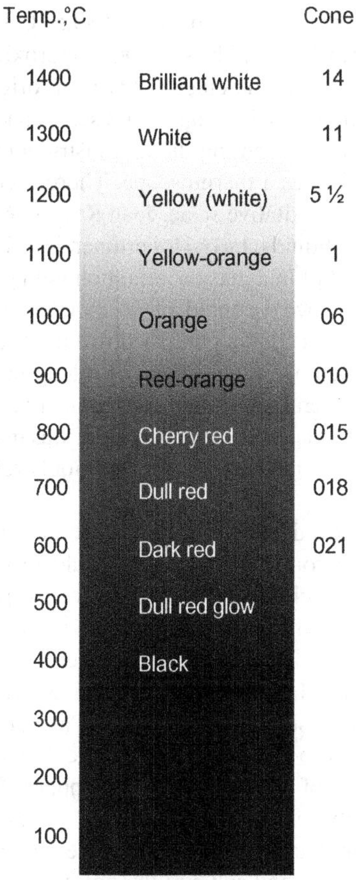

Temp.;°C		Cone
1400	Brilliant white	14
1300	White	11
1200	Yellow (white)	5 ½
1100	Yellow-orange	1
1000	Orange	06
900	Red-orange	010
800	Cherry red	015
700	Dull red	018
600	Dark red	021
500	Dull red glow	
400	Black	
300		
200		
100		

FIGURE 5.2 Kiln Firing Chart (For color version of this figure, please refer the color plate section at the end of the book.)

for incorrectly firing pottery or making cement. Throughout history, hominids have recognized the difference between hot and cold, freezing and boiling. Although temperature was not quantitatively measured, it must have been precisely controlled. The most obvious means of assessing the energy level of a fire is by looking at its color. Figure 5.2 shows the kiln firing chart that relates color to temperature in Kelvin and a scale used in the pottery industry based on the deformation of cones.

The first recorded temperature scale was developed by the Greco-Roman physician Galen around 170 AD. He proposed combining proportions of ice and boiling water to form a scale. The neutral point was equal masses of water and ice. There were four degrees above and below this neutral point. Coincidentally, while the temperature scale we now use was being developed, the English potter Josiah Wedgewood proposed a scale above the boiling point of mercury

(356 °C). The starting point was defined as 580.5 °C but the actual starting temperature was closer to 269 °C. The steps were approximately 16.9 °C, which is much lower than the 54 °C increments that were originally quoted. By the eighteenth century, as many as 35 temperature scales had been conceived.

Galileo is credited with inventing the first instrument to assess the change in temperature, referred to as a thermoscope. Thermoscopes show changes in temperature but lack a quantitative scale. Jean Rey in 1632 was the first to use the dilation properties of liquids. In 1641, Ferdinand II, Grand Duke of Tuscany, hermetically sealed alcohol in a glass contraption with graduation marks.

He, together with Prince Leopold of Medici, created the Academia del Cimento and in 1667, the academy published five treatises: the first instrument was that invented by Fredinand, consisting of a long necked-tube on top of a spherical bulb with 50 increments. The second instrument was similar to the first but with 100 increments. The third instrument was much larger and had 200 hundred increments. These instruments are commonly referred to as Florentine thermometers.

Hooke (1664) improved on the design with an expanding liquor (a red dyed alcohol) in a tube four feet long that would reach the top in the heat of the summer and approach the bottom at the coldest point in the winter. He graduated the stem of the thermomenter by marking zero at the height of the fluid in the stem when the bulb was placed in freezing water. One year later, the Dutch Mathematician Huyghens proposed freezing and boiling water as two reference points to designate a scale. Fabri, a French Jesuit, constructed an instrument with the zero reference equal to the freezing point of water and selected the highest heat of the summer as the upper reference point. He then divided the distance between the two points into eight equal parts in 1669. In 1701, the Danish Astronomer Romer defined the zero point equal to the freezing point of a mixture of water and salt and the boiling point of water was set to equal 60. Newton also developed a temperature scale but rejected using boiling water as a reference point because it begins to boil at one point and then boils vehemently at another point. In fact, the temperature of boiling water depends on atmospheric pressure, chemical purity, the shape of the vessel as well as the position of the thermometer (Sherwood-Taylor, 1942).

Fahrenheit was the first to construct a mercury thermometer (although at least ten others long before him had tested the use of Hg) and adapted a new scale from Romer's in 1714. Originally, he used the Florentine scale (−90, 0, 90). Then he developed his own scale (0, 12, 24) which was derived from foot measurements (Bolton, 1900). Because the graduations were imprecise, he added a factor of 4 to the scale so that it ranged from 0 to 96. The zero point was chosen to be the freezing point of brine; a mixture of ice and water was set to equal 32°; and the average body temperature was designated as 96°. (The scale was later refined such that the boiling point of water equaled 212° exactly and the temperature of melting water was 32°.)

Ferchault de Reaumur de Facheur (1731) used alcohol diluted with 20% water and showed that this mixture expanded from 1000 to 1080 volumes between the freezing point and boiling point of water. Thus, his scale varied between zero at the freezing point of water and 80 at the boiling point. A professor of astronomy in St. Petersburg, Joseph Nicolas de l'Isle, proposed a scale in which 0 represented the boiling point of water and 150 its freezing point. In 1742, A. Celsius—a professor of astronomy—adopted 100 as the freezing point of water and zero as its boiling point. Shortly thereafter, two scientists reversed the Celsius scale independently in Uppsala (Marten Stromer) and Lyons (Christin). Up to 18 different temperature scales were used in 1840. In 1900, three scales were in common use but no nation adopted the scale of its own citizen: English-speaking countries adopted the Fahrenheit scale (German), Germany chose the Reaumur scale, and the Celsius (Swedish) scale as modified by Christin of Lyon was popular in France, Belgium, and Switzerland.

Jacques Charles discovered in 1787 that at constant pressure, the volume of a gas drops by 1/273 per °C, which suggested that gases would disappear if cooled beyond −273 °C. In 1848, Lord Kelvin proposed that molecular translational energy and not volume becomes zero at this value. The Kelvin scale assigns the value of zero as this point and 273.15 K as the melting point of water. It was selected as the metric unit for temperature in 1954 and changed from °K to K in 1967. In 1859, William J. Rankin assigned 0 as the thermodynamic absolute zero and 459.67 as the melting point of water.

Table 5.2 summarizes some characteristic temperatures of important physical properties. The range is from absolute zero to the temperature of the outer surface of the sun at close to 6000 K

Although the Kelvin scale is the accepted international standard, it is infrequently used except in scientific articles. Even in journals the Celsius scale is often used. The Fahrenheit scale remains the official scale in the United States and Belize. To convert from °C to °F:

$$T_{°F} = 1.8 \, T_{°C} + 32.$$

Whereas standard pressure has recently been defined by the IUPAC as 1 bar, the conditions of the term "standard" have not as of yet been universally accepted. In the gas industry, standard conditions (STP) are 1 atm and 60 °F (15.6 °C). The IUPAC definition of standard temperature is 0 °C, whereas some assign 25 °C as standard temperature. Its definition of normal temperature and pressure is 20 °C and 1 bar, respectively. Whenever a new instrument—particularly flow meters—is brought into service, the manufacturer's definition of STP must be verified to avoid introducing a systematic error.

5.2.1 Wet-Bulb, Dry-Bulb Temperature, Dew Point

A class of process equipment (Unit Operations) of Chemical Engineering is concerned with either condensing a vapor from a gas stream or saturating a

TABLE 5.2 Characteristic Temperatures

°C	K	°F	
55067	5780	9944	Temperature on the surface of the sun
3410	3683	6170	Melting point of tungsten
1668	1941	3034	Melting point of titanium
1064	1337	1948	Melting point of gold
420	693	787	Melting point of zinc
100	373	212	Boiling point of water
58	331	136	Highest temperature recorded on earth
37	310	98.6	Body temperature
20	293.15	68	Ambient temperature
3.97	277	39	Temperature of water at its maximum density
0.01	273.16	32	Triple point of water
−17.78	255	0	Zero on the Fahrenheit scale
−89	184	−129	Coldest temperature recorded on earth
−219	54	−362	Triple point of oxygen
−259	13.8	−435	Triple point of hydrogen
−273.15	0	−459.67	Thermodynamic absolute zero

gas stream with a vapor—"gas" refers to the non-condensable component of the stream and vapor refers to the condensable component. In humidification and dehumidification operations, water is the vapor component and air is the gaseous component. A saturated gas is one in which the vapor concentration is in equilibrium with the gas. Relative humidity is the ratio of the water partial pressure to the saturated water vapor pressure at the same conditions (100% humidity). The Antoine equation may be used to calculate absolute humidity, as a function of temperature (see Table 4.1), but other more complex and accurate correlations have also been proposed.

Note that the partial pressure of water vapor in air at 100% humidity and 5 °C is much less than at 50% relative humidity and 30 °C. This has important implications with respect to drying operations; the driving force for evaporation is proportional to the difference between the water vapor pressure and the web-bulb temperature, which is equivalent to what the body would feel when the skin is wet and then exposed to a moving stream of air. It is measured by placing a damp wick over a liquid-in-glass thermometer and then spinning the thermometer at a high enough rate to simulate an air stream.

The dry-bulb temperature is the thermometer reading in the absence of humidity and radiation; it is the temperature most often reported by meteorological agencies. The dew point is the temperature at which the water vapor in the air first begins to condense: the dew point equals the measured temperature when air is 100% saturated—100% humidity.

5.2.2 Humidex, Heat Index

The sensation of heat is more pronounced when the humidity is high compared to a dry environment at the same dry-bulb temperature. Perspiration (sweat) cools the body when it evaporates. The evaporation rate depends on many factors of which the relative humidity is possibly the most important: at 100% relative humidity, air is saturated with water vapor and so at this condition there is no driving force to evaporate perspiration from the body. This effect is accounted for by two scales: in the United States, meteorologists report the Heat Index while in Canada they report the Humidex (Steadman, 1979). The Humidex is based on the dew point temperature, T_{dew} and the dry-bulb temperature, T, and is expressed as:

$$T_{HX} = T + 0.5555 \left(6.11 e^{5417.7530 \left(\frac{1}{273.15} - \frac{1}{T_{dew}} \right)} - 10 \right).$$

A Humidex of 30 is mildly uncomfortable—it feels like 30 °C with no humidity. A value of 40 is extremely uncomfortable and 45 is dangerous. Heat stroke is probable at values above 54.

This expression was developed when Canada was still officially using the Fahrenheit scale and could be simplified by multiplying the constants in the brackets by 0.5555. Furthermore, there is an inconsistency with respect to the coefficients—eight significant figures for the coefficient in the exponential and only two for the value "10". Clearly two significant figures would be sufficient to communicate the sensation of heat due to humidity.

The Heat Index (Rothfusz, 1990) is rather more complicated and is expressed as:

$$\begin{aligned} T_{HI} = &-42.379 + 2.04901523T + 10.14333127 H_R - 0.22475541 T H_R \\ &-6.83783 \times 10^{-3} T^2 - 5.481717 \times 10^{-2} H_R^2 \\ &+1.22874 \times 10^{-3} T^2 H_R + 8.5282 \times 10^{-4} T H_R^2 \\ &-1.99 \times 10^{-6} T^2 H_R^2. \end{aligned}$$

The variable T refers to the dry-bulb temperature in °F and H_R is the relative humidity in %. As with the expression for Humidex the coefficients with 10 significant figures are excessive. However, with higher order polynomial expressions many significant figures are often required: the difference in the calculated value of T_{HI} between the equation above and with coefficients with only four significant figures is 18 (at 90 °F and 80% H_R).

5.2.3 Wind Chill Factor

Whereas the Humidex accounts for the effect of humidity on how the body perceives temperatures above 20 °C, the wind chill factor accounts for the enhanced cooling effect caused by wind. The sensation of temperature depends on various phenomena including the evaporation rate, conduction, convection and, during sunny days, radiation. The heat lost from the body increases with increasing wind speed due to higher local convection rates on the skin and around the body. As in the seventeenth to nineteenth centuries, a world standard to account for the wind chill has yet to be adopted. Since 2001, the weather services in Canada and the US report Wind Chill according to the following relationship:

$$T_{WC} = 13.12 + 0.6215T - 11.37V^{0.16} + 0.3965T\,V^{0.16},$$

where T is the dry-bulb temperature in °C and V is the wind speed measured at a height of 10 m (km h^{-1}).

5.3 MECHANICAL INSTRUMENTS

According to the second law of thermodynamics, heat flows from a hot body to a cold body (unless work is done to reverse the direction). Temperature is an intensive thermodynamic property—unlike length and mass, whose values can only be compared to a standard—and represents the level of molecular energy of molecules.

Many instruments have been developed to measure the change in molecular energy and they can be broadly classified as mechanical, electrical, or based on the detection of electromagnetic waves (infrared and visible light). Although 10 000 yr ago the concept of temperature must have been well understood and quantitatively measured (most probably based on color), the first documented evidence of a device to measure it dates back to around 240 BC. Philo of Byzantium observed that bubbles would escape a tube immersed in water which was connected at the other end to a (sealed) hollow sphere when it was placed in the sun. When it was placed in the shade, water would rise into the tube. Hero of Alexandria (10–70 AD) further developed this concept, which later inspired Fludd (English physician, 1538), Galileo (1592), and Sanctorius (1612).

5.3.1 Gas Thermometers

Mechanical-based instruments used to measure temperature are based on the dilation properties of gases, liquids, and solids. The first thermometers were based on gases. The volume of an ideal gas will change in direct proportion to its temperature (under isobaric conditions). According to Charles' Law:

$$T = T_{ref}\frac{V}{V_{ref}}.$$

Figure 5.3 illustrates the thermoscope of Sanctorius, reproduced from an illustration by Biancani. The long tube is immersed at the bottom in a pool of water and at the top by a hollow sphere. When the top sphere is heated, the

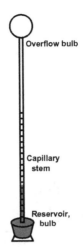

FIGURE 5.3 Thermoscope of Sanctorius

volume of gas expands and the liquid level drops. The liquid level increases when the sphere at the top is cooled. Temperature gradients along the tube length will make a precise measurement of the temperature difference.

The range of a constant-pressure gas thermometer is limited by the volume of the instrument of the graduated tube recording the change in height of the fluid. High pressure and low pressure ranges may be examined independently by pressurizing the tube. Higher ranges may be achieved with pressure-measuring constant-volume gas thermometers. For ideal gases, temperature increases proportionally with pressure at constant volume.

A special case of "constant-volume" gas thermometry is that based on the vapor pressure of a liquid. As the temperature of a solvent is increased, its vapor pressure increases proportionately over a wide range of conditions. The temperature range is limited to temperatures between the boiling point and freezing point of the solvent and it must also be lower than the critical point of the fluid. These thermometers are easy to adapt to common experimental equipment.

The pressure can be measured by a Bourdon gauge or a U-shaped manometer, as shown in Figure 5.4. In the case of a Bourdon gauge, the volume is constant and the pressure increases. In the case of a U-tube manometer, the volume increases with pressure and therefore, may not strictly be considered a "constant-volume" gas thermometer. Although the simplicity of the apparatus is attractive, it is extremely difficult to achieve a high degree of accuracy.

The volume of the solvent immersed in the medium to be measured must be much larger than all connecting lines between the pressure instrument and the solvent. In the case of Figure 5.4, the solvent will condense on all surfaces at temperatures lower than the bath temperature. In fact, the tubing of a U-tube

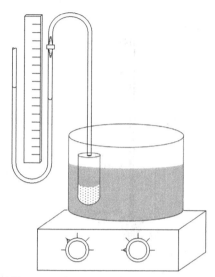

FIGURE 5.4 U-Shaped Manometer Configured to Measure Temperature

should be heated to a greater temperature than that of the bath to minimize condensation.

Example 5.1. The temperature of the water in a beaker on a hot plate is measured by a "constant-volume" thermometer. The cylindrical body of the thermometer is 200 ml. It is connected directly to a Bourdon gauge. The barometric pressure is 750 mmHg. Assume that the following experiments are conducted rapidly such that the tubing operates isothermally.

(a) What is the temperature of the water if the bourdon gauge reads 1.0 psig?
(b) In the second case, 50 ml of ethanol is placed in the body of the thermometer. What is the pressure recorded on the Bourdon gauge if the temperature increases to 55 °C?
(c) In the third case, the thermometer is connected to a U-tube manometer of which the total length of tubing (1/4 in. diameter) between the top and the zero point of the manometer is 1m. What is the temperature of the water if the differential height in the U-tube is 14 in.? (The cylindrical body contains only air.)

Solution 5.1a. The Bourdon gauge should be placed as close to the cylindrical body to minimize the total volume of gas that remains unheated. Assuming that this dead volume is negligible, Amontons' Law of Pressure-Temperature applies:

$$P_1/T_1 = P_2/T_2.$$

The absolute pressure must be calculated, from which the temperature rise is readily calculated:

$$T_2 = T_1 P_2/P_1 = (273 + 25)\frac{750 \text{ mmHg}\frac{101.325 \text{ kPa}}{760 \text{ mmHg}} + 1 \text{ psig}\frac{101.325 \text{ kPa}}{14.696 \text{ psig}}}{750 \text{ mmHg}\frac{101.325 \text{ kPa}}{760 \text{ mmHg}}}$$

$$= 298\frac{100 \text{ kPa} + 6.895}{100 \text{ kPa}} = 318.5 \text{ K} = 45.4 \,°\text{C}.$$

Solution 5.1b. In the case of placing an organic fluid in the cylindrical body of the thermometer immersed in the water, two factors affect the pressure gauge reading: the expansion of the air as well as the increased vapor pressure of the organic fluid. The vapor pressure of the fluid is calculated according to Antoine's equation, while the overall pressure is determined from the ideal gas law. Amontons' law is inapplicable because the total number of moles of gas in the vapor phase of the cylinder will increase. (We shall also assume that the change in volume of the liquid is inconsequential.)

From Antoine's equation, we first calculate the vapor pressure of ethanol from which we can then derive its initial mole fraction in the vapor phase (at 25 °C):

$$\ln P_1^{\circ} = A - \frac{B}{T_1 + C} = 16.8958 - \frac{3795.17}{25 + 230.918},$$

$$P_1^{\circ} = 7.9 \text{ kPa}.$$

The ethanol was charged to the cylinder at 25 °C and barometric pressure. The partial pressure of air equals the difference between the barometric pressure and the partial pressure of ethanol: $P_{\text{air}} = 100 - 7.9 = 92.1$ kPa. The total moles of ethanol and air are:

$$n_{\text{air}} = \frac{P_{\text{air}} V}{RT} = \frac{92.1 \text{ kPa}(0.200l - 0.050l)}{8.314 \text{ kPa l mol}^{-1} \text{ K}^{-1}(25 \text{ k} + 273 \text{ k})} = 0.00558 \text{ mol},$$

$$n_{\text{EtOH}} = n_{\text{air}}\frac{y_{\text{EtOH}}}{y_{\text{air}}} = 0.0056\frac{0.079}{0.921}0.00048 \text{ mol}.$$

The vapor pressure of ethanol at 55 °C is calculated next:

$$\ln P_2^{\circ} = A - \frac{B}{T_2 + C} = 16.8958 - \frac{3795.17}{55 + 230.918},$$

$$P_2^{\circ} = 37.4 \text{ kPa}.$$

Two equations describe the system with two unknowns—moles of ethanol in the vapor phase, $n_{\text{EtOH},2}$, and the total pressure, P_2:

$$\frac{P_2}{n_2 T_2} = \frac{P_1}{n_1 T_1},$$

$$P_2 = P_1\frac{n_2 T_2}{n_1 T_1},$$

$$P_2^{\circ} = y_{\text{EtOH}} P_2 = \frac{n_{\text{EtOH},2}}{n_2} P_2,$$

where

$$n_2 = n_{\text{EtOH},2} + n_{\text{air}}.$$

We can solve for the total moles of ethanol in the vapor phase by substituting the expression for the total pressure (from the ideal gas law) into the expression for the vapor pressure:

$$P_2^{\circ} = \frac{n_{\text{EtOH},2}}{n_2} P_1 \frac{n_2 T_2}{n_1 T_1},$$

from which

$$n_{\text{EtOH},2} = n_1 \frac{P_2^{\circ}}{P_1} \frac{T_1}{T_2} = 0.00606 \frac{37.4}{100} \frac{298}{328} = 0.00206 \text{ mol}.$$

With the total number of moles of ethanol in the vapor phase together with the moles of air, the pressure is now calculated:

$$P_2 = \frac{n_{\text{EtOH},2} + n_{\text{air}}}{n_{\text{EtOH},2}} P_2^{\circ} = \frac{0.00206 \text{ mol} + 0.00558}{0.00206} 37.4 \text{ kPa} = 139 \text{ kPa}.$$

This pressure is substantially greater than when only air is placed in the cylinder and thus the precision is superior. However, the design of the thermometer is much more difficult due to the possibility of vapor condensation in the Bourdon tube or in the tubing connecting the two. Furthermore, the lines connecting the tube to the gauge should also be completely immersed in the measured fluid to reduce ambiguity with respect to possible temperature variations.

Solution 5.1c. The situation with the U-tube manometer is a little more complicated. There are two systems that are coupled. The cylinder immersed in the water is a constant-volume vessel. The pressure and temperature will increase and there will be an outflow of moles to the tube. The tube is operated isothermally and the volume increases as the liquid level drops in the U-tube. The pressure in the cylinder equals the temperature in the tubing. The conditions in the cylinder before heating are:

$$P_1 = 750 \text{ mmHg} \frac{101.325 \text{ kPa}}{760 \text{ mmHg}} = 100 \text{ kPa},$$

$$V_c = 200 \text{ mm},$$

$$T_1 = 25 + 273 = 298 \text{ K},$$

$$n_{c,1} = \frac{P_1 V_c}{RT_1} = \frac{100 \text{ kPa} 200 \times 10^{-3} 1}{8.314 \text{ kPa l mol}^{-1} \text{ K}^{-1} 298 \text{ k}} = 0.008\,07 \text{ mol}.$$

The pressure and temperature in the tubing are the same as in the cylinder. The number of moles and volume are:

$$n_{t,1} = \frac{P_1 V_{t,1}}{RT_1} = \frac{100 \cdot 31.7 \times 10^{-3}}{8.314 \cdot 298} = 0.00128 \text{ mol}.$$

As the cylinder heats up, the pressure increases and the gas expands coincidentally such that the total number of moles in the cylinder decreases. The decrease in the number of moles in the cylinder corresponds to the increase in moles in the tubing, Δn, which equals the volume displacement of the manometer fluid. We assume that the tubing temperature is constant, but the pressure increases, which is calculated based on the displacement of the fluid in the manometer:

$$V_{t,2} = V_{t,1} + \Delta V = V_1 + \frac{\pi}{4}d^2h$$

$$= 31.7 + \frac{\pi}{4}\left(\frac{0.0254}{4}\right)^2 \cdot 0.0254 \cdot 14 \times 10^6 = 43.0 \text{ ml,}$$

$$P_2 = P_1 + \Delta P = P_1 + \rho g h = 100$$
$$+ 1000 \cdot 9.8 \cdot 0.0254 \cdot 14/1000 = 103.48 \text{ kPa.}$$

Again from the ideal gas equation, the total number of moles in the tube may be calculated including the volume expansion and the increased pressure:

$$n_{t,2} = \frac{P_2 V_{t,2}}{RT_1} = \frac{103.48 \cdot 43.0 \times 10^{-3}}{8.314 \cdot 298} = 0.00180 \text{ mol.}$$

So, the number of moles exiting the cylinder and entering the tubing is the difference between the moles in the tubing before and after heating:

$$\Delta n = n_{t,2} - n_{t,1} = 0.00180 - 0.00128 = 0.00052 \text{ mol.}$$

For the conditions in the cylinder for which we wish to calculate the temperature, we can again apply the ideal gas equation since all of the conditions are now defined:

$$n_{c,2} = n_{c,1} - \Delta n = 0.00807 - 0.00052 = 0.00755 \text{ mol,}$$

$$T_2 = \frac{P_2 V_2}{n_{c,2} R} = \frac{103.48 \cdot 200 \times 10^{-3}}{0.00755 \cdot 8.314} = 330 \text{ K} = 57 \,°\text{C.}$$

5.3.2 Liquid Thermometers

Liquid thermometers are the most common instrument employed to measure atmospheric temperature. They have an excellent range and can measure temperatures from as low as $-200\,°\text{C}$ to above $600\,°\text{C}$. Similarly to gas thermometers, they are based on the dilation properties of liquids when subject to changes in temperature. Many liquids have been used to measure temperature including water, alcohols, mercury, paraffins, and aromatics.

The liquid-in-glass thermometer consists of a bulb containing a reservoir of fluid. A stem, connected to the bulb, contains a capillary tube on the order of less than 0.5 mm in diameter. As the bulb is heated or cooled, the fluid in the capillary will rise or fall. A scale is etched along the stem to represent the temperature. Often, a small bulb at the other end of the capillary is added for safety: if the bulb is accidentally overheated, excess fluid will accumulate in this region, thus reducing the risk that the thermometer bursts. To improve accuracy, the stem is notched to indicate the height to which the thermometer should be immersed in the fluid. The fluid and glass expansion properties are different, which must be accounted for in the calibration of the thermometer.

Among the first commercial thermometers were those manufactured by Fahrenheit with mercury. The advantages of mercury over most other fluids include its very low vapor pressure, excellent temperature range—from $-38\,°C$ to $650\,°C$—and short response time. Short response time is a convenient, if not important, factor with respect to medical applications. Although Sanctorius' "thermoscope" was used for medical purposes up until 1866, clinical thermometers were 300 mm long and took 20 min to record a patient's temperature. In 1866, Albutt invented a thermometer half that length that required only 5 min (Adler, 1974).

Despite the advantages of Hg, alcohol thermometers are replacing mercury as the liquid of choice principally driven by safety considerations. Mercury is toxic. Moreover, when the thermometer breaks, mercury beads into small spheres making it difficult to clean, which increases both its environmental and health hazard. Disadvantages of solvents include their lower thermal conductivity, which are as much as 50 times lower than mercury and their high vapor pressure resulting in a lower level in the capillary and thus in the possible underestimation of the true temperature.

Amyl alcohols, as shown in Table 5.3, have a higher coefficient of expansion compared to mercury, which increases their sensitivity. By pressurizing the capillary with an inert gas (nitrogen or argon), the evaporation of the solvent can also be minimized and at the same time the higher pressure increases the effective temperature range. This technique is also used for Hg thermometers: the boiling point of mercury is $357\,°C$ but temperatures as high as $650\,°C$ are possible with pressure.

Eutectic alloys, such as galinstan, are also used as fluids for thermometers due to their excellent properties—high thermal conductivity, low vapor pressure, and excellent temperature range (Knoblauch et al., 1999).

The design parameters of a dilation thermometer depend mostly on the properties of the liquid—its melting point, boiling point, and thermal expansion coefficient—together with the range of operations. In many cases, when the temperature range is narrow, the thermal properties of the glass may be ignored and a linear expansion of fluid may be assumed.

In cases where a high precision is required, both these factors must be considered. A second-order polynomial characterizes the relationship between temperature and density very well, and thermal expansion is expressed in terms of volume as:

$$\Delta V = V_o[\alpha(T - T_o) + \beta(T - T_o)^2]. \tag{5.1}$$

TABLE 5.3 Physical Properties—Liquid-in-Glass Thermometers (Holman, 2001)

| | MP | BP | k | Therm. Exp. Coeff. | | Vap. Press. |
| | (°C) | (°C) | $\left(\frac{W}{mK}\right)$ | $(m^3 m^{-3} K^{-1})$ | | mmHg |
				$\alpha \times 10^3$	$\beta \times 10^6$	
Amyl alcohol	−117–8	102–138	0.15	0.900	0.657	1.5 @20 °C
Mercury	−38	3.7	8.34	0.182	0.0078	0.27 @100 °C
Ethanol	−110	1.0	0.171	0.750		40 @19.0 °C
n-Pentane	−200	20	0.136	1.58		4.0 @18.5 °C
Toluene	−95	1.1	0.151	1.07		40 @31.8 °C
Water	0	1.0	0.607	−0.064	8.505	101.3 @100 °C
Galinstan	−19	> 1300	16.5	0.115		$< 10^{-8}$ @500 °C
Glass	~ 1500		0.9–1.3	0.0255		
Pyrex	8.21		1.005	0.0099		

The height change in the capillary is related to its diameters and the expansion properties of the liquid as well as the initial volume, V_o:

$$h = \frac{\Delta V}{X_A} = \frac{V_o}{X_A}[\alpha(T - T_o) + \beta(T - T_o)^2]. \qquad (5.2)$$

Example 5.2. Consider a mercury thermometer fabricated for a custom experimental apparatus that reads from −20 °C to 300 °C with a 3 mm inside diameter.

(a) The density of mercury is 13596 kg m^{-3} at 0 °C. What is its density at −20 °C (the temperature at which the mercury is charged to the bulb)?

(b) If the bulb of the thermometer was charged with 100 mg of Hg, what is the length of the capillary at its maximum temperature?

(c) If the capillary is initially pressurized to 1 barg with argon and the spherical head space (bulb) at the top is 1mm in diameter, what would its pressure be at 300 °C when the thermometer is entirely immersed in the apparatus? Neglect the thermal expansion properties of the glass.

Solution 5.2a. The expansion coefficients shown in Table 5.3 relate the change in volume with respect to temperature. This may be rewritten in terms of density:

$$V_T = V_o[1 + 1.82 \times 10^{-4}(T - T_o) + 7.8 \times 10^{-9}(T - T_o)^2],$$

$$\rho = \frac{m}{V} = \frac{m}{V_o} \frac{1}{1 + 1.82 \times 10^{-4}(T - T_o) + 7.8 \times 10^{-9}(T - T_o)^2},$$

$$\rho_{-20} = \frac{\rho_0}{1 + 1.82 \times 10^{-4}(T - T_o) + 7.8 \times 10^{-9}(T - T_o)^2},$$

$$\rho_{-20} = \frac{13596}{1 + 1.82 \times 10^{-4}(-20 - 0) + 7.8 \times 10^{-9}(-20 - 0)^2},$$

$$\rho_{-20} = 13\ 646 \text{ kg m}^{-3} = 13\ 600 \text{ kg m}^{-3}.$$

Solution 5.2b. The initial volume of mercury charged to the thermometer, V_o, equals the quotient of the mass charged and density:

$$V_o = \frac{m}{\rho_{-20}} = \frac{0.1 \text{g}}{13.646 \text{ g l}^{-1}} \frac{1000 \text{ ml}}{1} = 7.32 \text{ ml}.$$

The minimum volume of mercury must be at least twice the volume that rises in the capillary. Since the glass will also expand, let us assume that we require 20% more Hg than V_T. The volume rising into the capillary, ΔV, equals:

$$\Delta V = X_A h = \frac{\pi}{4}(0.3 \text{ mm})^2\, h = 0.0707h,$$

$$V_T = 1.2 \cdot 2\Delta V = 0.170h,$$

$$\Delta V = V_o(1.82 \times 10^{-4}[300 - (-20)] + 7.8 \times 10^{-9}[300 - (-20)]^2)$$

$$= 7.32 \text{ ml} \cdot 0.0590,$$

$$h = \frac{V_T}{0.170} = \frac{7.32 \text{ ml}}{0.170 \text{ mm}^2} = 43 \text{ mm}.$$

Solution 5.2c. The total gas volume of the thermometer ($V_{g,o}$) is the sum of the capillary volume, ΔV, and the capillary bulb, V_{cb}:

$$V_{g,o} = \Delta V + V_{cb} = \frac{V_T}{1.2 \cdot 2} + \frac{\pi}{6}D_{cb}^3 = \frac{7.32 \text{ ml}}{2.4} + \frac{\pi}{6}.$$

The volume occupied by the gas at high temperature is simply V_{cb}. Applying Charles' law:

$$\frac{P_1 V_1}{T_1} = \frac{P_2 V_2}{T_2},$$

$$P_{300} = P_o \frac{T_{300}}{T_o} \frac{V_{g,o}}{V_{300}} = 1 \text{ barg} \frac{300 + 273 \text{ K}}{-20 + 273 \text{ K}} \frac{3.57 \text{ ml}}{0.52 \text{ ml}} = 13.3 \text{ barg (unsafe!)}.$$

Clinical thermometers are classified as partial immersion thermometers since only bulb and part of the stem enter the mouth. Optimally, the thermometer is calibrated and an immersion line is etched onto the side of the stem to indicate to what point the thermometer should be immersed. This type of thermometer is best for a narrow temperature range (such as for measuring body temperature) since the expected variation in temperature is much less than 10 °C. Total immersion thermometers are those in which it is immersed to the same level as the height of the liquid column of the thermometer. The entire thermometer—bulb, stem, and capillary bulb—is exposed to the temperature being measured in a complete immersion thermometer; these are often used in refrigerators, freezers, incubators and for ambient temperature measurements.

The level at which the thermometer is immersed in the medium to be measured is important to calculate the expansion of the glass. As shown in Table 5.3, the coefficient of thermal expansion is much less than that of organic compounds but is only five times lower than that of mercury. Over a narrow temperature range, it is negligible (when adequately calibrated) but, for accurate measurements over a wide temperature range, the change in volume of the glass must be considered.

The advantage of the complete immersion thermometer is that the entire glass casing is at the same temperature as the system to be measured: the correction factor is easily calculated. Calculation of volume expansion of the glass casing for partial and total immersion thermometers is less obvious since the bulb and part of the stem will measure the system temperature whereas the upper part of the stem will be at a different temperature. When the temperature to be measured is higher than the calibration temperature, the liquid level attained will be lower than anticipated since the glass will expand. When total immersion thermometers are only partially immersed in the medium to be measured, a correction factor is applied:

$$\Delta T_{im} = \alpha_{g,1} \Delta T_h (T_1 - T_2), \tag{5.3}$$

where ΔT_{im} is the correction factor (°C or °F), $\alpha_{g,1}$ is the expansion coefficient (~ 0.00016) for °C for Hg and 0.001 for organic fluids), ΔT_h is the differential temperature between the point at which the liquid emerges from the measured medium at the top of the liquid column (°C or °F), T_1 is the bulb temperature (°C or °F), and T_2 is the ambient temperature halfway between the point at which the liquid column exits the measured medium and the top of the liquid column (see Figure 5.5).

Example 5.3. The thermometer of the previous example is a total immersion thermometer with a resolution of 0.2 °C. The temperature indicated on the thermometer is 265.6 °C when the thermometer is immersed at the 50.8 °C mark. What is the correct temperature reading when the ambient temperature halfway up the column is -10 °C?

Solution 5.3. The calculation is an iterative procedure since the actual bulb temperature, T_1, is only known approximately:

$$\Delta T_{im,1} = 0.00016(265.6 - 50.8)(265.6 - (-10)) = 9.5 \ ^\circ C,$$
$$\Delta T_{im,1} = 0.00016(265.6 - 50.8)(265.6 + 9.5 - (-10)) = 9.8 \ ^\circ C,$$
$$T = 265.6 + 9.8 = 275.4 \ ^\circ C.$$

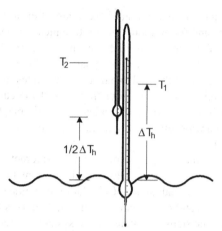

FIGURE 5.5 Thermocouple Immersion

5.3.3 Bimetallic Thermometers

Gas volume (or pressure) increases proportionately with temperature. The volume of organic liquids, aqueous solutions, mercury and eutectic alloys increases linearly over a narrow range of temperatures but their coefficient of expansion is several orders of magnitude lower than for gases. Thermal expansion coefficients of metals are even lower—three orders of magnitude lower than for liquids (Table 5.4). However, when two metals are bonded together, the difference in their thermal expansion coefficients causes the strip to curve. When the strip is heated beyond the temperature that they were welded together, the strip will bend away from the metal with the highest thermal expansion coefficient. When it is cooled, the curvature will be toward the metal with the highest heat transfer coefficient.

The strips may be bands or even coils. Coils are more sensitive to changes in temperature. Although they may operate up to temperatures as high as 1000 °C, they are more typically used as thermostats for applications near ambient temperature and in typical household appliances including refrigerators, freezers, irons, and hair and clothes dryers.

Thermostats may be used to indicate temperature but their main application is in the regulation of heating and cooling and as circuit breakers. They are much less sensitive than liquid-in-glass thermometers and achieving accuracy below 0.1 °C is unrealistic. The strip is fixed at one end which is attached to an electrical power supply. The other end is free to move up and down. For heating applications, the dial sets the position in which the free end completes an electrical circuit, which will then heat the strip. When the strip temperature rises beyond the set-point (the desired temperature set by the manual dial), the band will move in the opposite direction, thus breaking the contact. At this

TABLE 5.4 Mechanical and Thermal Properties of Metals Used in Bimetallic Strips

Metal	Therm. Exp. Coeff. $\times 10^6$ K, α	Modulus of Elast. GPa, E
316 Stainless Steel	15.3	214
Zinc	31.0	69.4
Chromium	6.5	279
Copper	17.0	138
Tin	23.5	58.2
Aluminium	23.5	75.2
Brass	20.2	96.5
Invar (Fe/Ni—64:36)	1.7	147
Nickel	13.3	177
Monel 400	13.5	179
Inconel 702	12.5	217

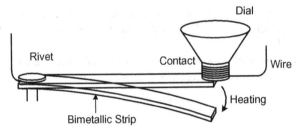

FIGURE 5.6 Thermostat Configuration with a Bimetallic Strip

point, the band will begin to cool and move back toward the contact point to complete the circuit. Figure 5.6 demonstrates the curvature of the band together with a typical configuration for a heating application.

The radius of curvature depends on the physical properties of the bonded metals, their thicknesses, as well as the temperature at which they were welded:

$$r = \frac{t[3(1+m)^2 + (1+mn)(m^2 + 1/mn)]}{6(\alpha_2 - \alpha_1)(T - T_o)(1+m)^2}, \qquad (5.4)$$

where r is the radius of curvature, t is the sum of the thickness of each strip, α_1 is the coefficient of expansion of the metal with the lowest coefficient, α_2 is the coefficient of expansion of the metal with the highest coefficient, m is the ratio of the thickness t_1/t_2 where t_1 corresponds to α_1 ($m \leqslant 1$), G_1 is the Young's modulus of the metal with the lowest modulus, G_2 is the Young's modulus of

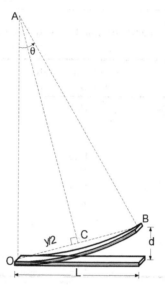

FIGURE 5.7 Deflection of a Bimetallic Strip

the metal with the highest modulus, n is the ratio of Youngs' moduli G_1/G_2 ($n < 1$), T is the temperature (K), and T_o is the temperature at which the metals were welded (K).

In practice, the thickness of both metal strips is the same and thus the ratio, m, equals 1. The equation for the curvature reduces to:

$$r = \frac{t[12 + \frac{1}{n}(1 + n)^2]}{24(\alpha_2 - \alpha_1)(T - T_o)}. \tag{5.5}$$

The deflection of the strip, d, is calculated based on geometry, as shown in Figure 5.7. The arc length, $r\theta$, equals the length of the strip, L. The line bisecting the angle θ is at a right angle mid-way between the arc and the straight line extending from the fixed point **O** and **B** equals y:

$$y = 2r \sin\left(\frac{\theta}{2}\right). \tag{5.6}$$

The deflection is calculated based on the angle of the segment **O** and **B** and the angle $\frac{\theta}{2}$:

$$d = y \sin\left(\frac{\theta}{2}\right). \tag{5.7}$$

Substituting the expression for the segment length, y, into that for the deflection, d, gives:

$$d = 2r \sin^2\left(\frac{\theta}{2}\right). \tag{5.8}$$

At low levels of deflection, typical of thermostats, $\sin\left(\frac{\theta}{2}\right)$ is approximately equal to $\left(\frac{\theta}{2}\right)$, so the expression for deflection may be simplified to:

$$d = 2r \sin^2\left(\frac{\theta}{2}\right) = 2r\left(\frac{\theta}{2}\right)^2 = 2r\left(\frac{L}{2r}\right)^2 = \frac{L^2}{2r}. \tag{5.9}$$

Example 5.4. Brass and Inconel strips 10.0 cm long are welded together at 50 °C. The thickness of each strip is 0.30 mm. Calculate the radius of curvature together with the deflection when the strip is heated to 150 °C.

Solution 5.4. The physical properties for the metals are given in Table 5.4:

$$n = 96.5/217 = 0.445,$$
$$\alpha_2 = 2.02 \times 10^{-5}\ \text{K}^{-1},$$
$$\alpha_1 = 1.25 \times 10^{-5}\ \text{K}^{-1},$$
$$m = 0.30\ \text{mm}/0.30\ \text{mm} = 1.0,$$
$$t = 0.30 + 0.30\ \text{mm} = 0.60\ \text{mm},$$
$$T = 150\ °\text{C},$$
$$T_o = 50\ °\text{C}.$$

Since the thickness of each strip is equal, the simplified form to calculate the curvature is applicable:

$$r = \frac{t[12 + \frac{1}{n}(1+n)^2]}{24(\alpha_2 - \alpha_1)(T - T_o)},$$
$$r = \frac{0.60[12 + \frac{1}{0.445}1.445^2]}{24(2.02 \times 10^{-5} - 1.25 \times 10^{-5})(150 - 50.)} = 542\ \text{mm} \approx 0.54\ \text{m}.$$

The simplified expression to calculate the deflection gives:

$$d = \frac{L^2}{2r} = \frac{100^2}{2 \cdot 542} = 9.2\ \text{mm}.$$

To verify, we calculate the complete expression for the deflection: the radius of curvature of the 100 mm strip is 542 mm and so the angle from the fixed point to the other end is:

$$\theta = L/r = 0.1/0.542 = 0.185,$$

and thus

$$d = 2r \sin^2\left(\frac{\theta}{2}\right) = 2 \cdot 542 \cdot \sin^2 0.185/2 = 9.2\ \text{mm}.$$

The simplified expression is an excellent approximation for the deflection at low angles, which is the case for most flat bimetallic strips.

5.4 ELECTRICAL INSTRUMENTS

Electrical instrumentation, including thermistors and resistance temperature detectors (RTDs), is virtually the only type of instrumentation used in chemical processes for monitoring and control due to their versatility, accuracy, repeatability, and rapid response time. Each relies on an electrical signal, which may then be amplified, compensated (for calibration) and filtered before a signal is relayed either to a database or displayed. Thermistors rely on the change of electrical resistance with temperature of semi-conductors whereas RTDs are fabricated from pure metals. Thermocouples are the most widely used device and they are based on measuring the change in electrical potential (emf) with temperature: when two dissimilar metals come into contact, emf is generated and this emf varies with temperature.

5.4.1 Thermistors

Common thermistors are based on semiconductors whose resistance decreases significantly as the temperature increases. Currently, thermistors are manufactured from many different materials including polymers and other ceramics and their resistance may either increase with temperature (positive temperature coefficient—PTC) or decrease with temperature (negative temperature coefficient—NTC). The conductivity of pure metals increases with increasing temperatures and instruments based on metals are referred to as RTDs.

Because of their the sensitivity to temperature, thermistors made with semi-conductors are among those with the highest precision of any of the electrical transducers with a resolution as good as $0.01\,°C$. Together with the high sensitivity, thermistors are inexpensive to manufacture and are compact. One substantial limitation is that they only work over a narrow temperature range, which is typically lower than $150\,°C$ and higher than $-80\,°C$.

Over a very narrow range of temperatures, the resistance, R, varies linearly with temperature:

$$\Delta R = k \Delta T, \tag{5.10}$$

but to calculate the temperature accurately, a third-order nonlinear polynomial expression is necessary—the Steinhart-Hart equation:

$$\frac{1}{T} = a + b \ln R + c \ln^3 R, \tag{5.11}$$

where a,b, and c are constants specific to each thermistor. The constant a may be eliminated from the equation but measuring the resistance, R_o, at a known reference temperature T_o, to give:

$$\frac{1}{T} = \frac{1}{T_o} + b \ln R/R_o + c \ln^3 R/R_o. \tag{5.12}$$

The value of c is often negligible for NTC resistors, thus the variation of resistance as a function of temperature may be written as:

$$R = R_o e^{\beta(1/T - 1/T_o)}, \tag{5.13}$$

where the factor β typically varies from 3500 K to 4600 K.

Example 5.5. The factor β has a value of 1420 K and its resistance equals $785 \pm 4 \ \Omega$ at 100 °F.

(a) What would the temperature reading be for a measured resistance of $2315\Omega \pm 6\Omega$.
(b) Estimate the uncertainty in the temperature.
(c) If the uncertainty in the reference temperature were 0.2, what would be the uncertainty in temperature?

Solution 5.5a. The first step is to convert °F into K:

$$T_o = (100 \ °F - 32 \ °F)/(1.8 \ °C) + 273.15 \ K = 310.9 \ K.$$

The next step is to express the temperature as a function of resistance:

$$R = R_o \exp\left(\beta\left(\frac{1}{T} - \frac{1}{T_o}\right)\right),$$

$$1/T = 1/T_o + 1/\beta \ln\frac{R}{R_o}.$$

Substituting the values of $T_o, R_o(311 \ K) = 785\Omega$ and $R(T) = 2315\Omega$ into the equation gives:

$$1/T = 1/310.9 + 1/1420 \ln 2315/785 = 0.00398,$$
$$T = 251.4 \ K = -21.8 \ °C.$$

Solution 5.5b. From Chapter 2, the uncertainty of a quantity is a sum of the squares of the uncertainties of each of factor:

$$\Delta_f^2 = \left(\frac{\partial f}{\partial x_1}\Delta_1\right)^2 + \left(\frac{\partial f}{\partial x_2}\Delta_2\right)^2 + \left(\frac{\partial f}{\partial x_3}\Delta_3\right)^2 + \cdots + \left(\frac{\partial f}{\partial x_n}\Delta_n\right)^2.$$

For the case of exponential functions, $f = x_1^{a_1} x_2^{a_2} x_3^{a_3} x_4^{a_4} \ldots x_n^{a_n}$, a simple expression was derived that gives:

$$\frac{\Delta_f}{f} = \sqrt{\sum_{i=1}^{n} \left(\frac{a_i}{x_i}\Delta_i\right)^2}.$$

However, in this case, the functional relationship involves a logarithm and therefore, the partial derivatives with respect to each of the factors must be derived as given below:

$$\frac{\partial T}{\partial R} = -\frac{\beta/R}{[\ln R/R_o + \beta/T_o]^2} = -\frac{1420/2315}{[\ln 2315/785 + 1420/311]^2} = 0.0192,$$

$$\frac{\partial T}{\partial R_o} = -\frac{\beta/R_o}{[\ln R/R_o + \beta/T_o]^2} = -\frac{1420/785}{[\ln 2315/785 + 1420/311]^2} = 0.0567.$$

The uncertainty in the resistance at the reference temperature, Δ_{R_o}, and the measured temperature, Δ_R, was 4Ω and 6Ω, respectively:

$$\Delta_T = \sqrt{\left(\frac{\partial T}{\partial R}\Delta_R\right)^2 + \left(\frac{\partial T}{\partial R_o}\Delta_{R_o}\right)^2}.$$

5.4.2 Resistance Temperature Devices (RTDs)

As with thermistors, in order to record a signal, a DC current must be applied across the filament to generate a voltage from which the resistance is then derived:

$$R = V/I. \tag{5.14}$$

RTDs are manufactured with pure metals as opposed to semi-conductors (for thermistors) and their resistance increases with temperature, which makes them among the most expensive instruments to measure temperature.

Over most temperature ranges the variation is linear:

$$\alpha_T = \frac{R - R_o}{R_o(T - T_o)} = \frac{1}{R_o}\frac{dR}{dT}, \tag{5.15}$$

where α_T is the temperature coefficient of resistance (K^{-1}), R_o is the reference resistance (Ω), and T_o is the reference temperature (K or °C).

The resistance is expressed as a function of temperature and the standard conditions by:

$$R = R_o[\alpha_T(T - T_o) + 1]. \tag{5.16}$$

Platinum wire is the metal of choice because of its linearity, its chemical inertness, and its accuracy, which increases with purity. Other metals are also used and their characteristics are given in Table 5.5. Temperature ranges for Pt RTDs range from $-260\,°C$ to $660\,°C$, which is significantly greater than for thermistors. Temperatures as high as $850\,°C$ are also possible but it is difficult to avoid contamination of the platinum by the thermometer metal sheath.

Commercial thermistors are manufacture to certain standards, the most common of which is the Pt100—a platinum metal element with a standard resistance of $100\,\Omega$ (R_o) at a standard temperature of $0\,°C$ (T_o).

TABLE 5.5 Resistance Properties of Selected Metals Used in RTDs
$T_O = 20°C$ (Haynes, 1996)

Metal	α_T (K^{-1})	R ($\mu\Omega$ cm)
Ni	0.0067	6.85
W	0.0048	5.65
Al	0.0045	2.65
Cu	0.0043	1.67
Pb	0.0042	20.6
Ag	0.0041	1.59
Au	0.004	2.35
Pt	0.00385	10.5
Semiconductor	−0.068 to 0.14	10^9

For a very large temperature range, the resistance no longer varies linearly with temperature and a quadratic expression becomes necessary:

$$R = R_o(1 + aT + bT^2). \tag{5.17}$$

In addition to the nonlinearity of the resistance with respect to temperature, another critical factor that may introduce error in the measurement is due to the heating caused by the electrical current required to generate the signal. The power (heat) generated, q_E, by an electrical current through a wire is related to the resistance and current:

$$q_E = I^2 R = IV = V^2/R. \tag{5.18}$$

The change in temperature induced by the current is related to the power produced and the rate of energy dissipated to the surroundings (q_{diss}), which is represented by a coefficient K_T:

$$q_{diss} = K_T(T_R - T), \tag{5.19}$$

where T_R is the temperature of the thermistor and T is the ambient temperature (quantity of interest).

Solving for the temperature as a function of resistance gives the temperature difference between the ambient conditions and the thermistor:

$$T = T_R - \frac{1}{K_T}\frac{V^2}{R}. \tag{5.20}$$

The dissipation coefficient, K_T, is on the order of 1.5 m W K^{-1} in still air. Values double that are realistic in flowing situations. The temperature of the thermistor, T_R, is calculated based on either the linear or quadratic expression for resistance as a function of temperature to calculate:

$$T_R = T_o - \frac{1}{\alpha_T}(R/R_o - 1). \qquad (5.21)$$

Example 5.6. Calculate the temperature of gas estimated by a Pt100 thermistor for which the resistance equals 200 Ω when a voltage of 40 mV is applied. What is the temperature difference between the thermistor and the gas?

Solution 5.6. For a Pt100 thermistor, the resistance, R_o, is 100 Ω at a temperature of 0°C(T_o). The temperature coefficient of resistance, α_T, for Pt equals 0.00385 K. In stagnant air, we assume that the dissipation coefficient, K_T, equals 1.5 m W k^{-1}. The temperature of the resistor, assuming a linear relationship, is:

$$T_R = T_o - \frac{1}{\alpha_T}(R/R_o - 1) = 0 - \frac{1}{0.00385}(100/200 - 1) = 129.9\ °C.$$

The temperature of air is calculated from:

$$T = T_R - \frac{1}{K_T}\frac{V^2}{R} = 129.9 - \frac{1}{1.5}\frac{40^2}{200} = 124.6\ °C = 125\ °C.$$

The temperature difference due to heating is simply $T - T_R = 5\ °C$.

5.4.3 Thermocouples

Thermocouples are the most common instrument to measure temperature in the chemical industry. They are versatile, compact, are applicable over a wide temperature range, and are relatively inexpensive. Not only are they used as a means to monitor the process, they are used to troubleshoot, identify areas of non-uniformity, and even determine the hydrodynamic pattern of fluids in motion. The technique of detecting flow maldistribution based on measuring the concentration is demonstrated in Figures 2.10 and 2.11. In Figure 5.8, both the concentration profile and temperature profile of an experimental methanation fluidized bed reactor are illustrated:

$$CO + 3H_2 \rightarrow CH_4 + H_2O.$$

The reaction is highly exothermic and to maximize performance and avoid overheating that might destroy the catalyst, it is important to avoid hot spots. Hot spots are regions in which the local temperature rises significantly above the desired operating temperature and can easily exceed 100 °C. They result when the heat generated by the reaction is much greater than the heat transfer rate.

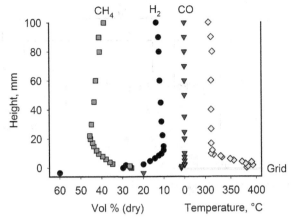

FIGURE 5.8 Fluidized Bed Methanation: Concentration and Temperature Axial Profiles (Kopyscinski, 2009)

The CO and H_2 are fed to the fluidized bed as a stoichiometric mixture (three moles of hydrogen for every one mole of carbon monoxide). Within the first 1 cm, the concentration of hydrogen drops from 60% to 10%. If the temperature were not measured simultaneously with the concentration, several hypotheses could adequately account for the drop—for example, one might consider that the mass transfer rate throughout the bed was extremely high. However, the temperature at the entrance drops from a maximum of approximately 400 °C to around 300 °C coincidentally with the drop in concentration.

These measurements are based on steady-state conditions. Thermocouples are also used to measure transients and are linked to safety alarms and interlocks to avoid hazardous conditions. For example, catalyst oxides can selectively convert alkanes and alkenes (paraffins and olefins) to oxygenated products. Oxygen is often co-fed to these processes and under some conditions may combust the hydrocarbon or form an explosive mixture (Hutchenson et al., 2010). When these conditions arise, it is critical for the instrumentation to detect changes in temperature in less than a second to change to non-hazardous operating conditions, i.e. reducing the oxygen or hydrocarbon feed rate or flooding the system with nitrogen, etc. Thermocouples are used for these conditions not only because of their rapid response times but also because they may be easily adapted to different geometries as well as being robust.

In 1821, T.J. Seebeck discovered that when two distinct metals were joined at a point as part of a circuit at different temperatures, a compass needle could be deflected. Later he determined that the temperature difference caused an electrical current and the voltage generated was close to proportional to the ΔT. An example of a circuit is shown in Figure 5.9, where T_h is referred to as the hot junction and T_c represents the cold junction and **A** and **B** represent two dissimilar metals (or alloys). The voltage resulting from joining two metals

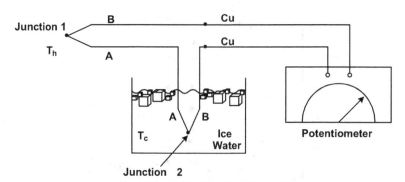

FIGURE 5.9 Thermocouple Configuration

at different temperatures is referred to as the Seebeck effect (although it was Nobili in 1829 who invented the first thermocouple).

Peltier showed in 1834 that by passing a current through two dissimilar metals, the junction would heat up in one direction and cool down in the opposite direction. This is known as the Peltier effect and can be used to liqueify nitrogen ($-196\,°C$).

Over a narrow range of temperatures, the electromotive force (ΔE or emf) is directly proportional to the temperature differential (ΔT):

$$\Delta E = \alpha_s \Delta T, \tag{5.22}$$

where α_s is the Seebeck coefficient ($\mu V\ K^{-1}$).

The Seebeck coefficient varies with the metal pair and many combinations are used. Table 5.6 lists common metal pairs together with their temperature range and corresponding voltage and the Seebeck coefficient. The most common thermocouples are the Types J and K. The Type J thermocouple is composed of iron and constantan, which is a copper-nickel alloy. It has a wide temperature range from $-210\,°C$ to $760\,°C$ and the second highest Seebeck coefficient of $50.2\mu\ VK^{-1}$, resulting in a higher degree of precision. The Type K thermocouple is made of two alloys—chromel (nickel and chromium) and alumel (nickel, manganese, and aluminum). Its Seebeck coefficient is lower but it can be used at temperatures as high as $1372\,°C$. A Type C thermocouple measures temperatures almost $1000\,°C$ higher than that of a Type K and contains tungsten and rhenium.

To achieve a better accuracy over a large temperature range, the emf is correlated to a third-order polynomial with respect to temperature:

$$E = AT + BT^2/2 + CT^3/3. \tag{5.23}$$

Alternatively, the emf may be read from NIST thermocouple reference tables that list the voltages as a function of temperature differential in $1\,°C$ increments

TABLE 5.6 Principal Thermocouple Types, Compositions, and Operating Range (Holman, 2001)

Type	Composition, Polarity	Temp. (°C)	Emf (mV)	Seebeck Coeff. (μV K^{-1})	T (°C)
T	(+) Copper	−270 to 400	−6.258 to 20.872	38	0
	(−) Constantan [55% Cu, 45% Ni]				
J	(+) Iron	−210 to 760	−8.095 to 42.919	50.2	0
	(−) Constantan [55% Cu, 45% Ni]				
K	(+) Chromel [90% Ni, 10% Cr]	−270–1372	−6.458 to 54.886	39.4	0
	(−) Alumel [94% Ni, 3% Mn, 2% Al, 1% Si]				
E	(+) Chromel [90% Ni, 10% Cr]	−270 to 1000	−9.835 to 76.373	58.8	0
	(−) Constantan [55% Cu, 45% Ni]				
S	(+) Platinum—10% Rhodium	−50 to 1768	−0.2356 to 18.693	10.3	600
	(−) Platinum				
R	(+) Platinum—13% Rhodium	−50 to 1768	−0.226 to 21.101	11.5	600
	(−) Platinum				
C	(+) Tungsten—5% Rhenium	0–2320	0–33.0	19.5	600
	(−) Tungsten—26% Rhenium				

(Burns et al., 1993). The voltages for Type J, K, and T thermocouples are given in Tables 5.7–5.9, respectively, at 10 °C increments (Burns et al., 1993). Note that since 0 °C is selected as the reference temperature, the temperature differential is simply equal to the temperature. Standardizing the voltage to 0 °C as a reference point (cold junction) has an advantage with respect to simplicity in reporting temperature but also experimentally: it is the most reproducible temperature.

The choice of thermocouple type is generally related to the temperature and the "bare wire environment" but accuracy can also be an important consideration; it depends on the chosen metals (or alloy) but the geometry of the application is also a factor. Besides the Seebeck and Peltier effects a third effect that might need consideration is the Thomson effect: a potential could be created due to a temperature gradient along the length of the thermocouple wires.

Together with these effects, multiple grades of thermocouples for each type are available. For Type J thermocouples, the error of the standard grade is 2.2 °C (or 0.75%, whichever is greater); the error of the special grade is 1.1 °C (or 0.4%, whichever is greater). The Type K thermocouple has the same error (for both grades) above 0 °C but below 0 °C, the error is 2.2 °C or 2.0% for the standard grade. The Type T thermocouple has a slightly lower error limit at 1.0 °C and 0.75% above 0 °C and 1.0 °C and 1.5% below zero. The special grade has an error of only 0.5 °C or 0.4%.

Example 5.7. The potential of a thermometer is approximated by the following third-order polynomial.

TABLE 5.7 Type J Thermocouple Reference Table—Recommended for Reducing, Vacuum, and Inert Environments; Limited Use for High Temperature and Oxidizing Conditions; Not Recommended for Low Temperatures

°C	mV	°C	mV	°C	mV	°C	mV	°C	mV
−210	−8.095	70	3.650	350	19.090	630	34.873	910	52.500
−200	−7.890	80	4.187	360	19.642	640	35.470	920	53.119
−190	−7.659	90	4.726	370	20.194	650	36.071	930	53.735
−180	−7.403	100	5.269	380	20.745	660	36.675	940	54.347
−170	−7.123	110	5.814	390	21.297	670	37.284	950	54.956
−160	−6.821	120	6.360	400	21.848	680	37.896	960	55.561
−150	−6.500	130	6.909	410	22.400	690	38.512	970	56.164
−140	−6.159	140	7.459	420	22.952	700	39.132	980	56.763
−130	−5.801	150	8.010	430	23.504	710	39.755	990	57.360
−120	−5.426	160	8.562	440	24.057	720	40.382	1000	57.953
−110	−5.037	170	9.115	450	24.610	730	41.012	1010	58.545
−100	−4.633	180	9.669	460	25.164	740	41.645	1020	59.134
−90	−4.215	190	10.224	470	25.720	750	42.281	1030	59.721
−80	−3.786	200	10.779	480	26.276	760	42.919	1040	60.307
−70	−3.344	210	11.334	490	26.834	770	43.559	1050	60.890
−60	−2.893	220	11.889	500	27.393	780	44.203	1060	61.473
−50	−2.431	230	12.445	510	27.953	790	44.848	1070	62.054
−40	−1.961	240	13.000	520	28.516	800	45.494	1080	62.634
−30	−1.482	250	13.555	530	29.080	810	46.141	1090	63.214
−20	−0.995	260	14.110	540	29.647	820	46.786	1100	63.792
−10	−0.501	270	14.665	550	30.216	830	47.431	1110	64.370
0	0.000	280	15.219	560	30.788	840	48.074	1120	64.948
10	0.507	290	15.773	570	31.362	850	48.715	1130	65.525
20	1.019	300	16.327	580	31.939	860	49.353	1140	66.102
30	1.537	310	16.881	590	32.519	870	49.989	1150	66.679
40	2.059	320	17.434	600	33.102	880	50.622	1160	67.255
50	2.585	330	17.986	610	33.689	890	51.251	1170	67.831
60	3.116	340	18.538	620	34.279	900	51.877	1180	68.406

(a) When the hot junction is exposed to a temperature of 1210 °C and the cold junction is at 31 °C, what will a potentiometer read?

$$E = 0.38T + 1.932 \times 10^{-5}\frac{T^2}{2} - 1.867 \times 10^{-7}\frac{T^3}{3},$$

where E is measured in V and the temperature in °C.

(b) If the voltage suddenly drops by 10%, what is the new temperature reading?

TABLE 5.8 Type K Thermocouple Reference Table—Recommended for Clean Oxidizing and Inert Environments; Limited Use for Vacuum or Reducing Conditions; Wide Temperature Range

°C	mV	°C	mV	°C	mV	°C	mV	°C	mV
−270	−6.458	60	2.436	390	15.975	720	29.965	1050	43.211
−260	−6.441	70	2.851	400	16.397	730	30.382	1060	43.595
−250	−6.404	80	3.267	410	16.820	740	30.798	1070	43.978
−240	−6.344	90	3.682	420	17.243	750	31.213	1080	44.359
−230	−6.262	100	4.096	430	17.667	760	31.628	1090	44.740
−220	−6.158	110	4.509	440	18.091	770	32.041	1100	45.119
−210	−6.035	120	4.920	450	18.516	780	32.453	1110	45.497
−200	−5.891	130	5.328	460	18.941	790	32.865	1120	45.873
−190	−5.730	140	5.735	470	19.366	800	33.275	1130	46.249
−180	−5.550	150	6.138	480	19.792	810	33.685	1140	46.623
−170	−5.354	160	6.540	490	20.218	820	34.093	1150	46.995
−160	−5.141	170	6.941	500	20.644	830	34.501	1160	47.367
−150	−4.913	180	7.340	510	21.071	840	34.908	1170	47.737
−140	−4.669	190	7.739	520	21.497	850	35.313	1180	48.105
−130	−4.411	200	8.138	530	21.924	860	35.718	1190	48.473
−120	−4.138	210	8.539	540	22.350	870	36.121	1200	48.838
−110	−3.852	220	8.940	550	22.776	880	36.524	1210	49.202
−100	−3.554	230	9.343	560	23.203	890	36.925	1220	49.565
−90	−3.243	240	9.747	570	23.629	900	37.326	1230	49.926
−80	−2.920	250	10.153	580	24.055	910	37.725	1240	50.286
−70	−2.587	260	10.561	590	24.480	920	38.124	1250	50.644
−60	−2.243	270	10.971	600	24.905	930	38.522	1260	51.000
−50	−1.889	280	11.382	610	25.330	940	38.918	1270	51.355
−40	−1.527	290	11.795	620	25.755	950	39.314	1280	51.708
−30	−1.156	300	12.209	630	26.179	960	39.708	1290	52.060
−20	−0.778	310	12.624	640	26.602	970	40.101	1300	52.410
−10	−0.392	320	13.040	650	27.025	980	40.494	1310	52.759
0	0.000	330	13.457	660	27.447	990	40.885	1320	53.106
10	0.397	340	13.874	670	27.869	1000	41.276	1330	53.451
20	0.798	350	14.293	680	28.289	1010	41.665	1340	53.795
30	1.203	360	14.713	690	28.710	1020	42.053	1350	54.138
40	1.612	370	15.133	700	29.129	1030	42.440	1360	54.479
50	2.023	380	15.554	710	29.548	1040	42.826	1370	54.819

TABLE 5.9 Type T Thermocouple Reference Table—Recommended for Mild Oxidizing, Reducing, Vacuum, Inert, and Humid Environments; Optimum for Low Temperature and Cryogenic Applications

°C	mV	°C	mV	°C	mV	°C	mV	°C	mV
−270	−6.258	−130	−4.177	10	0.391	150	6.704	290	14.283
−260	−6.232	−120	−3.923	20	0.790	160	7.209	300	14.862
−250	−6.180	−110	−3.657	30	1.196	170	7.720	310	15.445
−240	−6.105	−100	−3.379	40	1.612	180	8.237	320	16.032
−230	−6.007	−90	−3.089	50	2.036	190	8.759	330	16.624
−220	−5.888	−80	−2.788	60	2.468	200	9.288	340	17.219
−210	−5.753	−70	−2.476	70	2.909	210	9.822	350	17.819
−200	−5.603	−60	−2.153	80	3.358	220	10.362	360	18.422
−190	−5.439	−50	−1.819	90	3.814	230	10.907	370	19.030
−180	−5.261	−40	−1.475	100	4.279	240	11.458	380	19.641
−170	−5.070	−30	−1.121	110	4.750	250	12.013	390	20.255
−160	−4.865	−20	−0.757	120	5.228	260	12.574	400	20.872
−150	−4.648	−10	−0.383	130	5.714	270	13.139		
−140	−4.419	0	0.000	140	6.206	280	13.709		

Solution 5.7a. At a temperature of 1210 °C, the potential equals 363.7 mV. At 31 °C, it is 11.8 mV. The potentiometer reads the difference between these two values:

$$E = E_{T_h} - E_{T_c} = 363.7 - 11.8 = 351.9 \text{ mV}.$$

Solution 5.7b. Assuming that the drop is due to a change in the hot junction temperature, E will equal the potential of the cold junction temperature and the differential:

$$\Delta E = 0.1 \cdot 351.9 \text{ mV} = 35.2 \text{ mV},$$

$$E = 351.1 - 35.2 = 315.9 \text{ mV},$$

$$E_{T_h} = E_{T_c} + E = 11.8 + 315.9 \text{ mV} = 327.7 \text{ mV}.$$

The hot junction temperature now equals 327.7 mV and this value is used in the quadratic equation to solve for the temperature.

$$T = 1002 \text{ °C}.$$

The drop in 10% in the potential is equivalent to a drop of about 17% in temperature.

Tables 5.7–5.9 report the voltage as a function of temperature and third-order polynomials are generally sufficiently accurate to calculate the potential. However, to relate temperature as a function of thermocouple voltage, polynomials to as high as the ninth order are used:

$$T = \beta_0 + \beta_1 e + \beta_2 e^2 + \beta_3 e^3 + \beta_4 e^4 + \beta_5 e^5 + \beta_6 e^6 + \beta_7 e^7 + \beta_8 e^8 + \beta_9 e^9, \quad (5.24)$$

where T is the temperature (°C), e is the electrical potential (mV; cold junction reference temperature at 0 °C), and β_i are the polynomial coefficients.

5.4.4 Thermopile

The sensitivity of a thermocouple is related to the voltage generated. The voltage is greater for a larger temperature differential; it is 5.269 mV for a Type J thermocouple measuring a temperature of 100 °C and approximately double that at 200 °C. To increase the sensitivity, thermocouples may be connected in series and this series connection is referred to as a thermopile.

The emf generated is directly proportional to the number of junctions. A four-junction Type J thermopile generates 21.076 mV—or four times that of a single thermocouple. Greater precision may also be achieved by more advanced electronics with respect to potentiometers but for precise measurements of temperature differentials, a thermopile is very effective. Figure 5.10 illustrates the electrical circuit of a four-junction thermopile measuring the temperature differential between two points.

Example 5.8. Compare the uncertainty of measuring the temperature differential in the entrance region of the methanation reactor of Figure 5.8 with:

(a) Two Type J thermocouples (special grade).

(b) A differential thermopile with four-junction pairs illustrated in Figure 5.10.

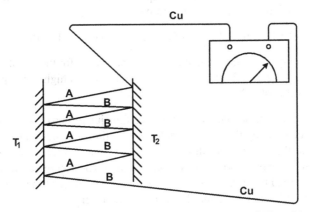

FIGURE 5.10 Four-Junction Pair Thermopile to Measure Temperature Differential

The uncertainty in the potentiometer reading is 0.1% FS for case (a) and it measures voltages up to 150 mV. For case (b), the full-scale reading is 15 mV with a 0.1% error at full scale.

Solution 5.8a. The simplest approach to this problem is to calculate the uncertainty based on the square root of the sum of the squares of the two temperature measurements. The uncertainty of the temperature reported for Type J thermocouples at 400 °C is 1.6 °C (0.4% × 400 °C):

$$\Delta T = T_2 - T_1,$$
$$\Delta_{\Delta T} = (\Delta_{T_2}^2 + \Delta_{T_1}^2)^{1/2} = (1.6^2 + 1.6^2)^{1/2} = 2.2 \,°\text{C}.$$

This approach ignores the uncertainty related to the potentiometer. To derive a more precise value, the sensitivity of the thermocouple should be taken into consideration either with the high-order polynomial or by deriving a lower-order polynomial in the vicinity of the temperature of interest. For a Type J thermocouple in the range of 0–1000 °C, a second-order polynomial accounts for 99.994% of the variance in the data (i.e. $R^2 = 0.99994$):

$$T_J = 19.1e - 0.0328e^2,$$

where T_J is the thermocouple temperature given in the NIST tables (Table 5.7) in °C and e is the voltage from Table 5.7 (mV).

The error in the measured temperature, Δ_{T_J} (from Chapter 2) is the partial derivative of the temperature function with respect to the voltage:

$$\Delta_{T_J}^2 = \left(\frac{\partial T_J}{\partial e}\Delta_e\right)^2 = \left(\frac{\partial}{\partial e}(19.1e - 0.0328e^2)\Delta_e\right)^2,$$
$$\Delta_{T_J} = (19.1 - 0.0656e)\Delta_e.$$

At 400 °C, the emf generated is approximately 21.8 mV. The uncertainty of the instrument is 0.1% of 150 mV, which equals ± 0.15 mV. So, the uncertainty of a single temperature measurement will be:

$$\Delta_{T_J} = (19.1 \,°\text{C mV}^{-1} - 0.0656 \cdot 21.8 \,°\text{C mV}^{-1})0.15 \,\text{mV} = 2.7 \,°\text{C}.$$

The uncertainty, $\Delta_{\Delta T}$, is calculated in the same way as for the simple approach and equals $\sqrt{2} \cdot 2.7 = 3.8 \,°\text{C}$, which is almost twice as high as the previously calculated value of 2.2 °C.

Solution 5.8b. The second case involves a four-junction pair thermopile. The total emf recorded by the potentiometer is four times greater than for a single junction but the temperature difference is very small—it could be as little as 1 °C, whereas in case (a) the emf recorded was based on a temperature of 400 °C. The voltage read from the potentiometer, E_t, is a product of the sensitivity, S, the number of junctions, n, and the temperature differential:

$$E_t = nS\Delta T.$$

The expression for the temperature differential becomes:

$$\Delta T = \frac{E_t}{nS}.$$

The sensitivity is calculated by differentiating the potential with respect to temperature:

$$S = \frac{dE}{dT}.$$

In the same way a polynomial expression was derived to express the temperature as a function of voltage, a polynomial may also be formulated that approximates the emf as a function of temperature:

$$E = \beta_1 T_J - \beta_2 T_J^2.$$

The constants for the polynomial expression are $\beta_1 = 0.05185$ and $\beta_2 = 6.20 \times 10^{-6}$. This relationship accounts for 99.92% of the variance in the data for a range of temperatures from 0 °C to 1000 °C. The sensitivity is then equal to:

$$S = \frac{dE}{dT_J} = 0.05185 - 12.4 \times 10^{-6} T_J,$$

$$S_{400\,°C} = 0.05185 - 12.4 \times 10^{-6}(400) = 0.469 \text{ mV °C}^{-1}.$$

Again from Chapter 2, since the expression for the temperature differential is a simple product, the expression for the uncertainty:

$$\Delta_{\Delta T_J}^2 = \left(\frac{\partial \Delta T_J}{\partial E_t}\Delta_{E_t}\right)^2 + \left(\frac{\partial \Delta T_J}{\partial S}\Delta_S\right)^2,$$

is simplified to:

$$\frac{\Delta_{\Delta T}}{\Delta T} = [(\Delta_{E_t}/E_t)^2 + (\Delta_S/S)^2]^{1/2},$$

$$\Delta_{\Delta T} = [(\Delta_{E_t}/nS\Delta T)^2 + (\Delta_S/S)^2]^{1/2}\Delta T.$$

The uncertainty of the potentiometer $\Delta_{E_t} = 0.00115$ mV $= 0.015$ mV and the sensitivity of the special grade thermocouple wire is 0.4% (Δ_S/S). The equation for the uncertainty in temperature is a function of the temperature level:

$$\Delta_{\Delta T} = \left[\left(\frac{0.015 \text{ mV}}{4 \cdot 0.0469}\right)^2 + (0.04\Delta T)^2\right]^{1/2} = [0.0800^2 + (0.04\Delta T)^2]^{1/2}.$$

The uncertainty in a 5 °C temperature differential equals 0.22 °C and it is about double that (0.41 °C) at a differential of 10 °C.

5.4.5 Radiation

The temperature measured by a thermocouple is related to several phenomena including convection, conduction, humidity, and radiation. In fluid applications, stagnant regions should be avoided. Probes should be placed in areas where the fluid is in motion—thus increasing the convective heat transfer rate. Radiation effects are often neglected but can be significant, particularly in laboratory applications with electrical heating and glass/quartz equipment. In order to minimize complications arising from radiation, it is best to purchase thermocouples that are shielded. In the case of a probe placed in flowing gas, conduction is negligible and a heat balance gives:

$$hA(T_g - T_p) = \sigma \varepsilon A(T_p^4 - T_w^4), \tag{5.25}$$

where h is the heat transfer coefficient (W m^{-2} K^{-1}, A is the surface area (m^2), T_g is the gas temperature (K), T_p is the probe temperature (K), T_w is the wall temperature (K), σ is the Stefan-Boltzmann constant (5.669 × 10^{-8} W m^{-2} K^{-4}), and ε is the surface emissivity of the probe ($0 < \varepsilon < 1$, i.e. 1 equals a black body).

When probes are immersed in opaque solids—catalytic reactors, for example—the solids temperature, probe temperature, and surrounding temperature are equal—$T_p = T_w$. Above the bed of solids, the reading will depend on the surrounding environment: it may remain unchanged if the vessel wall temperature is at the same temperature as the fluid (and the stream is highly convective); if the wall is cooler than the gas, the thermocouple will irradiate heat resulting in a lower than expected temperature; in the case where the walls are heated electrically, the thermocouple will be hotter than the fluid temperature. Regardless of whether or not the stream is highly convective, the temperature reading will be determined by the surroundings!

5.5 PYROMETRY

Co-author: Paul Patience

A pyrometer is a non-contacting temperature measurement instrument that is usually used for temperatures above 500 °C, although with some modifications it can measure temperatures below room temperature. The word pyrometry comes from the Greek words *pyro* (fire) and *meter* (measure). The basic principle relies on the notion that all bodies emit thermal radiation proportional to their temperature. Pyrometers detect this thermal radiation and through Planck's law the temperature can be determined.

5.5.1 Thermal Radiation

Thermal radiation is essentially a conversion of thermal energy into electromagnetic energy. All bodies at a temperature above absolute zero radiate thermal energy. However, they can also absorb, reflect, and transmit energy from other sources. Absorptivity (α), reflectivity (ρ), and transmissivity (τ) are related through the following equation:

$$\alpha + \rho + \tau = 1. \tag{5.26}$$

Gases approach ideal transmissivity, highly polished surfaces approach ideal reflectivity, and a small opening in a large hollow body approaches ideal absorptivity. An ideal absorber is called a black body, as opposed to a gray or non-gray body. Black bodies are perfect absorbers: they absorb all incoming energy. This makes them perfect emitters as well, because at constant temperature, a body emits the same amount of energy it absorbs; otherwise the temperature would either rise or drop. Kirchhoff's law states that the absorptivity of an object is equal to its emissivity (ϵ), or:

$$\epsilon = \alpha. \tag{5.27}$$

Planck's distribution law relates the energy emitted by a black body to the absolute temperature and the wavelength of the radiation:

$$E_{\lambda,b} = \frac{C_1 \lambda^{-5}}{\exp\left(C_2/\lambda T\right) - 1}, \tag{5.28}$$

where $E_{\lambda,b}$ is the energy emitted by a black body at wavelength $\lambda (\text{W m}^{-2}), T$ is the temperature (K), λ is the wavelength (μm), C_1 is 374.18 MW μm^4 m^{-2}, and C_2 is 14 388 μm K.

As temperature increases, the amount of energy emitted increases, and the peak radiation wavelength (the wavelength at which the radiation is the highest) decreases.

Non-black bodies emit less energy than black bodies since they are not perfect absorbers (they reflect and transmit incoming radiation as well as absorb it). Non-black bodies include gray bodies and non-gray bodies. Gray bodies have a constant emissivity with respect to wavelength, while the emissivity of non-gray bodies varies with wavelength. Planck's law can be adapted for non-ideal bodies by multiplying the emissivity by the black body equation:

$$E_{\lambda} = \frac{\epsilon(\lambda) C_1 \lambda^{-5}}{\exp\left(C_2/\lambda T\right) - 1}, \tag{5.29}$$

where $\epsilon(\lambda)$ is the emissivity.

5.5.2 Pyrometers

Pyrometers, also known as radiation thermometers (since they measure temperature through radiation), essentially consist of an optical system and a detector. The optical system detects energy given off by a body and focuses it onto the detector. The detector returns a value proportional to the energy radiated and is used to determine the object's temperature.

Radiation thermometers are non-contact instruments—they can be used at a distance from the object being measured. This is useful in situations where other methods of temperature measurement are difficult or impossible to use. They are classified based on the wavelength measured. Broadband pyrometers absorb energy over a very broad range of wavelengths, while narrowband pyrometers only absorb energy over a single wavelength. Ratio pyrometers measure radiated energy at two wavelengths and compare them to determine the temperature. Optical pyrometers, also known as disappearing filament pyrometers, compare the object's color to that of a heated filament inside the pyrometer.

Broadband pyrometers can measure thermal radiation from 0.3 μm to 20 μm, depending on the instrument. No filters are used to narrow the range of wavelengths detected, unlike other pyrometers. In the case of broadband pyrometers, the Stefan-Boltzmann law is used to calculate the temperature:

$$q = \epsilon \sigma T^4, \tag{5.30}$$

where q is the radiant heat flux emitted by source (W m^{-2}), ϵ is the emissivity of the source, σ is the Stefan-Boltzmann constant (which equals 5.6074×10^{-8} W m^{-2} K^{-4}), and T is the temperature (K).

It is important that the path to the object be free of particles such as dust, gases, vapor, smoke, etc., since these obstructions will cause error in the pyrometer's reading—these particles might absorb some of the body's radiation, causing the pyrometer to read a lower temperature than is correct.

Narrowband pyrometers operate at a single wavelength. This is accomplished through the use of filters. The choice of wavelength depends on the temperature range required. Since peak radiation is achieved at decreasing wavelengths with increasing temperature, a narrowband pyrometer measuring temperature over a shorter wavelength will tend to be used for higher temperatures. These radiation thermometers are generally used at above 500 °C, with a wavelength ranging from 0.65 μm to 0.85 μm.

Ratio pyrometers are a variation of narrowband pyrometers used to reduce the influence of the emissivity of the object being measured (they are especially useful in cases where the emissivity varies with time). The intensity of radiation is measured at two wavelengths (λ_1 and λ_2) and Planck's law is used:

$$\frac{E_{\lambda_1}}{E_{\lambda_2}} = \left(\frac{\lambda_1}{\lambda_2}\right)^5 \frac{e^{C_2/\lambda_2 T} - 1}{e^{C_2/\lambda_1 T} - 1}. \tag{5.31}$$

If the emissivity varies little with time and the two wavelengths are close, it will cancel out and thus not affect the result.

Disappearing filament pyrometers were among the first pyrometers used. The original optical pyrometers used red filters to limit the wavelength around 0.65 μm. A tungsten filament is located inside the pyrometer and will initially appear hot or cold compared to the source (the operator looks into the eyepiece to compare the filament and the source). A current is run through the filament to adjust its temperature, and therefore its color. This current is changed until the filament "disappears" inside the source: they will both have the same color. A calibration curve relates the current sent through the tungsten filament and the equivalent temperature of the source.

5.6 EXERCISES

5.1 Repeat the first example in Chapter 5 where the cylinder body is charged with 100 ml of acetone and mercury is the operating fluid in the U-tube manometer (differential height is 14 in.). Discuss the potential sources of error in this measurement and how to minimize them.

5.2 How did Sir Thomas Allbutt reduce the size of the clinical thermometer from 30 cm to 15 cm and reduce the response time from 20 min to 5 min?

5.3 A thermometer with a 0.4 mm ID capillary may be charged with either galinstan or toluene.

(a) Which of the two fluids will give a higher resolution?
(b) A higher temperature range?
(c) At 100 °C which fluid is higher in the column and by how much?

5.4 The temperature of water in a beaker on a hot plate is measured by a "constant-volume" thermometer. The cylindrical body of the thermometer is 250 ml. It is connected directly to a Bourdon gauge and the barometric pressure is 1.0 bar. One hundred grams of isopropanol is placed in the body of the thermometer. What is the temperature of the water if the pressure recorded on the Bourdon gauge is 10 kPa?

5.5 To prepare a dish of rice well, it must be cooked at a temperature of 90 °C. Room temperature is 25 °C and a Type T thermocouple is available. *N. Ly*

(a) Will the dish of rice be well prepared if the thermocouple indicates a value of 3.814 mV?
(b) Calculate the value that the potentiometer should indicate to obtain a temperature of 90 °C.

5.6 A total immersion Hg thermometer with a range from −20 °C to 200 °C is 40 cm in length. It is immersed in a boiling fluid in a beaker on a hot plate to a depth of 5 cm. If the thermometer reads 160 °C, what is the actual temperature?

5.7 (a) Derive an expression for the radius of curvature of a bimetallic strip when the thickness of each strip is equal and the Young's moduli are approximately the same.

(b) At what value of n does the error in the calculated value of the radius of curvature exceed 5%?

5.8 In the sixth example in Chapter 5, what would be the uncertainty in the temperature if the uncertainty in the reference temperature was 2 °F?

5.9 Derive an expression for the uncertainty of the resistance, R, of an RTD with respect to reference resistance and temperature for the case that the resistance varies with the following quadratic:

$$R = R_0(1 + aT + bT^2).$$

5.10 For the same conditions of the sixth example in Chapter 5, would the temperature differential between the thermistor and air be greater with Ni? How much lower would the temperature rise in the case of a flowing gas? Assume a Ni100 thermistor.

5.11 Derive a second-order polynomial expression relating the temperature and *emf* for a Type K thermocouple from 0 °C to 1000 °C.

5.12 The internal temperature of a camel can vary from 34 °C to 42 °C in order to adapt to the Sahara's temperature changes—usually from 10 °C to 50 °C during the summer. In the desert of southern Tunisia (Tozeur City), a camel's temperature was studied during the day with a Type T thermocouple. A voltmeter is used as a cold junction and its temperature is estimated at 25 °C, 50 °C, 10 °C at 9 AM, 3 PM and 9 PM. The other junction is placed in the camel's mouth. *B. Sana*

(a) If voltage drops are negligible, what is the temperature of the camel if the voltmeter shows a voltage of 0.486 mV, 0.959 mV, and −0.355 mV at 9 AM, 3 PM, and 9 PM?

(b) What would the voltage indicated by the voltmeter be if the reference temperature suddenly dropped from 10 to −1 °C overnight (at 9 PM)?

(c) Assuming that the voltmeter measuring the *emf* has an uncertainty of 0.003 mV and the temperature of the cold junction is set at a 1 °C resolution, calculate the accuracy in measuring the temperature of the camel for $T_{ref} = 25$ °C (9 PM).

5.13 Determine the *emf* indicated on three thermocouples of Types J, T, and K if the sample temperature is 290 K and the cold junction is in a refrigerator at 4 °C. Keeping the junction in the refrigerator, what would the temperature of a beer be if the *emf* indicated by the thermocouple is 0.203 mV? *P. Malo-Couture*

5.14 Repeat part (c) of the second example in Chapter 5 taking into account the thermal expansion properties of glass.

5.15 A copper-constantan thermocouple (Type T) is exposed to a temperature of 500 °F. The cold junction temperature is estimated to be 50 °F. *A. Benamer*

(a) Calculate the electromotive force *(emf)* indicated by the potentiometer.

(b) If the potentiometer indicates an *emf* of 3.723 mV, what temperature (in °C and °F) will the thermocouple be exposed to?

(c) If the temperature of the cold junction were to hit 5 °F, what would (a) be?

5.16 Determine the radius of curvature of a metallic strip thermometer's bimetallic strip, composed of yellow brass and chrome. The two metals are sealed at 60 °C and each band has a thickness of 0.20 mm. The strip has a length of 15.0 cm when exposed to a temperature of 250 °C. *J. Tran*

5.17 An expansion thermometer—with graduations of 1 °C—and two thermocouples—Types T and K—are used to observe the evolution of temperature in a heated water bath. The temperature is read directly off the expansion thermometer. The reference temperature for the two thermocouples is 25 °C and the data is summarized in Table Q5.17. Note that the thermocouples' precision is 0.05 mV. *C. Neagoe*

TABLE Q5.17 Thermocouple Reference Temperature

#	Exp. Therm. (°C)	Type T (mV)	Type K (mV)
1	21	−0.147	−0.144
2	25	0.039	0.030
3	29	0.170	0.190
4	33	0.344	0.408
5	39	0.550	0.550
6	42	0.720	0.725
7	47	0.910	0.880
8	51	1.063	1.020

(a) Calculate the calibration curves for both thermocouples.

(b) Calculate the absolute and relative errors for each device.

(c) Based on the temperature values and the error of each instrument, what can you conclude?

5.18 A butcher discovers some of his beef has putrefied from what appears to be poor temperature control in the freezer and thus decides to replace the controller and measuring device. The new system consists of a thermistor with a resitance of (800 ± 4) Ω at 80 °F and $\beta = 3400$ K. *Ludmilla*

(a) Determine the temperature of the freezer for a resistance of (2424 ± 5) Ω.

(b) Determine the uncertainty of the temperature.

REFERENCES

Adler, M. J. 1974. Albutt, Sir Thomas Clifford. Encyclopaedia Britiannica. Micropaedia 1, 251.

Berna, F., Behar, A., Shahack-Gross, R., Berg, J., Boaretto, E., Gilboa, A., Shaorn, I., Shalev, S., Shilstein, S., Yahalom-Mack, N., Zorn, J.R., Weiner, S., 2007. Sediments exposed to high temperatures: reconstructing pyrotechnological processes in Late Bronze and Iron Age Strata at Tel Dor (Israel). Journal of Archaeological Science 34, 358–373.

Bolton, H.C., 1900. Evolution of the thermometer. The Chemical Publishing Co., pp. 1592–1743.

Brown, K.S., Marean, C.W., Herries, A.I.R., Jacobs, Z., Tibolo, C., Braun, D., Roberts, D.L., Meyer, M.C., Bernatchez J., 2009. Fire as an engineering tool of early modern humans. Science 325, 859–862.

Burns, G. W. , Strouse, G. F., Croarkin, M. C., Guthrie, W.F., 1993. NIST Mono. 175 Thermocouple Reference Functions on ITS–90.

Goren, Y., Goring-Morris, A.N., 2008. Early pyrotechnology in the near east: experimental lime-plaster production at the pre-pottery neolithic B site of Kfar HaHoresh, Israel. Geoarchaelogy: An International Journal 23 (6), 779–798.

Goren-Inbar, N., Alperson, N., Kislev, M.E., Simchoni, O., Melamed, Y., Ben-Nun, A., Werker, E., 2004. Evidence of hominin control of fire at gesher benot ya'aqov, Israel. Science 304 (5671), 725–727.

Haynes, W.M. (Ed.), 1996. CRC Handbook of Chemistry and Physics, 92nd ed. CRC Press.

Holman, J.P., 2001. Experimental Methods for Engineers, seventh ed. McGraw-Hill Inc., New York.

Hummel, R.E., 2005. Understanding Materials Science: History, Properties, Applications, second ed. Springer.

Hutchenson, K .W., LaMarca, C., Patience, G.S., Laviolette, J.-P., Bockrath, R.E., 2010. Parametric study of Homogeneous Oxidation of Butane in a Circulating Fluidized Bed Reactor. Applications Catalysis A: General 376, 91–103.

Knoblauch, M., Hibberd, J.M., Gray, J.C., van Bel, A.J.E., 1999. A galinstan expansion femtosyringe for microinjection of eukaryotic organelles and prokaryotes. Nature Biotechnology 17, 906–909.

Kopyscinski, J., Schildhauer, T.J., Biollaz, S.M.A. 2009. "Employing Catalyst Fluidization to Enable Carbon Management in the Synthetic Natural Gas Production from Biomass", Chem. Eng. Technol., 32 (3) 343–347.

Temperature Measurement, Wiley.

Nicholson, P.T., Shaw, I., 2000. Ancient Egyptian materials and Technology, Cambridge University Press.

Rothfusz, L.P., 1990. The Heat Index "Equation" (or, More Than You Ever Wanted to Know About Heat Index), Scientific Services Division (NWS Southern Region Headquarters) SR 90–23.

Sherwood-Taylor, F., 1942. The origin of the thermometer. Annals of Science 5 (2), 129–156.

Steadman, R.G., 1979. The Assessment of sultriness. Part I: a temperature-humidity index based on human physiology and clothing science. Journal of Applied Meteorology 18, 861–873.

Thieme, H., 1997. Lower palaeolithic hunting spears from Germany. Nature 385, 807.

Weisberger, G., Willies, L., 2001. The use of fire in prehistoric and ancient mining: firesetting. Paleorient 26 (2), 121–149.

Chapter 6

Fluid Metering

6.1 OVERVIEW

Fluid flow is not only a critical aspect of the process industry: it had a tremendous importance in the development of ancient civilizations. The Sumerians in 4000 BC were the first to use canals for irrigation. Babylon is known to have had toilets and Sargon II's (700 BC) palace had drains connected to a 1 m high and 5 m long sewer that ran along the outer wall of the city. Sargon II's son, Sennacherib, was the first to build an aqueduct (65 km long), which was constructed for the Assyrian capital city Nineveh. In the Persian Achaemenian Empire, qanats were invented that tapped ground water and relied on gravity as the driving force for transport: underground tunnels were dug to the level of the ground water at one point and sloped downwards toward the exit. Nebuchadnezzar is credited for erecting the Hanging Gardens of Babylon: Water was fed to the gardens through the use of a noria—a water wheel, which could be considered one of the first automatic pumps. Norias were been designed to deliver water for irrigation at rates of up to 2500 l h^{-1}. The largest water wheel was 20 m tall and was built in Hama, Syria. Prior to water wheels, manual pumps called shadoofs were used to raise water from wells in Mesopotamia (2000 BC) and from the Nile in Egypt. Maximum rates are of the order of 2500 l d^{-1} at a maximum height of about 3 m. In 200 BC the reciprocating pump was invented by Ctesibius and Archimedes described the screw pump (some believe that the screw pump was used 500 years earlier in Babylon). Bellows may be considered the first air pump and were used by the Egyptians in 1490 BC. In the fifth century BC, the Chinese developed double-action piston bellows and Du Shi of China adapted a water wheel to power a bellows used for smelting iron.

Experimental Methods and Instrumentation for Chemical Engineers. http://dx.doi.org/10.1016/B978-0-444-53804-8.00006-X

The Romans mastered hydraulic engineering, which is principally concerned with the transport of water and sewage. Fresh water was supplied to Rome through 10 independent aqueducts that varied in length from 15 km to 91 km, the first of which was built in 312 BC. Estimates of the daily flow rate range from 300 kt d^{-1} to 1000 kt d^{-1}. Transporting water over these great distances required regulation basins, culverts, and energy dissipaters called stepped and dropshaft cascade chutes. The design tolerances for the aqueducts downward gradient were as low as 20 cm km^{-1}.

Together with the aqueducts, the Romans also advanced the technology of water distribution from the aqueducts to multiple sites—baths, residences, fountains, etc. Plumbing is derived from the Latin word *plumbum* which means lead. Piping used in Roman times included lead pipes, masonry channels as well as earthenware pipes. The water was delivered to baths and some public homes at a constant rate. The cost of the water was charged based on the pipe cross-sectional area, which served as a restriction orifice (Chanson, 2002, 2008).

Whereas the transport of water to major centers allowed civilizations to flourish, the measurement and control of fluid flow has been a critical aspect of the development of industrial processes. Not only is metering flow important to maintaining stable and safe operating conditions, it is the prime means to account for the raw materials consumed and the finished products manufactured. While pressure and temperature are critical operating parameters for plant safety, the measurement of flow rate has a direct impact on process economics. For basic chemicals (as opposed to specialty chemicals or pharmaceuticals) like ethylene, propylene, methanol, sulfuric acid, etc. profit margins are relatively low and volumes are large, so high precision instruments are required to ensure the economic viability of the process.

Flow meters are instruments that measure the quantity of movement of a fluid in a duct, pipe, or open space. The fluid can be water, a liquid solution, a chemical product or slurry, gas or vapor, and even solid—powders, for example. In everyday life, we use flow meters to pump gasoline into automobile fuel tanks, methane gas is metered to houses and metering water to houses is becoming more common. With respect to anatomy, the heart's pumping action ensures blood circulation and lung health is assessed by measuring the volume of air the lungs can hold. Trees are amazing for their ability to transport water in xylem and phloem for vertical distances exceeding 100 m!

Despite the importance of transporting water and its contribution to the rise of many great civilizations, most of the ancient technology used to build and maintain aqueducts, water distribution, and sewage systems was lost. Rome declined from a city of 1.6 million habitants at its zenith in 100 AD to less than 30 000 during the dark ages up until the Renaissance partly because of the destroyed aqueducts that remained in disrepair.

6.2 FLUID DYNAMICS

Modern fluid dynamics was pioneered by the Swiss physicist Bernoulli who published the book entitled "Hydrodynamica" in 1738. Bernoulli's work is based on the principle of the conservation of energy, which holds that mechanical energy along a streamline is constant. He demonstrated that increasing the potential energy of a flowing fluid (by raising the elevation of a pipe, for example) reduces the pressure of the fluid in the pipe (above that which would be expected based on wall friction); decreasing the cross-section of the pipe increases the velocity head and decreases the pressure head.

Consider the flow restriction of a pipe in Figure 6.1. The fluid accelerates from left to right as it passes through the restriction. Based on continuity—conservation of mass—the mass flow rate, $\dot{m}$, crossing point 1 equals that at point 2:

$$\dot{m}_1 = \dot{m}_2,$$
$$\dot{m} = \rho_1 X_{A,1} u_1 = \rho_2 X_{A,2} u_2, \tag{6.1}$$

where ρ is the fluid density (kg m^{-3}), X_A is the cross-sectional area (m^2), and u is the velocity (m s^{-1}).

For an incompressible fluid, the fluid accelerates in proportion to the ratio of the cross-sectional areas.

Bernoulli derived an equation based on an energy balance around a fluid flowing in a pipe and is expressed in a simplified form as:

$$\frac{\Delta P}{\rho} + \frac{1}{2}\Delta u^2 + g\Delta Z = h_f, \tag{6.2}$$

where h_f is the head loss due to friction.

In the case of Figure 6.1, besides neglecting friction losses (as well as some other simplifying assumptions), the change in elevation is equal to zero and thus the pressure drop going from point 1 to point 2 is simply calculated based

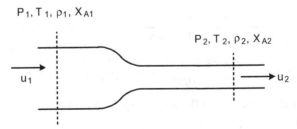

P_1, T_1, ρ_1, X_{A1}

P_2, T_2, ρ_2, X_{A2}

u_1

u_2

FIGURE 6.1 Fluid Flow Through a Constriction

on the change in fluid velocity:

$$P_1 - P_2 = \Delta P = -\frac{1}{2}\rho \Delta u^2 = -\frac{1}{2}\rho(u_1^2 - u_2^2). \tag{6.3}$$

Example 6.1. Light hydrocarbons are often stored in spheres and are transported by pipeline at moderate to high pressures in pipelines to chemical plants. To minimize pressure drop, pipeline diameter may be higher than that at the plant. Smaller pipe diameters are preferred in plants to minimize the cost of instrumentation and process control equipment such as valves and flow meters. Calculate the pressure drop, in mbar, resulting from a reduction of 6″ Sch40 pipe to 4″ Sch40 pipe transporting 10 000 kg h^{-1} of n-butane at 60 °F and 7 atm.

Solution 6.1. Pipe diameters are quoted in terms of their nominal pipe size as well as their schedule number that represents wall thickness—Table 6.1. The two most common pipe schedules are 40 and 80. Schedule 80 pipe is used for higher pressure applications and has thicker walls. Since the outside diameters (OD) of all nominal pipe sizes (NPS) are the same, higher schedule pipes have a smaller inner diameter (ID). The inside diameter of the 6 in. Sch40 pipe is 6.065 in. and it is 4.026 in. for the 4 in. Sch40 pipe.

Fluid properties as a function of temperature and pressure may be retrieved from the NIST database at http://webbook.nist.gov/chemistry/fluid. Butane has a density of 584 kg m^{-3} at a temperature of 60 °F and a pressure of 7 atm.

The cross-sectional area of each pipe is:

$$X_{A,1} = \frac{\pi}{4}d^2 = \frac{\pi}{4}(6.065 \text{ in.} \cdot 0.0254 \text{ m in.}^{-1})^2 = 0.0186 \text{ m}^2,$$

$$X_{A,2} = \frac{\pi}{4}d^2 = \frac{\pi}{4}(4.026 \text{ in.} \cdot 0.0254 \text{ m in.}^{-1})^2 = 0.00821 \text{ m}^2.$$

The velocity in each pipe is calculated based on continuity:

$$u = \frac{\dot{m}}{\rho X_A},$$

$$u = \frac{10\,000 \text{ kg h}^{-1}}{585 \text{ kg m}^{-3} \cdot 0.0186 \text{ m}^2} \cdot \frac{1 \text{ h}}{3600 \text{ s}} = 0.256 \text{ m s}^{-1},$$

$$u = \frac{10\,000 \text{ kg h}^{-1}}{585 \text{ kg m}^{-3} \cdot 0.00821 \text{ m}^2} \cdot \frac{1 \text{ h}}{3600 \text{ s}} = 0.579 \text{ m s}^{-1}.$$

Substituting the velocity into Bernoulli's equation gives:

$$\Delta P = -\frac{1}{2}\rho(u_1^2 - u_2^2) = -\frac{1}{2}584 \text{ kg m}^{-3}[(0.256 \text{ m s}^{-1})^2 - (0.579 \text{ m s}^{-1})^2]$$

$$= 78.8 \text{ Pa} \cdot \frac{1 \text{ bar}}{100\,000 \text{ Pa}} \cdot \frac{1000 \text{ mbar}}{1 \text{ bar}} = 0.788 \text{ mbar}.$$

TABLE 6.1 Nominal Pipe Sizes

NPS	OD (in.)	Sch. No.	Wall Thickness (in.)	ID (in.)
1/8	0.405	40	0.068	0.269
		80	0.095	0.215
1/4	0.54	40	0.088	0.364
		80	0.119	0.302
3/8	0.675	40	0.091	0.493
		80	0.126	0.423
1/2	0.84	40	0.109	0.622
		80	0.147	0.546
1	1.315	40	0.133	1.049
		80	0.179	0.957
2	2.375	40	0.154	2.067
		80	0.218	1.939
3	3.5	40	0.216	3.068
		80	0.300	2.900
4	4.5	40	0.237	4.026
		80	0.337	3.826
5	5.563	40	0.258	5.047
		80	0.375	4.813
6	6.625	40	0.280	6.065
		80	0.432	5.761
8	8.625	40	0.322	7.981
		80	0.500	7.625

The head loss due to friction has been neglected in the calculations but it is very important for long pipelines and in the case of elbows, valves, tees, and other restrictions. In the case of aqueduct design, restrictions were necessary to control flow rate. In the case of pipe flow, straight lines are preferred and all obstructions and changes in direction should be minimized to minimize pressure drop.

The pressure drop in a straight pipe is determined by factors including fluid velocity and viscosity, pipe surface roughness as well as flow regime—whether or not the flow is turbulent or laminar, a concept introduced by Stokes in 1851. In 1888, Osborne Reynolds conducted experiments that clearly delineated the difference between the different flow regimes. He injected a fine filament of colored water in a pipe together with water. At low flow rates, the filament

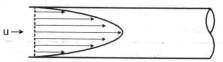

FIGURE 6.2 Laminar Velocity Profile, $N_{Re} < 2000$

maintained a parallel line from one end to the other. When the flow rate was increased beyond a certain value, the colored filament would break up forming eddies and vortices. The eddies and vortices promoted mixing of the colored water such that the filament disappeared and the color became uniformly distributed throughout the cross-section.

At the low velocities, before the filament broke up, the velocity profile was parabolic—the centerline velocity was twice as high as the average and the wall velocity was essentially equal to zero, as shown in Figure 6.2.

At higher velocities the filament would form eddies and the velocity profile became more flat at the center—the wall velocity remained zero (Figure 6.3).

Reynolds studied the conditions at which the flow pattern transitioned from laminar to turbulent and derived the following relationship that is now known as the Reynolds number, N_{Re}:

$$N_{Re} = \frac{\rho u D_h}{\mu}, \tag{6.4}$$

where μ is the fluid viscosity (Pa s or kg m^{-1} s^{-1}) and D_h is the hydraulic diameter (m).

This dimensionless number is a ratio of the inertial forces, $\rho u D_h$, to the viscous forces, μ. For circular pipes, the hydraulic diameter equals the pipe diameter. For square ducts, it is equal to four times the cross-sectional area divided by the perimeter. Subsequent experiments demonstrated that at about a Reynolds number of 2000, the flow regime was no longer laminar but neither was it entirely turbulent—this regime was designated as the intermediate regime. The turbulent regime is often considered to begin at a Reynolds number of 4000.

To calculate the Reynolds number in the previous example, all quantities are known except for the viscosity. From the NIST (2011) database, the viscosity of butane at 60 °F and 7 atm is reported as 0.000175 Pa s. The Reynolds number

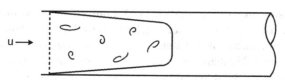

FIGURE 6.3 Turbulent Velocity Profile, $N_{Re} > 4000$

in the 6 in. pipe is 132 000 while it is 198 000 in the 4 in. pipe—the flow regime is turbulent.

Non-dimensional numbers are useful to size equipment but also as scaling parameters to design experiments. Engineers use scaled models to study phenomena that are difficult to study or prohibitively expensive: fluid flow experiments in a 6 in. pipe or larger would require large pumps or blowers, flow meters, valves, fittings, etc. The conditions of the experiments are chosen such that there is a similarity between the full-scale system and the model. Thus, the fluid, diameter, and velocities are chosen to maintain the same Reynolds number.

Example 6.2. Xylem vessels are the conduits through which water is transported from the roots throughout trees and other plants up to the leaves. They have a peculiar helical geometry, as shown in Figure E6.2 and vary substantially in diameter (and length). Design an experimental model to study the hydrodynamics of the fluid flow using glycerol whose viscosity is 2500 cP and density equals 1250 kg m^{-3}.

Solution 6.2. The density and viscosity of water are nominally 1000 kg m^{-3} and 0.001 Pa s. So, neglecting the helical inner construction and assuming a circular geometry and a linear velocity of 1 mm s^{-1}, the Reynolds number in the xylem is:

$$N_{Re} = \frac{\rho u D}{\mu} = \frac{1000 \cdot 0.001 \cdot 150 \times 10^{-6}}{0.001} = 0.15.$$

Since we have chosen to use glycerol as the model fluid because of its high viscosity and transparency, the only choice left to make is the diameter of the apparatus (which will then determine the operating velocity):

$$uD = \frac{\mu N_{Re}}{\rho} = \frac{2500 \text{ cP} \cdot 1 \times 10^{-3} \text{ Pa s/cP} \cdot 0.15}{1250 \text{ kg m}^{-3}} = 0.00030 \text{ m}^2 \text{ s}^{-1}.$$

Operating in small diameter tubes is inconvenient not only to assemble but also because it results in high fluid velocities. With a 5 cm diameter tube

Cell Wall　　　　　　　　　　　　　　　Lignin

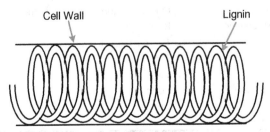

FIGURE E6.2 Schematic of a Xylem Vessel

and to maintain hydrodynamic similarity, the operating fluid velocity would be 6 mm s^{-1}, which is a reasonable value to conduct visualization experiments.

6.3 FLOW METER SELECTION

Just as with the measurement of temperature and pressure, many technologies have been invented to quantify the flow rate of gases and liquids. Crabtree (2009) identified 33 distinct technologies and divided them into eight categories. Table 6.2 includes an additional category for rotameters, which are also known as variable area meters. The classification has lumped thermal mass flow meters together with Coriolis meters as "mass flow" meters. Although they do measure mass flow (as opposed to volumetric flow), the operating principles are entirely unrelated: thermal mass flow meters measure heat transfer rates to deduce mass flow while the Coriolis meter relies on the force induced by a fluid passing through a moving pipe that has a semicircular (or arch) shape.

The types of instruments used in an industrial setting are often different from those used in the laboratory. The most common high precision laboratory instrument is the mass flow controller and rotameters are frequent for applications for both gases and liquids but are much less accurate. In industry, obstruction flow meters, Coriolis meters, and vortex shedders are more standard.

Selecting a flow meter for a given application depends on several criteria including:

- Process conditions.
- Required precision.
- Robustness.
- Size.
- Response time.

TABLE 6.2 Flow Meter Categories

Category	Example
Positive displacement	Wet test flow meters, pumps, gears, and impellers
Differential pressure	Orifice, Venturi, Tuyère, Pitot tube
Variable area meter	Rotameter
Mass flow	Thermal mass, Coriolis
Inferential	Turbine
Oscillatory	Vortex
Electromagnetic	Magmeters
Ultrasonic	Doppler
Open channel	Weir, flume

- Resolution.
- Facility of installation.
- Price.
- Maintenance frequency.
- Operator preference (familiarity).
- Repeatability.

Crabtree (2009) detailed most of the flow meters used in industrial plants. His classification for selecting measuring technology with respect to process application is reproduced in Table 6.3. All flow meters are suitable for clean liquids except for the Ultrasonic-Doppler instrument and only electromagnetic instruments are unsuitable for low conductivity fluids. Most instruments are suitable for high temperature operation or application under certain conditions except for the ultrasonic instruments. Many flow meters are suitable for gases. Few instruments can be used for open channel flow or pipes that are semifilled with the exception of weirs and flumes.

The more common flow meters are discussed in some detail further on. The following is a brief discussion of the open channel, ultrasonic, and electromagnetic flow meters. Open channel meters are found in irrigation applications as well as in waste water treatment, mining beneficiation, and sewage. The principle is to change the level of the water by introducing a hydraulic structure in the flow then inferring the flow rate by the change in level.

Electromagnetic flow meters (magmeters) are considered to be the ideal flow meter for conductive fluids—they are unsuitable for hydrocarbons, gases, steam, or ultra-pure water. They have a high range of operability, an accuracy of $\pm 0.1\%$, low space requirements, and are non-intrusive (that is, they do not affect the flow field). The principle is based on measuring the voltage induced when a conductive object passes a magnetic field. The voltage is proportional to the velocity.

Ultrasonic meters have been available for several decades and are an alternative to magmeters for measuring flow non-intrusively. They are suitable for a large range of pipe sizes and are particularly economic versus other methods for large diameter pipe with an accuracy of about 1%. Three types of meters are manufactured: Doppler, transit time, and frequency difference. The Doppler-effect meter relies on measuring the change of frequency when an ultrasonic beam is directed toward or away from a moving fluid. The time-of-flight meter measures the difference in time it takes an ultrasonic beam to reach one detector upstream of the beam (in the counter-current direction of flow) and another downstream of the beam (in the co-current direction). Finally, the frequency method sends an ultrasonic pulse to each detector and after each successive pulse reception, another signal is transmitted. The difference in the frequency of the downstream and upstream pointing beams is related to the velocity of the fluid. These meters require as much as 25 ppm of particles or

TABLE 6.3 Instrument Suitability ("+": Suitable; "0": Limited; "−": Unsuitable)

Technology	Clean Liquids	Dirty Liquids	Corrosive Liquids	Low Conductivity	High Temperature	Low Temperature	Low Velocity	High Viscosity	Non-Newtonian	Abrasive Slurries	Fibrous Slurries	Gas	Steam	Semifilled Pipe	Open Channel
Coriolis	+	+	+	+	0	+	+	+	+	0	+	+	−	−	−
Electromagnetic	+	+	+	−	0	−	+	0	0	+	+	−	−	0	0
Flow nozzles	+	0	0	+	0	0	0	0	0	0	0	+	+	−	−
Fluidic	+	0	+	+	0	0	−	+	−	−	−	+	+	−	−
Flumes	+	+	+	+	0	−	+	−	−	0	0	+	−	+	+
Orifice plate	+	0	+	+	+	+	+	0	0	−	−	+	+	−	−
Pitot tube	+	0	+	+	0	0	0	0	−	−	−	+	+	−	−
Positive displacement	+	−	0	+	0	+	+	+	0	−	−	+	0	−	−
Target	+	+	+	+	0	0	0	+	0	0	−	+	+	−	−
Thermal mass	+	0	0	+	0	−	+	0	0	0	−	+	+	−	−
Turbine	+	−	0	+	0	+	0	0	−	−	0	+	+	−	−
Ultrasonic-Doppler	0	+	0	+	−	−	0	0	0	0	−	−	−	0	0
Ultrasonic-transit time	+	0	0	+	−	0	0	0	−	−	0	+	0	−	−
Variable area	+	0	+	+	+	−	0	0	−	−	−	+	0	−	0
Venturi tubes	+	+	0	+	0	0	0	0	0	0	0	+	+	−	−
Vortex shedding	+	+	+	+	+	+	−	0	−	−	−	+	+	−	−
Vortex precession	+	0	0	+	0	−	−	0	−	−	−	+	+	−	−
Weirs	+	+	+	+	0	−	+	−	−	0	0	−	−	+	+

bubbles with a minimum 30 μm diameter: the amplitude of the Doppler signal depends on the number of particles or discontinuities.

6.4 POSITIVE DISPLACEMENT

Positive displacement meters—direct volumetric totalizers—are often used in laboratories for calibration. The principle is based on delivering a discrete volume of fluid in aliquots. In fact, water wheels and shadoofs could be considered as positive displacement meters. Pumps may also be considered as meters.

In the medical profession, infusion pumps introduce fluids at rates as low as 0.1 ml h^{-1}—much lower than a drip. Peristaltic pumps force larger volumes of fluids—solutions sufficient to feed a patient—at higher rates: a roller rotates around a cylinder on which flexible tubing is mounted. At each rotation, the roller pinches the tube around the cylinder thereby trapping the fluid and moving it forward. Infusion pumps consist of a motor turning a screw that advances the plunger of a syringe, as shown in Figure 6.4, and can deliver aliquots of 500 nl. High pressure syringe pumps such as Isco can deliver fluids at rates of nl min^{-1} and pressures of 1380 bar. These types of pumps are popular for analytical equipment such as High Performance Liquid Chromatography (HPLC).

Bubble meters are used calibrate laboratory meters that have low flow rates. The gas passes through a tee at the bottom connected to a balloon filled with soapy water. The balloon is squeezed to force the water into the neck and then a bubble forms as the gas passes through the liquid. The bubble ascends in the body of the vessel and passes the graduated marks. The volumetric flow rate is determined by recording the time the bubble takes to pass two graduated marks. Remember that reference conditions to calculate the volume displacement correspond to the pressure and temperature of the bubble meter.

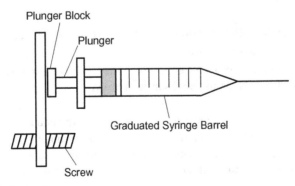

FIGURE 6.4 Syringe Pump

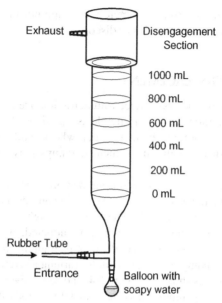

FIGURE E6.3 Bubble Flow Meter

Often meters report flow rates at standard conditions. Standard conditions must also be precisely defined (as mentioned in Chapter 1).

Example 6.3. The bubble flow meter, shown in Figure E6.3, consists of a 1 l cylinder with graduation marks every 200 ml. The time a bubble takes to cross two graduation marks (400 ml) is recorded with a precision stopwatch with a resolution of 0.01 s. Ten measurements are made, and the bubble meter recording at 400 ml intervals is (in s): 4.83, 4.95, 4.78, 5.01, 5.12, 4.85, 5.09, 4.70, 4.99, and 5.30.

(a) Calculate the volumetric flow rate and uncertainty assuming STP.
(b) Calculate the absolute error.
(c) How would the error and uncertainty change if the time was recorded between four graduation marks instead of two?

Solution 6.3a. The mean time equals 4.96 s, which corresponds to a volumetric flow rate of 80.6 ml s^{-1} at ambient conditions. The volumetric flow rate at standard conditions (sc) is calculated using the ideal gas law by converting from ambient temperature and pressure (T_a and P_a, respectively):

$$Q = Q_a \frac{T_a}{T_{sc}} \frac{P_{sc}}{P_a}.$$

If the laboratory was operating at 1 atm and 25 °C, then the volumetric flow rate at STP is:

$$Q = Q_a \frac{25 + 273.15}{273.15} \frac{1}{1.01325} = 86.8 \text{ ml s}^{-1} = 86.8 \text{ sccm.}$$

The units of volumetric flow rate in the laboratory are often quoted as sccm—standard cubic centimeters per minute.

The uncertainty in the measured time, Δ_t, is the product of the sample standard deviation and the Student's t-statistic for 10 sample points at a 95% confidence interval (Table 2.2):

$$\Delta_t = \pm t(\alpha, n - 1) \cdot s_{t,\bar{x}} = \pm t(95, 9) \frac{s_t}{\sqrt{n}},$$

$$\Delta_t = \pm 2.262 \frac{0.2}{\sqrt{10}} = 0.14 \text{ s} \cong 0.1 \text{ s.}$$

The uncertainty in the time measurement as a fraction is 0.02% or 2%. Note that although the resolution of the stop watch is 0.01 s, it is generally accepted that the accuracy (including the response time of an individual) is only about 0.1 s.

The uncertainty in the volumetric flow rate is a function of the uncertainty of measuring both time and volume:

$$\Delta_Q = \sqrt{\sum_{i=1}^{n} (a_i \Delta_i)^2} = \sqrt{(a_V \Delta_V)^2 + (a_t \Delta_t)^2}.$$

Assuming that the uncertainty in identifying the point at which the meniscus passes each graduation mark is less than 1% (4 ml), the uncertainty in the volume, Δ_V, equals the sum of the errors due to each graduation mark:

$$\Delta_V = \sqrt{0.01^2 + 0.01^2} = 0.014.$$

The uncertainty in the volumetric flow rate is then:

$$\Delta_Q = \sqrt{0.014^2 + 0.02^2} = 0.025 = 2.1 \text{ sccm.}$$

Therefore, the volumetric flow rate should be expressed as 87 ± 2 sccm.

Solution 6.3b. A systematic error of the volume is introduced by neglecting to correct for the pressure and temperature. It equals the difference between the true value and the measured value. In this case, it equals the difference between the true value and the reported value:

$$e_Q = Q_{sc} - Q_a = 87 \text{ sccm} - 81 \text{ sccm} = 6 \text{ sccm.}$$

At an atmospheric pressure of 1.03 atm (the highest recorded pressure in Figure 2.2), the ambient flow rate would equal 79 sccm and the error would be 8 sccm. The uncertainty is the same at standard and ambient conditions.

Solution 6.3c. The uncertainty in the volume at 400 ml is 4 ml (1%) and it is 4 ml at 800 ml (0.5%). The uncertainty in the time measurements should also be equal to 0.1 s. Therefore, the uncertainty in doubling the volume is:

$$\Delta_Q = \sqrt{(0.014/2)^2 + (0.1/10)^2} = 0.012 = 1 \text{ sccm.}$$

The uncertainty in the volumetric flow rate is twice as high when measuring the time at 400 ml versus 800 ml. Note that the effect of humidity has been ignored in these calculations. If a bone dry gas is fed to the bubble meter, the soapy water will have a tendency to humidify the gas: the volumetric flow rate will increase in proportion to the humidity and could be as large as 2%.

6.5 DIFFERENTIAL PRESSURE

6.5.1 Obstruction Meters—Orifice

The three most common obstruction meters include the orifice, Venturi, and Tuyère. The operating principle is based on reducing the cross-section of the pipe normal to the flow field and measuring the increase in pressure drop: the velocity head ($u^2/2$) increases at the expense of the pressure head (P/ρ). The reduction in pressure head is measured at two points in the pipe—one immediately downstream and the other upstream—Figure 6.5.

The orifice meter is the simplest to manufacture and occupies the least space. It consists of a thin plate with a round hole perforated such that the hole is in the center of the pipe. Upstream, the hole has a sharp edge; downstream, the edge may be bevelled. The position of the taps is somewhat arbitrary—the

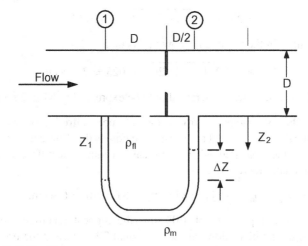

FIGURE 6.5 Orifice Meter

upstream tap is often placed at a distance equivalent to one pipe diameter. For the maximum pressure differential reading, the downstream tap should be located at the vena contracta (the point at which the flow converges to its narrowest point and at which the velocity head is maximum) at between 0.3 and 0.8 pipe diameters. For pipes below 150 mm ID, the flanges from the tap may overlap the optimum tap location and thus larger distances will be required. For applications where compression costs are considerable, this meter is not recommended.

The equation relating volumetric flow rate and pressure drop is derived from Bernoulli's equation and is expressed as:

$$Q = \frac{C_o X_{A2}}{\sqrt{1 - \beta^4}} \sqrt{\frac{2}{\rho}(P_1 - P_2)}, \tag{6.5}$$

where Q is the volumetric flow rate (m^3 s^{-1}), P_1 is the upstream pressure (Pa), P_2 is the downstream pressure (Pa), ρ is the fluid density (kg m^{-3}), X_{A2} is the cross-sectional area of the orifice (m^2), β is the ratio of orifice diameter to pipe diameter (d/D), and C_o is the discharge coefficient (about 0.61 for $N_{Re} > 20\,000$).

For a given flow rate, larger diameter orifices result in a lower pressure drop and, therefore, lower precision. Recent advances of the single orifice meter include introducing multiple perforations, eccentric holes, and hemispherical type geometries that may increase operability, accuracy, and/or reduce pressure drop. Eccentric holes located near the bottom of the pipe are for applications in liquids containing solids or dense fluids or gases that carry liquid droplets. The dense phase could become trapped and accumulate at the bottom of the pipe behind the orifice; positioning the hole near the bottom reduces this tendency.

When properly designed, an orifice plate can have an precision as low as 0.6% (2–3% is more typical) but it generally rises with time due to changes in the orifice bore caused by corrosion or other wearing mechanisms. The precision is best when operated at a turndown ratio greater than four to one—when the flow rate drops below 25% of the design full-scale rate, precision suffers. The accuracy depends on the length of straight pipe upstream and downstream of the orifice. Upstream pipe length is more critical than downstream length. The ISO 5167 standard specifies 60 pipe diameters as the minimum length upstream of the orifice and seven pipe diameters downstream. Crabtree (2009) suggests that the minimum pipe length upstream and downstream of the orifice depends on the β ratio: for β equal to 0.5, the minimum upstream is 25 D and 4 D downstream; for β equal to 0.7, the minimum upstream is 40 D and 5 D downstream.

Another parameter in the design of the orifice plate is its thickness: plate thickness must increase with increasing pipe diameter in order to minimize deflection (ISO 5167, ASME-MFC-3M). For pipes less than 150 mm in diameter, the plate should be around 3.2 mm thick; from 200 to 400 mm ID, the plate should be about 6.1 mm thick; and, for pipes greater than 450 mm in diameter, the plate should be 9.5 mm.

Example 6.4. An orifice meter is installed in a 4 in. Sch40 pipeline to measure the mass flow rate of butane up to $10\,000$ kg h^{-1}. Consider that butane is an incompressible fluid with a density of 600 kg m^{-3} and a viscosity of 0.230 cP. The orifice diameter is exactly half of the pipe diameter:

(a) Calculate the orifice Reynolds number.
(b) Determine the pressure drop across the orifice at the maximum flow rate.
(c) What is the uncertainty in the mass flow rate if the uncertainty in the pressure drop is ± 0.05 psi?

Solution 6.4a. The Reynolds number is a straightforward calculation and is required to ensure that it is at least greater than $20\,000$ (in order to assume that $C_o = 0.61$). The inside diameter of a 4 in. Sch40 pipe equals 4.026 in. and the orifice diameter is one-half the pipe diameter:

$$D = 4.026 \text{ in.} \cdot 0.0254 \text{ m in.}^{-1} = 0.1023 \text{ m},$$
$$d = \beta D = 0.0511 \text{ m}.$$

The Reynolds number may be expressed as a function of mass flow rate, diameter, and viscosity:

$$N_{Re} = \frac{\rho u d}{\mu} = \frac{4}{\pi} \frac{\dot{m}}{d\mu} = \frac{4}{\pi} \frac{10\,000 \text{ kg h}^{-1}}{0.0511 \text{ m} \cdot 0.23 \times 10^{-3}} \frac{1 \text{ h}}{3600 \text{ s}} = 300\,900.$$

Solution 6.4b. By multiplying the equation for volumetric flow by density, the equation relating the mass flow rate and pressure drop can be derived:

$$\dot{m} = \frac{C_o X_{A2}}{\sqrt{1 - \beta^4}} \sqrt{2\rho(P_1 - P_2)}.$$

Rearranging this equation to relate pressure drop as a function of the given conditions gives:

$$\Delta P = \frac{1 - \beta^4}{2\rho} \left(\frac{\dot{m}}{C_o X_{A2}} \right)^2 = \frac{1 - 0.5^4}{2 \cdot 600}$$
$$\times \left(\frac{10\,000 \text{ kg h}^{-1}}{0.61 \cdot \frac{\pi}{4}(0.0511)^2} \frac{1 \text{ h}}{3600 \text{ s}} \right)^2 = 3860 \text{ Pa}.$$

Solution 6.4c. Since the equation relating mass flux to pressure drop and geometry can be expressed as a simple power law, the simplified form for uncertainty is applicable:

$$\frac{\Delta_f}{f} = \sqrt{\sum_{i=1}^{n} \left(a_i \frac{\Delta_i}{x_i} \right)^2}.$$

Since only the uncertainty in the measured pressure drop was specified, the equation for uncertainty reduces to:

$$\frac{\Delta \dot{m}}{\dot{m}} = \sqrt{\sum_{i=1}^{n} \left(a_{\Delta P} \frac{\Delta_{\Delta P}}{\Delta P} \right)^2} = \frac{1}{2} \frac{\Delta_{\Delta P}}{\Delta P}.$$

The uncertainty in the pressure drop equals ± 0.05 psi, which is 345 Pa. Therefore, the uncertainty in the mass flux equals:

$$\Delta \dot{m} = \frac{1}{2} \frac{\Delta_{\Delta P}}{\Delta P} \dot{m} = \frac{1}{2} \frac{345}{3860} 10\,000 = 447 \text{ kg h}^{-1} \cong 500 \text{ kg h}^{-1}.$$

6.5.2 Obstruction Meters—Venturi

A Venturi meter consists of a converging truncated cone that leads to a straight cylindrical throat followed by a divergent cone, as shown in Figure 6.6. Pressure taps are located upstream of the converging cone and at the middle of the cylindrical neck, which corresponds to the vena contracta and will result in the highest pressure drop. Both the precision and accuracy of Venturi meters are good and at the same time the permanent pressure loss is lower compared to an orifice meter. Other advantages include minimal wear, lower tendency to become contaminated, and a higher turndown ratio—eight to one. However, Venturi meters are much more expensive to install and they are significantly larger, which limits their use for applications in which space is a constraining factor.

The equation relating the volumetric flow rate to pressure drop is similar to that for the orifice. The only difference is that the coefficient C_o is replaced by the Venturi coefficient, C_v. This coefficient approaches 0.99 in. large pipes at high Reynolds numbers. Figure 6.7 illustrates the variation of C_v as a function of pipe diameter and Reynolds number at the throat.

Other variants of the standard Venturi geometry include the Dall tube, Venturi nozzle, and Tuyère (or simply flow nozzle). The Venturi nozzle lacks the converging inlet but retains the flared outlet. A Tuyère (or simply flow nozzle) is an extreme variant of a Venturi in that it is essentially only the throat section—the converging and diverging sections are absent. It resembles an orifice because

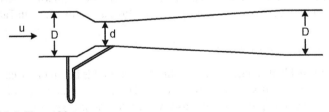

FIGURE 6.6 Venturi Meter

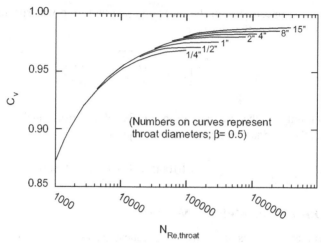

FIGURE 6.7 Variation of the Venturi Coefficient, C_v, with N_{Re} and Pipe Diameter (Holman, 2001)

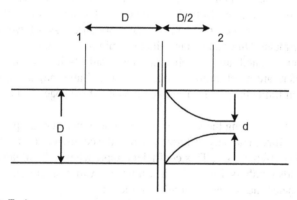

FIGURE 6.8 Tuyère

the fluid discharges from a single point, as illustrated in Figure 6.8. One pressure tap is located one pipe diameter upstream of the face of the converging cone and the second pressure tap is located at half a pipe diameter downstream of the face of the converging cone. It has the highest permanent pressure loss, while the Venturi has the lowest permanent pressure loss, as summarized in Table 6.4.

6.5.3 Compressible Flow

The compressible flow equations are suitable for liquids but in many cases also for gases—particularly at high pressure or when the pressure drop is less than 1% of the operating pressure. For cases where the compressibility of the gas is significant, an additional non-dimensional term, Y, is included in the expression

---

TABLE 6.4 Permanent Pressure Loss of Obstruction Meters as a Function of β (Holman, 2001)

β	Square-Edged Orifice	Tuyère	Venturi
0.4	0.86	0.8	0.1
0.5	0.78	0.7	0.1
0.6	0.67	0.55	0.1

---

relating mass flow rate and operating conditions:

$$\dot{m} = \frac{CYX_{A2}}{\sqrt{1 - \beta^4}}\sqrt{2\rho_1(P_1 - P_2)}. \tag{6.6}$$

The value of the term Y—the expansion factor—can be derived for a Venturi meter by assuming that the fluid is an ideal gas and the flow is isentropic:

$$Y_v = \left(\frac{P_2}{P_1}\right)^{1/\gamma}\sqrt{\frac{\gamma(1 - \beta^2)(1 - P_2/P_1)^{1-1/\gamma}}{(\gamma - 1)(1 - P_2/P_1)(1 - \beta^4(P_2/P_1)^{2/\gamma})}}, \tag{6.7}$$

where γ is the specific heat ratio C_p/C_v (1.4 for air).

When the pressure drop is low, Y equals 1 and it decreases as the pressure ratio P_2/P_1 decreases (i.e. increasing pressure drop). It also decreases with increasing β.

For a standard sharp-edged orifice, the expansion factor is calculated based on an empirical relationship:

$$Y_0 = 1 - \frac{0.41 + 0.35\beta^4}{\gamma}\left(1 - \frac{P_2}{P_1}\right). \tag{6.8}$$

The tendencies for the orifice expansion factor are the same as those for the Venturi but the values are higher—at a pressure ratio of 0.6 and β equal to 0.7, Y_v is 0.7 while Y_0 is 0.86. The critical pressure ratio for air is 0.53 at which point the flow is sonic and these equations are inapplicable.

6.5.4 Restriction Orifice

The first practical restriction orifice—also known as "choked flow"—was introduced by the Romans who were able to deliver a prescribed flow of water to a fountain, bath, or residence by selecting the appropriate pipe diameter. The advantage of a restriction orifice is that it functions unattended with no additional instrumentation. It is popular in the process industry in which low flow rates of gases can be delivered at a steady rate. In gas-solids systems, restriction orifices are used as "blowbacks" for pressure measurements. Instead of filters, pressure taps are purged with a steady flow of gas delivered through a restriction orifice.

In gas systems, the flow becomes choked when the exit velocity through the orifice plate reaches sonic velocity. The mass flow rate is essentially independent of downstream conditions but can be increased by increasing the upstream pressure or decreasing the temperature. For an ideal gas and isentropic flow, the pressure ratio to calculate the onset of sonic conditions depends on the ratio of the specific heats, g, and is often known as the isentropic expansion factor:

$$\frac{P_2}{P_1} = \left(\frac{2}{\gamma + 1}\right)^{\gamma/(\gamma-1)}. \tag{6.9}$$

At 20 °C, the value of γ is 1.67 for monatomic gases (Ar, He, etc.), 1.4 for diatomic gases (H_2, N_2, O_2, CO, etc.), and approximately 1.3 for triatomic gases. The following relationship correlates the value of γ as a function of the number of atoms in the molecule (n_{mol}) for many inorganic gases (excluding NH_3, for which γ equals 1.31):

$$\gamma = \frac{1 + 2(n_{mol} + 1)}{1 + 2n_{mol}}. \tag{6.10}$$

The values of γ for low molecular weight alkanes are: $CH_4 = 1.32, C_2H_6 = 1.22, C_3H_8 = 1.13, C_4H_{10} = 1.09$. The isentropic expansion coefficient generally decreases with increasing temperature.

Under choked flow conditions, the mass flow rate is calculated based on the following conditions:

$$\dot{m} = C_o A_2 \sqrt{\rho_1 P_1 \gamma \left(\frac{2}{\gamma + 1}\right)^{(\gamma+1)/(\gamma-1)}}, \tag{6.11}$$

where C_o is the discharge coefficient, A_2 is the cross-sectional area of the orifice (m^2), ρ_1 is the fluid density upstream of the orifice (kg m^{-3}), and P_1 is the pressure upstream of the orifice (Pa).

The volumetric flow rate is reasonably constant as the flow reaches sonic conditions, but the mass flow rate increases with increasing pressure and decreases with increasing temperature. It is proportional to the open area of the orifice.

6.5.5 Pitot Tube

The Pitot tube measures pressure, from which the speed of the fluid can be deduced by application of Bernoulli's equation. It consists of a tube whose tip face is perpendicular to the direction of the fluid flow, as shown in Figure 6.9, and an additional pressure tap to measure the static pressure. The kinetic energy

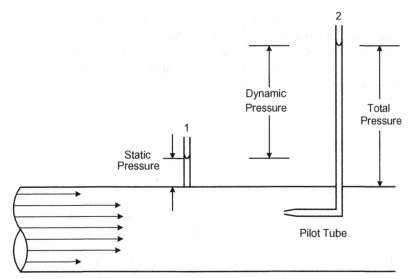

FIGURE 6.9 Pitot Tube

of the gas impinging on the opening of the tube is converted to potential energy. The total pressure of a fluid in motion is composed of the static and dynamic pressures. The static pressure is the force exercised perpendicular to the flow direction while the dynamic pressure is due to the motion of the fluid. For an incompressible fluid, Bernoulli's equation for a Pitot tube is given by:

$$\frac{P_1}{\rho} + \frac{1}{2}u_1^2 = \frac{P_2}{\rho}. \tag{6.12}$$

The total pressure (also known as the stagnation pressure) equals P_2 and the static pressure equals P_1. The difference between the two is the dynamic pressure:

$$P_{dyn} = P_{tot} - P_{stat} = P_1 - P_2 = \Delta P. \tag{6.13}$$

The equation to calculate the velocity of the fluid is simply:

$$u_1 = \sqrt{\frac{2(P_1 - P_2)}{\rho}}. \tag{6.14}$$

Note that this relationship applies to ideal Pitot tubes. The instrument should be calibrated and the deviation from ideal conditions is accounted for by including a constant factor K_p:

$$u_1 = K_p \sqrt{\frac{2(P_1 - P_2)}{\rho}}. \tag{6.15}$$

The factor depends on the geometry including the position of the static pressure tap in relation to the Pitot tube. Each design requires a calibration.

Note that, whereas obstruction devices give an average flow rate, Pitot tubes only give the local velocity. Several measurements must be made across the flow field in order to calculate the average volumetric flow rate.

For incompressible flow, the following simplification can be used to calculate velocity:

$$u_1 = \sqrt{\frac{2(P_1 - P_2)}{\rho_1 \left(1 + N_{\text{Ma}}^2/4\right)}},\qquad (6.16)$$

where N_{Ma} is the Mach number (u_1/u_s) and u_s is the speed of sound (m s^{-1}; $\sqrt{\gamma P_1/\rho_1}$).

The calculation of the velocity, u_1, requires an iterative procedure in which N_{Ma} is first estimated to give a value of u_1. This value is subsequently used to update the estimate of N_{Ma}.

One of the common applications for Pitot tubes is to calculate the speed of aircraft. In fact, contamination of the Pitot tube or freezing of the line has been cited as a probable cause of several aviation disasters including Austral Líneas Aéreas Flight 2553, Birgenair Flight 301, Northwest Airlines Flight 6231, AeroPeru Flight 603, and even Air France Flight 447 (although this claim has not been substantiated).

These instruments are also being developed to assess solids, velocity and mass flux in two-phase systems. The tip of the probe is placed perpendicular to the flow field and a vacuum is applied to the tube so that the gas (fluid) is withdrawn isokinetically with respect to the gas in the process.

6.6 ROTAMETERS

Rotameters consist of a vertically oriented tapered tube with a graduated scale along the length. As shown in Figure 6.10, a float (also known as a "bob") is placed in the tube and when a fluid enters from the bottom, it rises to a point at which the inertial and buoyancy forces acting upward equal the gravitation force acting downward. Since the tube is tapered, the velocity changes with height. Very large ranges may be achieved with the same rotameter by changing the float—steel versus glass, for example, as shown in Table 6.5.

At equilibrium, the force balance is given by:

$$F_g = F_D + F_A,\qquad (6.17)$$

where F_g is the gravitational force acting downward (N; $m_b g$), F_D is the drag force acting upward (N), and F_A is the buoyancy (Archimedes' force; N; $\rho_f V_b g$).

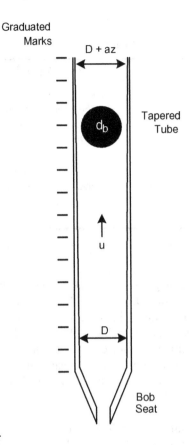

FIGURE 6.10 Rotameter

The gravitational force, F_g, depends on the mass of the bob, m_b, which is the product of its density and volume $\rho_b V_b$. Increasing the mass of the bob by increasing its density will increase the range of measurable volumetric flow rates of the rotameter. Increasing the volume of the bob increases the gravitational force acting downward but at the same time it increases the buoyancy force acting upward. Since $\rho_b > \rho_f$, increasing the bob volume will increase the range of the rotameter but proportionately less than increasing its density.

The buoyancy force was discovered by Archimedes in 212 BC in Syracuse. He found that an object that is partially or wholly immersed in a fluid is buoyed by a force equal to the weight of fluid displaced by the object (up to the point at which it is submerged). In the case of a bob in a rotameter, since it is entirely submerged, the buoyancy force is simply the product of the fluid density, ρ_f, the bob volume, V_b, and the gravitational constant.

TABLE 6.5 Density of Float Materials

Float Material	Density (kg m^{-3})
Teflon	2200
Glass	2500 (varies depending on type)
Sapphire	3970
Titanium	4510
Carbon Steel	7850
316 SS	8000 (Mo-based steel)
Hastelloy C	8890 (Ni-based steel)
Carboloy	15 000 (Tungsten-based alloy with Co or Ni)
Tantalum	16 600

We experience the drag force on a daily basis—walking on a windy day, for example. It is the result of a moving fluid across an object (or a moving object through a stationary fluid). The drag force increases with velocity and this contributes to the increase in automobile fuel consumption with increased speed. Minimizing drag on airplanes, race cars, boats, etc., directly affects both fuel consumption as well as maximum speed because it increases proportionately with surface area. The force exerted by a fluid on a body perpendicular to the direction of flow is called drag while the force exerted parallel to the direction of flow is called shear.

Because of the complexity of the fluid dynamics, correlations are used to estimate the drag coefficient, C_D, from which the drag force is calculated. The drag coefficient is defined as the ratio of the force per area ($F_D/A_{p,b}$, perpendicular to the fluid flow and where $A_{p,b} = \pi d_b^2/4$ is the projected surface area of the bob) to the product of the fluid density and the velocity head:

$$C_D = \frac{F_D/A_{p,b}}{\rho_f u_m^2/2}. \tag{6.18}$$

For spheres at low flow rates, the drag coefficient is inversely proportional to the particle Reynolds number:

$$C_D = \frac{24}{N_{Re,p}}, \tag{6.19}$$

where $N_{Re,p} = \rho u_m d_p/\mu$, and d_p is the particle diameter (and not the tube diameter; m).

Spherical floats are often used in rotameters but more complex geometries are also common.

Combining the expressions for the drag, buoyancy, and gravitational forces, the volumetric flow rate as a function of conditions gives:

$$Q = A_a u_m = A_a \sqrt{\frac{1}{C_D} \frac{2gV_{fl}}{A_{p,b}} \left(\frac{\rho_b}{\rho_f} - 1\right)}, \qquad (6.20)$$

where A_a is the annular area between the bob and the tube.

Because of the taper in the tube, the annular region increases with height and is approximated by the following relationship:

$$A_a = \frac{\pi}{4}((D + az)^2 - d_b^2), \qquad (6.21)$$

where d_b is the bob diameter at its widest point (m), D is the tube diameter at the inlet (m), and az is the taper variation with height (z; m).

The tube can be designed such that the quadratic relationship between area and height is nearly linear. When the rotameter is calibrated for the operating conditions, the variation of the physical properties related to the Reynolds number, and hence the drag coefficient, may be lumped together to give the following relationship:

$$\dot{m} = C_R z \sqrt{\rho_f(\rho_b - \rho_f)}, \qquad (6.22)$$

where C_R is the characteristic constant of the rotameter at calibration conditions.

For gas applications, the manufacturer will calibrate the rotameter with air. Since its density is three orders of magnitude lower than the bob, the relationship may be simplified to:

$$\dot{m} = C_{R,a} z \sqrt{\rho_f}, \qquad (6.23)$$

where $C_{R,a}$ is the rotameter constant that includes the density of the bob.

Note that this constant is specific to each type of bob: changing the bob of the rotameter will necessarily change the constant by the square root of the ratio of the bob densities. For liquid applications, the density of the fluid is non-negligible compared to the float and this simplified form is inapplicable.

Example 6.5. A rotameter measures the flow of air with a maximum of 1.00 std $m^3 min^{-1}$ at standard conditions ($0\,°C$ and 1 bar):

(a) Calculate the mass and volumetric flow rate of air when the float is at 50% of its maximum value and the air is at $40\,°C$ and 5.0 atm.
(b) What is the volumetric flow rate at standard conditions?

Solution 6.5a. The conditions at the entrance of the tube are different from the standard conditions at which the tube was calibrated. We assume that the rotameter constant, $C_{R,IG}$, remains unchanged, which may be a reasonable assumption since the viscosity only changes marginally from $25\,°C$ to $40\,°C$.

At standard conditions, 1 bar and 0 °C, the density of air is:

$$\rho_{f,std} = \frac{M_w P}{RT} = \frac{29 \text{ kg kmol}^{-1} \cdot 1 \text{ bar}}{0.08314 \text{ m}^3 \text{ bar kmol}^{-1}\text{K}^{-1} \cdot 273 \text{ K}} = 1.28 \text{ kg m}^{-3}.$$

At operating conditions, 5 atm and 40 °C, the density of air is:

$$\rho_f = \frac{M_w P}{RT} = \frac{29 \text{ kg kmol}^{-1} \cdot 5 \text{ atm}}{0.082056 \text{ m}^3 \text{ atm kmol}^{-1} \text{ K}^{-1} \cdot 313 \text{ K}} = 5.65 \text{ kg m}^{-3}.$$

To convert from standard conditions to operating conditions, the following relationship may be derived for the volumetric flow rate and mass flow rate, respectively:

$$Q_2 = Q_1 \frac{z_2}{z_1} \sqrt{\frac{\rho_{f,1}}{\rho_{f,2}}},$$

$$\dot{m}_2 = \dot{m}_1 \frac{z_2}{z_1} \sqrt{\frac{\rho_{f,2}}{\rho_{f,1}}}.$$

In this example, the calibration and operating gas are both air, so these equations can be simplified to:

$$Q_2 = Q_1 \frac{z_2}{z_1} \sqrt{\frac{P_1}{P_2} \frac{T_2}{T_1}},$$

$$\dot{m}_2 = \dot{m}_1 \frac{z_2}{z_1} \sqrt{\frac{P_2}{P_1} \frac{T_1}{T_2}}.$$

The volumetric flow rate with the bob at 50% of the maximum is:

$$Q_2 = 1 \text{ m}^3 \text{ min}^{-1} \frac{0.5}{1} \sqrt{\frac{1 \text{ bar}}{5.0 \text{ atm} \cdot 1.01325 \text{ bar atm}^{-1}} \frac{313 \text{ K}}{273 \text{ K}}}$$
$$= 0.238 \text{ m}^3 \text{ min}^{-1} \cong 0.24 \text{ m}^3 \text{ min}.$$

The mass flow rate is simply:

$$\dot{m}_2 = \rho_f, \; 2Q_2 = 5.65 \text{ kg m}^{-3} \cdot 0.228 \text{ m}^3 \text{ min}^{-1}$$
$$= 1.35 \text{ kg min}^{-1} \cong 1.4 \text{ kg min}^{-1}.$$

Solution 6.5b. The volumetric flow converted to standard conditions is:

$$Q_{2,std} = \frac{\dot{m}}{\rho_{f,1}} = \frac{1.35 \text{ kg min}^{-1}}{1.28 \text{ kg m}^{-3}} = 1.08 \text{ m}^3 \text{ min}^{-1} \cong 1.1 \text{ m}^3 \text{ min}^{-1}.$$

Rotameters are most common in laboratories and for instrumentation in industrial processes. Their precision is perhaps no better than 2–5%.

The precision of a float with sharp edges may be superior to spherical floats. Often, the floats will "bob"—rise and fall at a regular frequency, which decreases the precision. They are very sensitive to contamination—vapor, dirt, or oil—which can cover the surface and thereby change the drag coefficient. At times, when there is a sudden surge of fluid, the bob gets stuck at the top of the tube. Gentle tapping at the top of the tube can dislodge the float. Often rotameters may be installed in line with more precise instruments because they offer an easy means of verifying if there is flow. This is a major plus with respect to troubleshooting a process.

6.7 THERMAL MASS FLOW METERS

The most common high precision laboratory instruments are the thermal mass flow meters. (They are among the most expensive and can cost over 2000 $.) The principle is based on an energy balance: either the entire flow or a slipstream is heated at a constant rate and the temperature rise is recorded. The temperature (or temperature difference) of the fluid is measured and amplified; it is proportional to the mass flow rate:

$$S = KC_p\dot{m}, \tag{6.24}$$

where C_p is the specific heat of the fluid and K is a constant that includes heat conductivity, viscosity, and density.

Generally, the meters are calibrated for a specific fluid at the manufacturer. Careful calibration is recommended if a different fluid is to be measured. Thermal mass flow meters are generally instrumented with a controller to be able to set the flow rate at a desired level. These instruments are known as mass flow controllers (MFCs).

One of the major limitations of gas MFCs is that there is susceptibility to contamination by liquids. In many cases, when a fluid enters the MFC, it becomes entirely blocked. It is recommended to send it back to the manufacturer for conditioning. Alternatively, heating the block to a modest temperature of 50 °C, for example, can often unblock the sensing element.

6.7.1 Hot Wire Anemometry

Anemometry is a general term to represent the measurement of wind speed—*anemos* is the Greek word for wind. The earliest anemometer for meteorology is credited to Alberti in 1450. Hooke reinvented the device, which relied on cups or disks mounted on a pole that would rotate by the force of wind. Modern instruments to measure wind speed rely on laser Doppler shift, ultrasonic waves, propellers, and hot wire anemometers. The hot wire anemometer is commonly used for fluid flow measurements and in particular for research applications that require a detailed analysis of velocity in localized areas or for conditions that vary rapidly. They are used to measure fluctuations in turbulent flow at a frequency of 1000 Hz in air with a 0.0001 in. diameter Pt or W wire.

The concept of hot wire anemometry is similar to that of thermal mass flow meters as well as the Pirani gauge to measure pressure: a fine wire is placed in a flow stream and then heated electrically. The heat transfer rate from the fluid to the wire equals the rate heat is generated by the wire.

The heat, q, generated by the wire is the product of the square of the electrical current, i, and the resistance of the wire R at the fluid temperature, T_w:

$$q = i^2 R(T_w). \tag{6.25}$$

The resistance of the wire varies linearly with temperature and is calculated with respect to a reference temperature, T_0:

$$R(T_w) = R(T_0)(1 + \alpha(T_w - T_0)), \tag{6.26}$$

where α is the temperature coefficient of resistance.

The coefficient for platinum wires is 0.003729 K^{-1} and that for tungsten is 0.004403 K^{-1}. This heat must be carried away by the fluid. King (1914) expressed this rate as a function of fluid velocity, u_f, and the temperature differential between the wire and fluid temperatures, $T_w - T_f$:

$$q = a + b u_f^2 (T_w - T_f), \tag{6.27}$$

where a and b are constants determined through calibration.

Together with thin wires, anemometers have been made by coating a metallic film with a thickness of about 5 μm over an insulated cylinder. These devices are extremely sensitive and rapid and can measure frequencies of 50 000 Hz.

6.8 CORIOLIS

Electromagnetic and ultrasonic meters are truly non-intrusive meters since they do not alter the flow pattern or perturb the flow by introducing a probe. A major limitation of hot wire anemometry is that, although they are small, the probes can perturb the flow stream, which introduces error in the measurement. Coriolis meters have no intrusive probes but they rely on diverting the flow through tubes that are vibrated by an actuator in an angular harmonic oscillation. The vibration causes the tube to deflect due to the Coriolis effect. The instruments consist of a straight single tube or a dual curved tube. Depending on the geometry of the tube, the vibration ranges from 80 Hz to 1000 Hz. The accuracy is as good as $\pm 0.1\%$ for liquids and 0.5% for gases (although it might be as high as 2%) with a turndown ratio of 100:1. Small Coriolis meters are available with flow ranges from as low as 20 g h^{-1} to as high as 350 t h^{-1}.

The mass flow rate is calculated by the following equation:

$$\dot{m} = \frac{K_u - I_u \omega^2}{2K d^2} \tau, \tag{6.28}$$

where K_u is the stiffness of the tube (temperature dependent), I_u is the inertia of the tube, K is the shape factor, d is the width (m), ω is the vibration frequency, and τ is the time lag.

6.9 INFERENTIAL—TURBINE

Inferential meters include instruments in which the volumetric flow rate is inferred by the movement of a turbine, propeller, or impeller. The fluid impinging on a blade causes it to rotate at an angular velocity that is proportional to the flow rate. The early anemometers made with plates and cups are examples. These types of meters are becoming less and less common due to the need to calibrate and compensate for effects like viscosity. According to Crabtree (2009), the Coriolis and ultrasonic meters are replacing the turbine meters in most industrial applications.

6.10 OSCILLATORY—VORTEX

Vortex meters are intrusive because they rely on disturbing the flow regime by placing an object in the fluid stream to produce an oscillatory motion downstream. The object can take many shapes but often a thin wire is used, as shown in Figure 6.11, which minimizes the pressure drop. The oscillatory motion is referred to as a vortex and may be detected by piezoelectric transducers, or magnetic or optical sensors. The number of vortices present is proportional to the volumetric flow rate.

The frequency of vortex shedding (f) is proportional to the product of the Strouhal number (N_{St}) and the diameter of the wire (d_w) or some other characteristic dimension of the object used to generate the vortices and is inversely proportional to the fluid velocity (u_f):

$$f = \frac{N_{St} u_f}{d_w}. \tag{6.29}$$

The Strouhal number varies as a function of the Reynolds number: as shown in Figure 6.12, its value is close to 0.2 over a large range of Reynolds numbers— from 200 to 200 000.

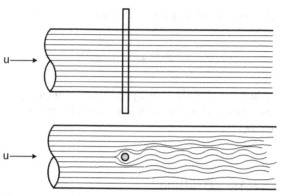

FIGURE 6.11 Vortices Induced by a Cylinder

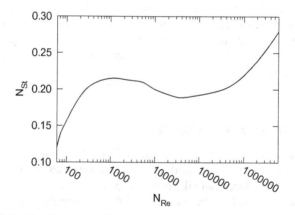

FIGURE 6.12 Strouhal Number as a Function of Reynolds Number

6.11 FLOW METERS IN AN INDUSTRIAL SETTING

Most chemical processes require many different types of flow meters to monitor gases and liquids as well as for control purposes. Crabtree (2009) has itemized the important considerations when installing flow meters including:

- Geometrical considerations:

    - Position (vertical, horizontal).
    - Provision for sufficient straight pipe upstream and downstream.
    - Allowance for piping expansion.
    - Sufficient clearance for installation and maintenance.
    - Provision of bypass lines for servicing.

- Process considerations:

    - Minimize, eliminate gas or vapor in liquid lines (and pressure taps).
    - Minimize, eliminate dust, vapor, or liquids in gas/vapor lines (and pressure taps).
    - Filtration upstream.
    - Maintain pressure tap lines full.

- Mechanical considerations:

    - Avoid, minimize vibration.
    - Avoid, minimize strong electromagnetic fields in the vicinity.
    - Avoid, minimize pressure, flow surges.
    - Design and implement a maintenance schedule.

To illustrate the extent of the use of flow meters in an industrial context, we will use the process to manufacture maleic anhydride from butane in a circulating fluidized bed reactor over a vanadium phosphorous catalyst. A schematic of the plant is shown in Figure 6.13. The catalyst is transported between two zones in the riser-fast bed a butane-rich stream reacts with the catalyst to form maleic anhydride and in the stripper-regenerator air reacts with the catalyst to re-oxidize the catalyst. The average diameter of the catalyst powder is 80 μm. To maintain the catalyst in the process, cyclones are installed at the exit of each reactor. Because of the high toxicity of the catalyst, the effluent gas passes through filters as well. The stream with maleic anhydride goes to an absorber where it hydrolyzes to maleic acid, which is then pumped to another process. After the regenerator filters, the effluent stream goes through a CO converter before leaving a stack to the atmosphere.

The process requires flow meters for the gas phase—air, butane, recycled gas, nitrogen, oxygen, steam—the liquid phase—condensed water, maleic acid—and the solids—catalyst recirculation from one reactor to the other. In total, there are 248 flow meters and, as shown in Table 6.6, most of these flow meters are dedicated to nitrogen and water. Only four flow meters monitor the product of interest—maleic acid. Figure 6.13 shows some of the major flow

TABLE 6.6 Number of Flow Meters for Each "Fluid" Type

Fluid	No.	Fluid	No.
Nitrogen	89	Dual service (air/N_2)	11
Water	74	Butane	7
Air	31	Oxygen	5
Recycled gas	25	Maleic acid	4

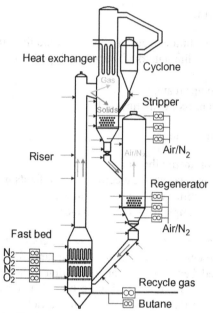

FIGURE 6.13 Circulating Fluidized Bed reactor to produce maleic anhydride from n-butane over a (VO)2PO4 catalyst.

meters. The horizontal arrows represent blow backs in which 1 kg/s of nitrogen was fed to each port. These ports were used to measure pressure drop.

Of the 248 flow meters, only 53 are controlled remotely by operators via the distributive control system (DCS). In the field, the instrument signal is first converted to a mass flow reading based on the pressure, temperature, and molecular weight for which they were calibrated, $\dot{m}_{field}$. The reading is then sent to the DCS where the values are compensated to account for the actual operating pressure, temperature, and molecular weight. The compensation factor not only includes the operating conditions but also the flow meter type: volumetric flow meters require a square root compensation, $\lambda_{comp,sr}$, whereas the mass flow meters require a linear compensation factor, $\lambda_{comp,l}$:

$$\dot{m}_{comp} = \dot{m}_{field}\lambda_{comp}, \tag{6.30}$$

$$\lambda_{comp,l} = \frac{P + P_0}{P_R} \frac{M_W}{M_{W,R}} \frac{T_R}{T + T_0}, \tag{6.31}$$

$$\lambda_{comp,sr} = \sqrt{\frac{P + P_0}{P_R} \frac{M_W}{M_{W,R}} \frac{T_R}{T + T_0}}, \tag{6.32}$$

where P is the actual pressure (barg), P_0 is the atmospheric pressure (atm; 1.01325 bar), P_R is the reference pressure (design basis; bara), T is the operating temperature (°C), T_0 is the absolute temperature (K), and T_R is the reference temperature (design basis; K).

Vortex, turbine, and thermal mass flow meters require linear compensation whereas ΔP meters require square root compensation.

Common instrument errors detected during the commissioning of the plant include incorrect reference conditions and assigning a square root compensation for the mass flow meters (and vice versa). In some cases, the error introduced is as much as 50%.

Example 6.6. During the construction of a chemical plant, a Venturi meter was originally chosen for a hydrocarbon stream but it was changed to a vortex due to space constraints. This change was not communicated to the instrumentation engineers. The DCS reading was 2100 kg h^{-1} and the design pressure and temperature in the DCS were reported as 5 atma and 140 °C. The operating pressure and temperature were 3.04 barg and 165 °C. In the field, the meter specified the operating pressure as 5.07 barg:

(a) Calculate the compensation factor and the true mass flow rate if it were a Venturi meter.
(b) What is the compensation term for the vortex shedder and what is the true mass flow rate?

Solution 6.6a. Two errors were made communicating the data from the field instrument to the DCS: the reference pressure was 5.07 barg (which equals 6 atm) instead of 5 atm and the compensation factor should be linear since the instrument is a vortex shedder and not a Venturi. The DCS readout is compensated reading, therefore, first we must correct the DCS reading by correcting for the compensation factor to determine what was the actual field reading:

$$\dot{m}_{field} = \frac{\dot{m}_{comp}}{\lambda_{comp}},$$

$$\lambda_{comp} = \sqrt{\frac{P + P_0}{P_R} \frac{T_R}{T + T_0}}$$

$$= \sqrt{\frac{3.04 \text{ barg} + 1.01325 \text{ bar}}{5 \text{ atm} \cdot 1.01325 \text{ bar atm}^{-1}} \frac{140 + 273}{165 + 273}} = 0.869.$$

The compensation factor to get the 2100 kg h^{-1} reading at the DCS was 0.869. Thus the field reading was 2420 kg h^{-1}. The correct compensation factor with the reference temperature is:

$$\lambda_{comp,sr} = \sqrt{\frac{3.04 \text{ barg} + 1.01325 \text{ bar}}{5.07 \text{ barg} + 1.01325 \text{ bar}} \frac{140 + 273}{165 + 273}} = 0.793.$$

Therefore, the DCS measurement for a Venturi meter for the mass flow rate would be 1920 kg m^{-3} (2420 · 0.793).

Solution 6.6b. The compensation term for a vortex shedder is simply the square of the compensation term for the Venturi meter, or 0.631:

$$\lambda_{comp,1} = \frac{3.04\ \text{barg} + 1.01325\ \text{bar}}{5.07\ \text{barg} + 1.01325\ \text{bar}} \cdot \frac{140 + 273}{165 + 273} = 0.631.$$

The actual flow rate is then the product of the field reading, 2420 kg m^{-3} and the linear compensation factor, which is equal to about 1500 kg h^{-1} (500 kg h^{-1} lower than originally reported).

6.12 EXERCISES

6.1 Syrup produced from sugar beets or sugarcane is concentrated using multi-effect evaporators. The flow rate of a partially heated stream of syrup is measured by a Venturi meter with a throat diameter of (3.00 ± 0.01) in. The piping upstream and downstream is 5 in. Sch40. At 50 °C 10 °Bx, the density of the sugar water is (1027 ± 3) kg m^{-3}. The pressure drop is measured with a liquid manometer. The density of the fluid equals (1250 ± 5) kg m^{-3}. For a differential height of 14 in. in the capillary: *I. Bouvier*

   (a) Calculate the uncertainty in the pressure drop.
   (b) What is the volumetric flow rate (m^3 s^{-1})?
   (c) What is the relative uncertainty in the flow rate (%).

6.2 A rotameter fabricated with a tantalum float measures the flow rate of a waste stream from a water treatment unit. The rotameter is calibrated such that the center of the float is at 30 for a flow of pure water at a rate of 15 000 l h^{-1}, as shown in Figure Q6.2. The waste water stream contains 10 wt% motor oil ($\rho_{oil} = 0.875$ g cm^{-3}): *M. Sayad*

   (a) When the float reaches a value of 30 with waste water, is its flow rate higher or lower than 15 000 l h? Why?
   (b) Calculate the mass flow of the waste water for the float position shown in Figure Q6.2.

6.3 Orifice meters in two pipes indicate a flow rate of 100 l min^{-1} of air. The first pipe operates at 100 °C and 1100 mmHg while the second operates at 85 °F and 8 psig. Calculate the mass flow rate in each pipe. *Naud*

6.4 The velocity of air at 20 °C and 1 atm measured by a Pitot tube equals 25 m s^{-1}. The precision of the instrument is 2%. What is the dynamic pressure and its uncertainty?

Graduated
Marks

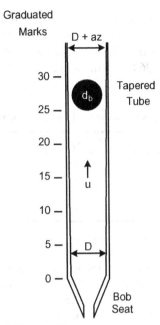

FIGURE Q6.2 Rotameter with Tantalum Float

6.5 Methane is shipped through an 8 in. Sch80 pipe at a rate of 5 t h^{-1} and a pressure such that its density equals 25 kg m^{-3}. The pipe is reduced to 4 in. Sch80 at the plant. The viscosity of methane at 40 bar equals 1.189×10^{-5} Pa s: *A.-M. de Beaumont-Boisvert*

(a) What is the pressure drop resulting from this pipe reduction?
(b) Is it in laminar or turbulent flow in the 8 in. pipe? In the 4 in. pipe?
(c) What is the uncertainty in the flow rate if the uncertainty in the measured pressure drop is 2%?

6.6 Ethanol at 85% w/w circulates in a pipe with a cross-sectional area of 0.5 m^2. The mass flow rate in the pipe equals 5000 kg h^{-1} and is measured with an orifice meter: $\beta = 0.5$ and $C_o = 0.61$. The density of ethanol is 0.789 and its viscosity is 1.2 cP: *M. Ménard*

(a) Calculate the Reynolds number at the orifice.
(b) What is the pressure drop across the orifice?

6.7 A rotameter is used to measure the flow rate of air at 25 °C and 1 bara. A 2 mm sapphire bead rises to a point in which the tube diameter is equal to 6 mm. Calculate the mass flow rate and the volumetric flow rate if the drag coefficient equals 0.1. *M. Lessard*

6.8 An unknown hydrocarbon flows through a 10 cm diameter pipe at a volumetric flow rate of $0.09 \text{ m}^3 \text{ s}^{-1}$. The flow regime is barely turbulent. Based on the equation for the Reynolds number and the data in Table Q6.8, determine the most likely hydrocarbon.

TABLE Q6.8 Properties of Suspected Hydrocarbons

Elements	ρ (kg m^{-3})	μ (cP)
Pentane	626	0.240
Hexane	655	0.294
Heptane	685	0.386
Octane	703	0.542

6.9 A vortex shedder is used to measure the flow rate of bio-ethanol in a newly constructed plant. The diameter of the wire traversing the 6 in. Sch40 pipe generating the vortices equals 10 mm. What is the mass flow rate of the ethanol for a measured vortex frequency of 120 Hz? Assume the density of ethanol equals 789 kg m^{-3} and its viscosity is 1.2 cP.

6.10 A rotameter is calibrated to measure the flow rate of air at a maximum of 1 m^3 min^{-1} STP (0 °C and 1 bar). Calculate the mass flow rate of methane when the bob is at 40% of its maximum height and the pressure and temperature are 25 °C and 4 barg.

6.11 A Venturi tube is installed on a 150 mm diameter pipe. What is the diameter of the constriction at a maximum flow rate of 17 l s^{-1} and a pressure differential of 34.5 kPa for water at 30 °C? $\rho = 995.7$ kg m^{-3} and $\mu = 0.801 \times 10^{-3}$ Pa s. *E.M. Benaissa*

6.12 Calculate the compensation term of a Tuyère when the design conditions are 3 barg and 0 °C and the operating conditions are 5 atm and 75 °C. What is the DCS reading when the field reports a value of 3500 kg h^{-1} as the flow rate. If the design conditions were incorrectly reported and they were actually 3 atma and 25 °C, what would DCS report?

6.13 A rotameter measures the flow of nitrogen at a temperature of 800.5 °F and a pressure of 58.01 psig. The volumetric flow rate equals 5 m^3 min^{-1} when the carbon steel ball is at full scale:

(a) Calculate the float height of a tantalum float at 2 m^3 min^{-1}.

(b) What would the height (at 2 m^3 min^{-1}) be if the pressure of the gas were increased by 20%, with the tantalum float and the steel ball?

6.14 Repeat the first example in Chapter 5 where the cylinder body is charged with 100 ml of acetone and mercury is the operating fluid in the U-tube manometer (the differential height is 14 in.).

6.15 Calculate the minimum pressure ratio required to achieve choked flow for carbon monoxide.

6.16 To measure the pressure drop in a two-phase gas-solids catalytic reactor, 1 in. diameter sonic orifices are installed. The supply nitrogen pressure is 9 atm and it is at a temperature of 20 °C:

(a) To maintain the mass flow rate at less than 1 kg s^{-1}, what is the maximum bore size of the orifice?

(b) At what downstream pressure is the orifice no longer at sonic conditions?

6.17 Calculate the dynamic pressure (in mbar) measured by a Pitot tube in a water stream moving at a speed of 0.3 m s^{-1}. If the uncertainty of measurement of dynamic pressure is 5 N m^{-2}, what is the uncertainty of the speed?

6.18 What is the speed of sound at sea level? What is it at a altitude of 40 000 ft?

6.19 Calculate the error of ignoring the N_{Ma} number correction for an aircraft flying at 10 000 m at an estimated speed of 800 km h^{-1}. Note that on average the atmospheric temperature drops by 6 °C per 1000 m.

REFERENCES

Chanson, H., 2002. Certains Aspects de la Conception Hydraulique des Aqueducs Romains. Journal La Houille Blanche (6–7), 43–57.

Chanson, H., 2008. The hydraulics of Roman aqueducts: what do we know? Why should we learn? In: Badcock Jr., R.W., Walton, R. (Eds.), Proceedings of World Environmental and Water Resources Congress 2008 Ahupua'a, ASCE-EWRI Education, Research and History Symposium, Hawaii.

Crabtree, M.A., 2009. Industrial flow measurement. M.Sc. Thesis, University of Huddersfield.

Holman, J.P., 2001. Experimental Methods for Engineers, 7th ed. McGraw-Hill Inc., New York. pp. 297. with permission.

King, L.V., 1914. On the convection of heat from small cylinders in a stream of fluid, with applications to hot-wire anemometry. Philosophical Transactions of the Royal Society of London 214 (14), 373–433.

McCabe, W.L., Smith, J.C., 1976. Unit Operations of Chemical Engineering. McGraw-Hill Chemical Engineering Series, 3rd ed. McGraw-Hill.

NIST, 2011. Propriétés thermophysiques des systèmes fluides. Retrieved from: <http://webbook. nist.gov/chemistry/fluid/>.

Chapter 7

Physicochemical Analysis

7.1 OVERVIEW

Material science interprets and predicts the behavior of materials, from their atomic-level properties (vibration of electrons and atomic networks) to their macroscopic-scale properties—fatigue, corrosion, roughness, strength, appearance, etc. This science applies the theoretical knowledge of chemistry and physics, hence the origin of the term "physicochemical." It allows us to study and characterize materials in their different states: solids and powders, such as metals, catalysts, construction materials (glass, ceramics, concrete, etc.), liquids—molten plastics, paints, resins, hydrocarbons—and gases. Also included are not only synthetic polymers but natural products such as pulp and paper, pharmaceuticals, and agricultural produce. The classes of materials are biomaterials, ceramics, composites, metal alloys, polymers, and semiconductors. In the last 50 yr, among the most significant advances in material sciences include: semiconductors, light-emitting diodes, carbon-reinforced plastics, nanomaterials, lithium-ion batteries and the materials to make them, scanning probe microscopes, lithography, and metamaterials (Elsevier, 2007). Besides the new applications with respect to electronics, replacing metals and natural products—wood, cotton, wool, etc.—with polymer composites continues to attract considerable industrial research. In fact, elastic polymers for electronic applications—to replace copper wire—are also under development (Karayianni et al., 2008). In the textile and footwear industries, material research focuses on introducing comfort into all articles: comfort means the control of heat, humidity, and compression (in the case of shoes).

The Boeing 787 Dreamliner was the first aircraft in which the structure was made out of 50% composite materials (Morazain, 2011). One of the drivers for composites is to use an inexpensive base material and a layer over with a thin

Experimental Methods and Instrumentation for Chemical Engineers. http://dx.doi.org/10.1016/B978-0-444-53804-8.00007-1

TABLE 7.1 Summary of the Three Laws

Phenomenon	Law
Thermal transfer	Fourier's law: $q = -kX_A \frac{dT}{dz}$
Mass transfer	Newton's law: $\tau = -\mu \frac{du}{dz}$
Binary diffusion of gases	Fick's law: $j_A = -D_{AB} \frac{dC_A}{dz}$

second (or third) more functional material—another example of composites is adding graphite particles to polystyrene to improve the heat reflection properties.

This chapter addresses the three fundamental transport properties characteristic of Chemical Engineering: heat transfer, momentum transfer, and mass transfer. The underlying physical properties that represent each of these phenomena are thermal conductivity, viscosity, and diffusivity and the equations describing them have a similar form. Heat flux through conduction is expressed as a temperature gradient with units of W m^{-2}. Note that heat flux, mass flux, etc. are physical measures expressed with respect to a surface (m^2). Momentum flux in laminar flow conditions is known as shear stress and has units of Pa (or N m^{-2}): it equals the product of viscosity and a velocity gradient. Finally, molar flux (or mass flux) equals the product of diffusivity and a concentration gradient with units of mol m^{-2} s^{-1}. These phenomena are expressed mathematically as shown in Table 7.1.

7.2 THERMAL CONDUCTIVITY

Several thermal properties are critical in establishing the energy balance for heat transfer applications such as the design of heat exchangers, manufacturing insulating materials, semiconductors, or highly conductive materials. A good practical example is in the construction field where energy savings and comfort are now two inseparable concepts. The choice of materials for the walls of a house must take into account thermal conductivity and diffusion of moisture.

Energy dissipation in semiconductors is important to avoid overheating, which leads to failure, and thus a profound understanding and control of thermal conductivity is important. For example, power transistors and solar cells can be exposed to intense heat. Diodes or semiconductor lasers require a heat sink with a substantial internal energy and a high thermal conductivity, which allows the rapid transfer of the excess energy to a heat sink (absorbing heat) (McCoy, 2011).

The thermal properties of many materials are listed in tables but precise measurements are still necessary for new products—polymers, composites, alloys—and new configurations, particularly at high temperatures (> 1000 °C).

7.2.1 Definition

Certain materials, particularly metals, conduct heat rapidly, while others, such as wood and most plastics and textiles, are poor conductors of heat. The physical property that describes the rate at which heat is conducted (transported) by the material is called thermal conductivity and is represented by the symbol k, or sometimes λ.

It is defined by Fourier's law: the heat flux is a product of the thermal conductivity and a temperature gradient

$$q = -k \frac{dT}{dz},\qquad (7.1)$$

where q is the heat flux in W m^{-2}, T is the temperature in K, z is the distance in m, k is the thermal conductivity in Wm^{-1}K, $Q = q X_A$, and X_A is the cross-sectional area in m^2.

In gases and liquids, transport of heat is through the movement of molecules. In a solid, however, energy is transported by the movement of electrons and the change in the vibration of the atoms in the lattice. In metals, the movement of free electrons is dominant, whereas in non-metals, the vibration of the ions is the more important. As shown in Table 7.2, the thermal conductivity of gases is an order of magnitude lower than that of liquids and some metals are as much as three orders of magnitude larger than that of water. Silver has a very high thermal conductivity—a 429 W m^{-1} K^{-1}—while, surprisingly, two non-metals (diamond and graphene) have an even higher thermal conductivity—as much as an order of magnitude higher.

7.2.2 Measurement of Solids

The principle behind the determination of the thermal conductivity of a material is based on the relationship between the heat flow through the material and the resulting temperature gradient.

Consider a heat flux through surface "A" of a solid plate of thickness Δz (see Figure 7.1). Following Fourier's law, the thermal conductivity of the plate is calculated:

$$k = -\frac{Q\Delta z}{X_A(T_2 - T_1)}.\qquad (7.2)$$

One could easily imagine a heat source on one side of the plate and a sink on the other that absorbs the energy traversing the object. A practical example is a combustion chamber in which the interior is maintained at temperatures above 800 °C, while the exterior might be close to ambient. Instrumentation measuring the thermal conductivity relies on thermocouple readings on each side of the vessel (or plate, as shown in Figure 7.1). Heat losses from the sides of the plate should be minimized in to limit heat losses. Thus, the sides may be insulated or a heat gradient might be added to mimic the axial gradient.

TABLE 7.2 Thermal Conductivity, k, of Various Gases, Liquids, and Solids

	k (@ 32 °F) (Btu h^{-1}ft^{-1}°F^{-1})	k (@ 0 °C) (W m^{-1}K^{-1})
Helium	0.0818	0.1416
Hydrogen	0.0966	0.167
Methane	0.0176	0.0305
Oxygen	0.0142	0.0246
Air	0.0140	0.024
Carbon dioxide	0.0052	0.0090
Acetone (20 °C)	0.102	0.177
Ethanol	0.105	0.182
Ethylene glycol	0.1530	0.265
Glycerine (20 °C)	0.164	0.284
Water	0.32	0.55
Graphene	2890	5000
Diamond	520–1340	900–2320
Silver	248	429
Copper	232	401
Aluminum	137	237
Iron	32	55
Stainless Steel	7–26	12–45
Lead	20	35

McCabe and Smith (1976).

To control the heat transfer rate, a cooling fluid may be circulated at a constant rate across the surface of the cool side of the plate (T_2 side). By measuring the temperature of the cooling fluid entering and exiting, the net heat rate may also be estimated. Figure 7.2 demonstrates a practical geometry to measure thermal conductivity. A heated guard plate is sandwiched between the sample of interest. The temperature of the guard plate as well as the heat load supplied must be measured precisely as well as the temperature at the surface of the sample at the point at which the coolant contacts the sample.

Example 7.1. The thermal conductivity of a 30 cm by 30 cm insulation insulator is measured in a guarded hot plate. The uncertainty of a differential thermocouple measuring the temperature is ±0.3 K. The power applied to the sample is 5 kW ± 1% and the temperature differential is 55 K across the 2.0 mm thick sample. What is the thermal conductivity of the insulation and what is the measurement uncertainty?

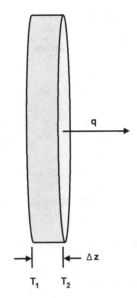

FIGURE 7.1 Heat Conduction Through a Plate

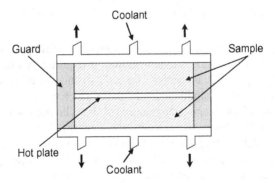

FIGURE 7.2 Guarded Hot Plate to Measure Thermal Conductivity

Solution 7.1. The surface area of the sample, X_A, is 0.09 m^2 and the heat rate across the surface, Q, is 5 kW $\pm$ 1%, so $\Delta_Q/Q = 0.01$. Introducing the numerical values in the relationship for thermal conductivity gives:

$$k = -\frac{Q}{X_A}\frac{\Delta z}{\Delta T} = -\frac{5 \text{ kW}}{0.090 \text{ m}^2}\frac{0.002 \text{ m}}{-55 \text{ K}} = 2.02 \text{ kW m}^{-1} \text{ K}^{-1}.$$

Because the relationship between the thermal conductivity and the measured values can be expressed as a power law, the uncertainty of the conductivity is given by:

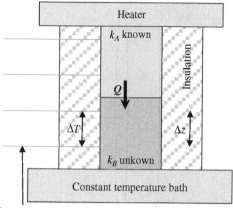

FIGURE 7.3 Configuration to Measure Metals' Thermal Conductivity

$$\frac{W_k}{k} = \sqrt{\left(\frac{W_{\Delta T}}{\Delta T}\right)^2 + \left(\frac{W_Q}{Q}\right)^2},$$

$$W_k = 2.02 \text{ W m K}^{-1}\sqrt{(0.3/55)^2 + 0.01^2} = \pm 0.02 \text{ W m K}^{-1}.$$

For highly conductive materials, such as metals and alloys, a very accurate measurement of small changes in temperature is required. A simple apparatus is illustrated in Figure 7.3. It consists of a metal cylinder with a known thermal conductivity joined to a cylindrical specimen for which the thermal conductivity is to be measured. Heat is applied to one end of the cylindrical assembly and the other is immersed in a thermal bath to absorb the energy. As with the guarded hot plate, the sides of the cylinder are insulated in order to minimize heat losses and ensure that the heat propagates in only one direction. Thermocouples, separated by a distance Δz, are placed along the length of the tube.

The thermal conductivity of sample A is given by:

$$k_A = -\frac{Q\Delta z}{X_A(T_2 - T_1)}. \tag{7.3}$$

The measurement principle is based on the assumption that the total heat flows through the sample; its precision depends on the ability to eliminate heat loss.

To ensure the best possible accuracy, the sample may be placed under vacuum to minimize convection. The measuring chamber is wrapped in thermal blankets to maintain the temperature and minimize radiation heat losses. Finally, the thermocouples are chosen to minimize heat losses. Using this method, the thermal conductivity of the metal heated to 600 °C can be measured.

7.2.3 Measurement of Fluids

The measurement of the conductivity properties of liquids and gases is less straightforward than for solids. Measuring the characteristics of saturated mixtures presents a greater level of difficulty. The first place to look for thermal conductivity data is in the NIST library: http://webbook.nist.gov/chemistry/fluid/. However, the values of many fluids remain under investigation. For example, (Marsh et al., 2002) published new experimental measurements for propane from 86 K to 600 K at pressures as high as 70 MPa in 2002.

The method relies on anemometry techniques (hot-wire anemometry): a thin wire, r_w, is suspended in a cell with a significantly larger radius, r_c. A prescribed power is applied over the length of the wire and the surrounding fluid in the cell begins to rise (King, 1914). The ideal temperature rise in the cell is given by the following equation (assuming a line energy source and an infinite medium with constant physical properties):

$$\Delta T_w = \frac{1}{4\pi k}\left(\ln t + \ln\left(\frac{4\alpha}{r_w^2 C}\right)\right), \qquad (7.4)$$

where t is the time in s, α is the thermal diffusivity of the fluid ($k/\rho C_p$), and $C = 1.7811$ is the exponent from Euler's constant.

The accuracy of this technique is to within $\pm 0.3\%$. Both tungsten and platinum wires are used with diameters of below 7 μm, more typically 4 μm (for low-pressure gas measurements). Small diameters reduce errors introduced by assuming the wire is a line source (infinitely thin). The effect becomes significant for gases and increases with decreasing pressure. The thermal diffusivity of gases is inversely proportional to the pressure and thus thermal waves may extend to the cell wall. For these cases, a steady state hot-wire technique is used for which the design equation is:

$$k = \frac{\frac{Q}{L}\ln\left(r_c/r_w\right)}{2\pi(T_w - T_c)}. \qquad (7.5)$$

7.2.4 Pressure, Temperature Effects

When the values of thermal conductivity of some simple compounds cannot be determined at the operating conditions desired, it is always possible to refer to the diagram in Figure 7.4. This diagram reports the reduced thermal conductivity, $k_r = k/k_c$, as a function of a reduced pressure, $P_r = P/P_c$, and reduced temperature $T_r = T/T_c$, which represent the operating pressure and temperature divided by the critical pressure and temperature, respectively.

The diagram in Figure 7.4 is unreliable for complex fluids but it does represent simple fluids quite well. Increasing the temperature increases the thermal conductivity for gases while it decreases it for many liquids. Critical properties of some gases are given in Table 7.3.

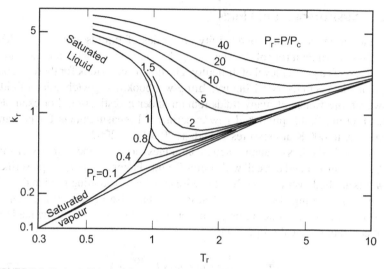

FIGURE 7.4   Reduced Thermal Conductivity as a Function of Reduced Pressure and Temperature (Bird et al., 1960, p. 250)

TABLE 7.3 Critical Properties of Some Common Gases

Gas	T_c (°C)	P_c (atm)	$T_{boiling}$ (°C)
He	−268	2.26	−269
H_2	−240	12.8	−253
Ne	−229	26.9	−246
N_2	−147	33.5	−196
CO	−140	34.5	−191
Ar	−122	48	−186
O_2	−118	50.1	−183
CH_4	−83	45.4	−161
CO_2	31	72.9	−78
NH_3	132	111	−33
Cl_2	144	78.1	−34

7.2.5 Insulation Design

Many processes operate at temperatures exceeding 700 °C and require both specialized metals—stainless steel, Inconel, Hastelloy, etc.—as well as refractory materials to insulate the metal from the extreme temperatures. Refractory linings and bricks are the most common insulating materials for industrial vessels. In selecting a refractory lining, its resistance to thermal shock

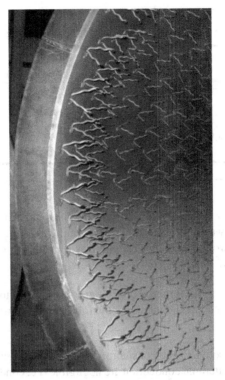

FIGURE 7.5 Refractory Anchors

and mechanical wear, chemical inertness as well as thermal conductivity and the coefficient of thermal expansion must be considered. Alumina (Al_2O_3), silica (SiO_2), and magnesia (MgO) are the most common insulators. For service at very high temperatures silicon carbide or graphite can be used but for oxidizing conditions, zirconia is preferred because carbon compounds will oxidize or burn. Figure 7.5 is a photograph of the interior of a 2 m diameter vessel wall illustrating the anchors on which the refractory material is mounted.

New technologies are being developed for ever higher pressure and temperature operation. Designing experimental equipment is often as challenging as designing the commercial equipment. The following example illustrates the design sequence to design a high pressure, high temperature experimental reactor.

Example 7.2. Propose a refractory and insulation layer design for a 0.0762 m ID reactor with an overall length of 0.914 m (36 in.) that is to operate at a maximum operating pressure and temperature of 35 bar and 1100 °C. The vessel is to be made of 304SS (stainless steel) with a maximum external diameter of 0.36 m (14 in.). Furthermore, the outer wall temperature should be maintained

TABLE E7.2 Physical Properties of Refractory/Insulating Materials

	k (W m^{-1} K^{-1})	Max Op. Temp. (°C)	Thick. Inc. Avail. (in.)
Alumina refractory	3.00	1930	0.0625
Insulating wool	0.30	1000	0.75
Ceramic paper	0.09	810	0.0625
Microporous insulation	0.04	640	0.25
Stainless steel	16.2	120	0.75

less than 120 °C in order to maintain its structural integrity. The reactor is to be installed in a well-ventilated room at a temperature of 25 °C. Table E7.2 summarizes the properties of a select number of insulating materials available for the project. *J.R. Tavares*

Solution 7.2. Due to the imposed constraints, the solution to this design problem will be iterative in nature including the following simplifying assumptions:

- Steady state.
- Thermal conductivities are constant (no variation with temperature).
- Axial and longitudinal conduction are negligible (i.e. conduction only in the radial direction).
- Heat transfer coefficient between the reactor wall and room is 10 W m^{-2} K^{-1} (natural convection).
- Heat losses due to radiation are negligible.

A common way of approaching problems with multi-layered systems is to consider each of the layers as a resistance to heat transfer (analogous to electrical circuits). In this case, because each refractory/insulating material is layered onto the previous material, these thermal "resistances" can be considered as being "in series" and are thus additive. In heat transfer, the resistance typically has units of temperature over heat rate (such as K W^{-1}).

The heat flow, Q, can be calculated using Fourier's law of heat transfer in cylindrical coordinates:

$$Q = -k(2\pi rl)\frac{dT}{dr},$$

where k is the thermal conductivity of the pipe material, r is the radius, l is the length of the pipe, and T is the temperature at a given r.

Integrating from the inner radius, r_i, to the outer radius, r_o, of a given layer gives:

$$Q = \frac{k(2\pi l)(T_i - T_o)}{\ln\left(r_i/r_o\right)}.$$

This equation can be expressed in terms of thermal resistance as follows:

$$Q = \frac{1}{R_i}(T_i - T_o),$$

in which:

$$R_i = \frac{\ln\left(r_i/r_o\right)}{k_i(2l)}.$$

Thus, it is possible to obtain the global heat flow out of the reactor by adding up the resistances contributed by each of the layers and using the resulting total resistance in the heat flow equation.

The challenge in solving the problem lies in the fact that each of the layers is constrained on both the maximum continuous operating temperature and thickness increments. These constraints make it such that the only way to solve this problem is through trial and error, validating whether or not the operational constraints are met for one particular combination and adjusting accordingly.

The available radial space for the refractory/insulating materials is 4.75 in. (7 in. outer radius minus 0.75 in. stainless steel shell minus 1.5 in. inner reactor radius).

As a first guess, we assume the following combination:

- Alumina refractory: 2 in. (32 layers, 0.0508 m).
- Insulating wool: 2.25 in. (three layers, 0.05715 m).
- Ceramic paper: 0.25 in. (four layers, 0.00635 m).
- Microporous insulation: 0.25 in. (one layer, 0.00635 m).

The first step is to calculate the global heat flow. The thermal resistance for each layer is:

$$R_{\text{refractory}} = \frac{\ln\left(r_2/r_1\right)}{k_{\text{refractory}}(2\pi l)},$$

$$R_{\text{wool}} = \frac{\ln\left(r_3/r_2\right)}{k_{\text{wool}}(2\pi l)},$$

$$R_{\text{paper}} = \frac{\ln\left(r_4/r_3\right)}{k_{\text{paper}}(2\pi l)},$$

$$R_{\text{microporous}} = \frac{\ln\left(r_5/r_4\right)}{k_{\text{microporous}}(2\pi l)},$$

where $r_1 = 0.03810$ m (inner reaction space), $r_2 = 0.08890$ m (refractory/wool interface), $r_3 = 0.14605$ m (wool/paper interface), $r_4 = 0.15240$ m (paper/microporous interface), $r_5 = 0.15875$ m (microporous/steel interface), and $l = 0.914$ m.

The resistance from the outer stainless steel shell will be:

$$R_{\text{steel}} = \frac{\ln\left(r_6/r_5\right)}{k_{\text{steel}}(2\pi l)},$$

where $r_6 = 0.17785$ m (outer vessel radius). And finally, the resistance due to natural convection will be:

$$R_{\text{convection}} = \frac{1}{hA} = \frac{1}{h(2\pi r_6 l)}.$$

By substituting these values into the heat flow equation, the global heat flow from the inner reactor conducted through the various layer and convected out of the system through natural convection is found:

$$Q = \frac{1}{R_{\text{refractory}} + R_{\text{wool}} + R_{\text{paper}} + R_{\text{microporous}} + R_{\text{steel}} + R_{\text{convection}}}$$
$$\times (T_i - T_o)$$

with $T_i = T_1 = 1373$ K and $T_o = T_6 = 298$ K (1100 °C and 25 °C, respectively).

For the proposed combination of materials, the global heat flow is $Q = 1762$ W.

Conservation of energy dictates that the heat will pass through each of the layers. Thus, it is possible to calculate the interfacial temperature between each layer by applying the same heat flow equation, but focusing around a single thermal resistance. For example, in the case of the alumina refractory layer:

$$Q = \frac{1}{R_{\text{refractory}}}(T_1 - T_2),$$

which can be rearranged to:

$$T_2 = T_1 Q R_{\text{refractory}}.$$

The same procedure can be repeated for each layer in order to measure the interface temperature. The values calculated for the proposed combination are:

- Alumina refractory inner temperature T_1: 1100 °C.
- Insulating blanket inner temperature or alumina refractory outer temperature T_2: 1013 °C.
- Ceramic paper inner temperature or insulating blanket outer temperature T_3: 498 °C.
- Microporous insulation inner temperature or ceramic blanket outer temperature T_4: 354 °C.

- Stainless steel inner temperature or microporous insulation outer temperature T_5: 54 °C.
- Stainless steel outer temperature T_6: 52 °C.

In the present configuration, it appears that all temperature constraints are met except the constraint on the insulating blanket (over by 13 °C). Thus the procedure must be repeated.

If three layers of alumina refractory are added (bringing the refractory thickness up to 2.1875 in.) at the expense of three layers of ceramic paper (bringing the ceramic paper thickness down to 0.0625 in.), the global heat flow Q increases to 1992.1 W and the new temperature profile is:

- Alumina refractory inner temperature T_1: 1100 °C.
- Insulating blanket inner temperature or alumina refractory outer temperature T_2: 996 °C.
- Ceramic paper inner temperature or insulating blanket outer temperature T_3: 437 °C.
- Microporous insulation inner temperature or ceramic blanket outer temperature T_4: 397 °C.
- Stainless steel inner temperature or microporous insulation outer temperature T_5: 58 °C.
- Stainless steel outer temperature T_6: 56 °C.

This combination respects all the required temperature constraints and, while it is not the only solution, it is noteworthy in that the stainless steel temperature is far below the maximum safety limit. This is of particular importance considering the fact that the stainless steel shell will be the component withstanding the strain of the high operating pressure.

To further refine the design, it would be pertinent to remove the simplifying assumption that the thermal conductivities of the various layers are constant with respect to temperature, as this is unrealistic.

Moreover, the proposed solution does not take into account any economic considerations. For example, the microporous insulation and the ceramic paper are significantly more expensive than the insulating wool. Also, the solution proposed uses the maximum allowable steel shell diameter, which may be prohibitively costly. Alternative solutions may be possible and less costly, particularly if forced convection (such as a fan) is used to draw additional heat away from the outer reactor wall.

Finally, the proposed design does not take into account the thermal expansion or the compressibility of the various layers, particularly the alumina refractory for the former and the insulating wool for the latter. Typical commercial designs make allowances for such phenomena.

7.3 VISCOSITY

The word "rheology" comes from the Greek "Rheos" which means flow. Rheology studies the deformation and flow of materials in response to the action of a force. It describes the interrelationship between force, deformation, and time and provides information on the consistency of materials, by their viscosity—which is simply the resistance to flow—and elasticity, i.e. structure and/or adhesion.

As an example of an application of rheology, one could cite the measurement of the hydrodynamics of submarine avalanches caused by earthquakes. The speed of the avalanche is related to the rheology of the sediments: viscous sediments flow more slowly compared to less viscous avalanches. The speed at which the avalanche proceeds has a direct relationship with the severity of ensuing tidal wave—tsunamis.

The invention of the water clock in 1650 BC in Egypt is arguably the first quantitative application of rheological principles to a practical application: the viscosity of fluids depends on temperature and Amenemhet took this into consideration in his design. He added a $7°$ angle modification to his water clock to account for the viscosity at night versus the day. The first detailed study of rheology emerged in the late seventeenth century, with the empirical law of deformation of Robert Hooke, which describes the behavior of solids. They are often represented by an elastic spring, whose elongation, ϵ, is proportional to the force, σ, applied at its extremity:

$$\sigma = E\epsilon. \tag{7.6}$$

Hooke was seeking a theory of springs, by subjecting them to successively increasing force. Two important aspects of the law are the linearity and elasticity. Linearity considers that the extension is proportional to the force, while the elasticity considers that this effect is reversible and there is a return to the initial state, such as a spring subject to weak forces. Hooke's law is valid for steels in most engineering applications, but it is limited to other materials, like aluminum only in their purely elastic region.

In 1687, the empirical law of Newton related the flow stress of a fluid to its velocity gradient. The constant of proportionality was the viscosity. In fact, the viscosity describes the internal resistance of the fluid to flow and deformation. For example, water has a low resistance to flow, hence its viscosity is lower compared to most oils, for example, that have a higher resistance (at room temperature). Fluids like water and most gases are ideal fluids that satisfy Newton's theory. Many real fluids have a much more complex relationship between stress, τ, and the velocity gradient, $(d\gamma/dt)$ (γ is the deformation). The viscosity of Newtonian fluids is represented by the symbol "μ" whereas for real fluids, it is represented by the symbol "η", which is also known as the dynamic viscosity:

$$\tau = \eta\dot{\gamma}. \tag{7.7}$$

TABLE 7.4 Different Types of Viscosity

Type of Viscosity	Symbol and Definition
Relative viscosity	$\eta_{rel} = (\eta/\eta_0) = (t/t_0)$
Specific viscosity	$\eta_{sp} = \eta_{rel} - 1 = (\eta - \eta_0/\eta_0) = (t - t_0/t_0)$
Reduced viscosity ($cm^3\ g^{-1}$)	$\eta_{red} = (\eta_{sp}/c)$
Inherent viscosity ($cm^3\ g^{-1}$)	$\eta_{inh} = (\ln \eta_{rel}/c)$
Intrinsic viscosity ($cm^3\ g^{-1}$)	$[\eta] = (\eta_{sp}/c)_{c=0} = (\ln \eta_{rel}/c)_{c=0}$

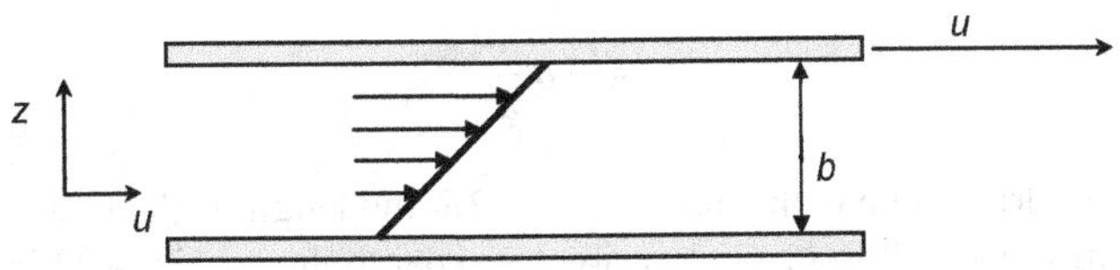

FIGURE 7.6 Laminar Flow Velocity Profile Between Two Large Plates

For Newtonian fluids, viscosity depends only on pressure and temperature, whereas for non-Newtonian fluids, viscosity also depends on shear rate.

Two hundred years after the early contributions of Newton and Hooke, various laws of real fluids emerged as well as a quantitative description of flow and the measurement of viscosity, including the work of Euler, Cauchy, Coulomb, Poiseuille, Hagen, Couette, Reynolds, and Bingham. In 1890, the first rotational rheometer was invented by Couette. In 1929, Reiner and Bingham founded the first rheological society.

Terminology and definitions to characterize viscosity are summarized in Table 7.4.

7.3.1 Single Phase Flow

Consider a fluid contained between two large parallel plates separated by a distance b, as shown in Figure 7.6 and assume that the lower plate is stationary while the top plate is driven with a constant velocity u.

When the flow is fully developed (meaning that it is far enough from the entrance—and exit—such that entrance effects have disappeared), the fluid velocity profile between the two plates is linear and can be defined by the following relationship:

$$\frac{du}{dz} = \frac{u}{b}. \tag{7.8}$$

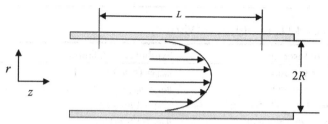

FIGURE 7.7 Laminar Flow Velocity Distribution in a Circular Tube

The fluid shear stress, τ, is proportional to the velocity gradient (for Newtonian fluids) and is given by the following equation:

$$\tau = -\mu \frac{du}{dz}. \tag{7.9}$$

In a cylindrical tube with diameter $D = 2R$ and length L, the flow due to the pressure difference ($P_0 - P_L$) is described by a parabolic profile (see Figure 7.7). The velocity gradient is given by the Hagen-Poiseuille relation:

$$\frac{du}{dr} = \frac{P_0 - P_L}{2\mu L} r. \tag{7.10}$$

At $r = 0$, the flow velocity is maximum and will be linked to the viscosity by the following equation:

$$u_{max} = \frac{P_0 - P_L}{4\mu L} R^2. \tag{7.11}$$

For incompressible fluid flow (fully developed), the volumetric flow rate is readily calculated as:

$$\dot{V} = \frac{P_0 - P_L}{8\mu L} \pi R^4. \tag{7.12}$$

7.3.2 Reynolds Number

Dimensionless numbers are pure numbers that are expressed as ratios of physical properties and are used to classify or understand a system. Already, in Chapter 6, we defined the Reynolds number as the ratio of the fluid's inertial forces, $\rho u d$, to its viscous force and used it to differentiate between the laminar, intermediate, and turbulent flow regimes. In the laminar flow regime, viscous forces dominate but, as the velocity increases, the Reynolds number increases and inertial forces predominate in the turbulent regime.

The Reynolds number has been applied to systems other than fluids flowing through tubes and pipes; it is used to characterize flow through irregular shaped

ducts and even flow of particles through fluids. The form of the equation, for a
Newtonian or non-Newtonian fluid, is:

$$N_{Re} = \frac{\rho u L}{\eta} = \frac{u L}{\nu},\tag{7.13}$$

where ν is the kinematic viscosity in $m^2 \ s^{-1}$ ($\nu = \eta/\rho$) and L is the
characteristic dimension in m.

The characteristic dimension of fluid flow in a tube is the diameter, d. It is the
particle diameter in fluid-powder systems, d_p. In square ducts, it is the hydraulic
radius, r_H. Note that the notion of turbulence and laminar flow for particles as
well as the values of the Reynolds number are quite different compared to flow
through pipes and ducts.

The units of viscosity both dynamic, η, and kinematic, ν, are given next.
Dynamic viscosity:

$$\begin{aligned}
1 \ N \ s \ m^{-2} &= 10 \ P \\
&= 1000 \ cP \\
&= 1 \ kg \ m^{-1} \ s^{-1}.
\end{aligned}$$

$$\begin{aligned}
1 \ P &= 100 \ cP \\
&= 1 \ dyn \ s \ cm^{-2} \\
&= 0.1 \ N \ s \ m^{-2} \\
&= 0.1 \ kg \ m^{-1} \ s^{-1}.
\end{aligned}$$

Kinematic viscosity:

$$\begin{aligned}
1 \ St &= 1 \ cm^2 \ s^{-1} \\
&= 100 \ cSt \\
&= 10^{-4} \ m^2 \ s^{-1}.
\end{aligned}$$

Example 7.3. Calculate the Reynolds number in a pipeline 1 m in diameter
(d) that transports 100 000 oil barrels per day with a specific gravity of 0.8 and
a viscosity 0.2 cP.

Solution 7.3. Pipelines should always be operated in the turbulent flow regime
because the friction losses are lowest, thus pumping costs are minimized.
Furthermore, investment costs will be lower because the cost of pipe per
volumetric flow rate is lower. The volume of one barrel of oil is 159 l. So,
the volumetric flow rate of the oil is 15 900 $m^3 \ d^{-1}$ or 0.184 $m^3 \ s^{-1}$.

The flow regime is highly turbulent in this example. One risk is that in the
winter the oil could become very viscous if the temperature of the line dropped
significantly. There exist crude oils with high wax content that are solid at
temperatures as high as 50 °C. Pipelines carrying these crude oils need to be
insulated to avoid solidification.

TABLE 7.5 Prandtl Numbers and Viscosity of Various Gases and Liquids

Gas/Liquid	N_{Pr} at 100 °C	μ, cP at 100 °C
Helium	0.71	0.02
Hydrogen	0.69	0.01
Methane	0.75	0.04
Oxygen	0.70	0.023
Air	0.69	0.02
Carbon dioxide	0.75	0.017
Acetone	2.4	0.17
Ethanol	10.1	0.32
Ethylene glycol	125	2.0
n-Octane	3.6	0.26
Water	1.5	0.26

McCabe and Smith (1976).

7.3.3 Prandtl Number

The Prandtl number is a dimensionless number that provides a measure of the efficiency of transport by momentum diffusivity to thermal diffusion:

$$N_{Pr} = \frac{C_p \mu}{k}, \tag{7.14}$$

where C_p is the specific heat in J kg^{-1} K^{-1} and μ is the viscosity in kg m^{-1} s^{-1}.

It is also written as the ratio of the kinematic viscosity, ν, which represents the momentum diffusivity, to the thermal diffusivity, α:

$$N_{Pr} = \frac{\nu}{\alpha} = \frac{\text{viscous diffusion rate}}{\text{thermal diffusion rate}}. \tag{7.15}$$

The viscous diffusion rate equals the viscosity divided by the density, while the thermal diffusion rate is the ratio of the thermal conductivity to the product of fluid density and heat capacity. While the Reynolds number includes a length scale, the Prandtl number depends only on fluid properties. The Prandtl number of many inorganic gases is approximately 0.70, as summarized in Table 7.5. It is much higher for liquids and the variation from one liquid to another is also greater: the Prandtl number of acetone is 4.5 at 20 °C, while it is 350 for ethyl ether. It equals 0.015 for mercury. Thermal diffusivity dominates in the case of mercury. Engine oils have a high Prandtl number and so the convective component of heat transfer is higher.

For simultaneous heat and mass transfer applications, the Prandtl number indicates the relative thickness of the momentum boundary compared to the

TABLE 7.6 Viscometry Techniques

Viscometer	Range (cP)	Fluid
Falling ball	$0.5–70 \times 10^3$	Newtonian
Capillary	$0.5–10^5$	Newtonian and non-Newtonian
Flow cup	$8–700$	Newtonian
Rotational	$10–10^9$	Newtonian and non-Newtonian
Rolling ball	$0.5–10^5$	Newtonian
Drawing ball	$0.5–10^7$	Newtonian

thermal boundary layer. (The laminar boundary layer is the region next to a solid surface—temperature, velocity, and concentration gradients are highest in the immediate vicinity of the boundary layer and tend to be zero in the fully developed region.) For low values of the N_{Pr}, the momentum boundary layer is smaller than the thermal boundary layer—heat is dissipated more rapidly than the momentum. When the viscosity is low compared to the ratio of the thermal conductivity to the heat capacity, the momentum boundary layer will be smaller compared to the thermal boundary layer, $N_{Pr} > 1$. The thickness of the momentum boundary layer equals the thermal boundary layers when $N_{Pr} = 1$.

7.3.4 Viscosity Instrumentation

Experimental methods to measure viscosity vary depending on the fluid type: capillary flow through a vertical tube is the most common method for transparent, low viscosity fluids (water, organic solvents, etc.); for viscous fluids such as oils, molten polymers, gels, etc. rotating concentric cylinders are used. The range of the different techniques is given in Table 7.6.

7.3.4.1 Newtonian Fluids

The falling ball viscometer consists of a tube that may be rotated about a horizontal axis, as shown in Figure 7.8. The tube is marked with two lines a and b and contains the fluid of interest maintained at a given temperature. A sphere (steel and glass are the most common materials) with a finely calibrated diameter is inserted into the tube. At the beginning of the test, the ball lies at the bottom. The tube is rotated by 180°, which brings the ball (sphere) to the top and then it drops through the fluid. The time it takes to traverse a prescribed distance L between the lines a and b is measured. The velocity of the ball is the distance between the two lines divided by the time.

The principle of the falling ball viscometer is similar to that of the rotameter (discussed in Chapter 5)—instead of suspending a bob in the flowing gas, the sphere moves at a constant velocity and the fluid is immobile. As with the rotameter, the gravitational force on the sphere, $V_s \rho_s g$, acts downward and the

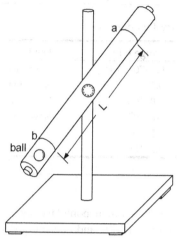

FIGURE 7.8 Falling Ball Viscometer

buoyancy force, $V_f \rho_f g$, and inertial force act upward. So, based on the force balance we have:

$$V \rho_f g + C_D A_{p,s} \rho_f \frac{u_s^2}{2} = V \rho_s g, \qquad (7.16)$$

where V is the volume of the sphere $\left(\frac{4}{3} \pi R^3\right)$, $A_{p,s}$ is the projected area of the sphere in m^2 (πR^2), ρ_s and ρ_f are the densities of the sphere (s) and the fluid (f) in kg m^{-3}, g is the gravitational constant in m s^{-2}, and C_D is the drag coefficient.

In the particle laminar flow regime—also known as the Stokes regime—the drag coefficient equals $24/N_{Re}$. So, the above equations can be simplified to express the time of descent, t, as a function of geometry and physical properties:

$$t = \frac{6\pi R L}{(\rho_s - \rho_f) V g} \mu = \frac{9}{2} \frac{L}{(\rho_s - \rho_f) R^2 g} \mu, \qquad (7.17)$$

where L is the distance between points a and b in Figure 7.8.

Example 7.4. The viscosity of a new polymer formulation is measured in a falling ball viscometer. A total of nine readings are made with a stopwatch for a 2.00 cm ball to traverse a distance of 0.300 m: 23.5 s, 22.9 s, 24.1 s, 21.5 s, 24.5 s, 23.3 s, 22.9 s, 23.7 s, and 23.3 s. The density of the steel ball is 7850 kg m^{-3} and the polymer density is 852 ± 8 kg m^{-3}. (a) What is its viscosity? (b) What is the uncertainty?

Solution 7.4.a. Based on the nine experiments, the average time of descent is 23.3 s with a standard deviation of 0.9 s. The radius, R, of the steel ball is 0.01 m, and so the viscosity may be calculated directly from the equation:

$$\mu = \frac{9}{2L} (\rho_s - \rho_f) R^2 g t = \frac{9}{2 \cdot 0.30} (7850 - 852)(0.01)^2 \cdot 9.82 \cdot 23.3 = 160 \text{ Pa s}.$$

Solution 7.4.b. The uncertainty in the calculated viscosity is due to the contributions of the uncertainty of the density of the polymer as well as the uncertainty in the measurement of time:

$$\frac{W_\mu}{\mu} = \sqrt{\left(\frac{W_{\Delta\rho}}{\Delta\rho}\right)^2 + \left(\frac{W_t}{t}\right)^2}.$$

The relationship between the polymer density, ρ_{poly}, and the difference in density, $\Delta\rho$, is linear and since only the uncertainty in the polymer density is quoted, $W_{\Delta\rho} = W_{\rho,poly} = \pm 8 \text{ kg m}^{-3}$. The uncertainty with respect to time is calculated based on the 95% confidence interval:

$$W_t = \pm t(\alpha, n-1)s_i = \pm t(95\%, 8)\frac{s}{\sqrt{n}} = 2.306\frac{0.85}{3} = 0.65,$$

$$W_\mu = \sqrt{\left(\frac{8}{7000}\right)^2 + \left(\frac{0.65}{23.3}\right)^2} \cdot 160 \text{ Pa s} \cong 4 \text{ Pa s}.$$

7.3.4.2 The Saybolt Viscometer

Calibrated glass capillary columns are commonly used to measure the viscosity of petroleum fluids. The procedures are described in ASTM methods D88, D445, D2170, and D2171 (the latter two are for bitumen, which is an extremely viscous opaque fluid). The procedure for D445 involves first loading the fluid to the calibrated tube and maintaining it at a constant temperature overnight. The fluid is then brought to just above the highest line in the tube and then it is allowed to drain. The timer is initiated when the bottom of the meniscus just touches the top line and it is halted when the meniscus touches the bottom line. This procedure is repeated three times. In method D445, the temperature of the bath is kept at 40 °C. The bath is then heated to 100 °C and the procedure is repeated.

The Saybolt viscometer works on a similar principle (ASTM D88)—the fluid is loaded to a tube calibrated to 60 ml. A cork is removed from the bottom of a narrow capillary and a timer is initiated. When 60 ml of the fluid is drained, the timer is halted. The time to drain the 60 ml volume is known as "Saybolt" seconds. The instrument operates at temperatures from ambient to as much as 99 °C. Because of the high sensitivity of viscosity to temperature, substantial efforts are devoted to maintaining isothermal conditions: quoted uncertainties around the temperature are on the order of ±0.03 °C. The accuracy of Saybolt viscometers is better than 0.2%. In order to ensure accuracy, they should be calibrated at regular intervals. For fully developed laminar flows, the drain time through the capillary at the base is directly proportional to viscosity. However, since the capillary section of the tube is short, a correction factor to the time is

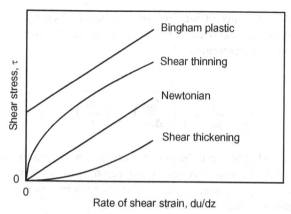

FIGURE 7.9 Classification of Non-Newtonian Fluids

introduced to account for the fact that the flow profile is not entirely developed:

$$v = 0.255t - \frac{208}{t}. \qquad (7.18)$$

Note that the units of the kinematic viscosity for this equation are in St—$cm^2\ s^{-1}$. The importance of the correction term decreases with increasing time. At 100 s, the second term on the right-hand side of the equation represents 8% of the total. Below a value of 28 s, the kinematic velocity equals zero.

7.3.4.3 Non-Newtonian Fluids

Most fluids exhibit non-Newtonian behavior—blood, household products like toothpaste, mayonnaise, ketchup, paint, and molten polymers. As shown in Figure 7.9, shear stress, τ, increases linearly with strain rate, $\dot{\gamma}$, for Newtonian fluids. Non-Newtonian fluids may be classified into those that are time dependent or time independent and include viscoelastic fluids. Shear thinning (pseudoplastic) and shear thickening (dilatant) fluids are time independent while rheopectic and thixotropic are time dependent. The shear stress (viscosity) of shear thinning fluids decreases with increasing shear rate and examples include blood and syrup. The viscosity of dilatant fluids increases with shear rate. The viscosity of rheopectic fluids—whipping cream, egg whites—increases with time while thixotropic fluids—paints (other than latex) and drilling muds— decrease their viscosity with the duration of the shear.

7.3.4.4 The Rotational Rheometer

The first cylindrical rotary rheometer was invented in 1890 by Couette. Rotational rheometers consist of two concentric cylinders, as shown in Figure 7.10.

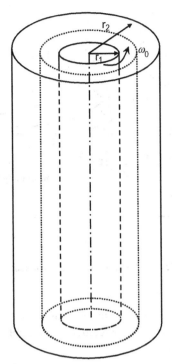

FIGURE 7.10 Schematic of a Coaxial Rotational Rheometer (Patience et al., 2011)

The test fluid is placed between the cylinders and the inner cylinder is rotated at a constant velocity ω while the outer cylinder remains fixed. A no-slip condition is assumed for the fluid in contact with each wall; therefore, the fluid in the outer wall is stationary while it rotates at an angular velocity ω_0 at the inner wall. Because of the rotational motion, the shear rate, $\dot{\gamma}$, varies across the gap and shear stress.

The torque of the shaft is related to the shear stress τ by the following equation:

$$\tau = \frac{M}{2\pi r^2 h},$$ (7.19)

where M is the torque of the shaft and $2\pi r^2 h$ is the wetted surface area in m^2.

7.3.5 Influence of Temperature and Pressure on Viscosity

Figure 7.11 shows the effect of temperature and pressure changes on viscosity. The reduced viscosity $\mu_r = \mu/\mu_c$ is the viscosity at pressure P and temperature T divided by the viscosity at the critical point, μ_c. This quantity is shown as a function of reduced temperature $T_r = T/T_c$ and reduced pressure $P_r = P/P_c$.

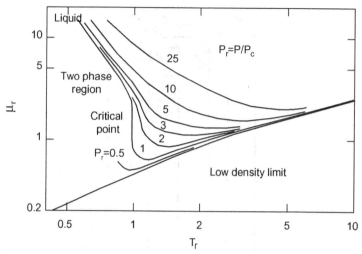

FIGURE 7.11 Diagram of Reduced Viscosity as a Function of Reduced Pressure and Temperature (Bird et al., 1960, p. 16)

For a gas of low density, viscosity increases with increasing temperature while it decreases for a liquid. The relationship is applicable to low molecular weight Newtonian fluids.

7.4 BINARY GAS DIFFUSION

The transport of chemical species from a region of high concentration to a region of low concentration can be demonstrated experimentally by the diffusion of a crystal of potassium permanganate ($KMnO_4$) with water in a beaker. The $KMnO_4$ begins to dissolve in water, a dark purple halo first forms around the crystal (region of high concentration) and then diffuses over time due to the concentration gradient, until the equilibrium concentration is reached.

Diffusion is the spontaneous movement of a species under the effect of a concentration difference, or in other cases, a gradient of temperature or force (electrostatic force in the case of charged species or chemical potential).

Diffusion is described mathematically by Fick's first law (1855), demonstrating that the net movement of a species through a unit area (flux) is proportional to the concentration gradient. The constant of proportionality is called the diffusion coefficient D, or diffusivity.

The applications of this law are numerous, but only a few examples will be given here. Take first our body as an example. Our lungs are made up of very thin cells (less than a millionth of a meter) and have a large effective area (about 100 m^2) that facilitates the exchange of gases, which depends on the diffusion and the solubility of the fluid in the membrane of the lungs. The transport of molecules through the pores and channels in the membrane of our cells as well as the transport of oxygen and carbon dioxide in the blood both proceed through diffusion.

A second example concerns packaging in the food industry. The effectiveness of thin transparent wrapping (polymers) in protecting food depends on their ability to prevent the spread of pathogenic microorganisms, external moisture causing mold, and oxygen responsible for the oxidation of food.

Finally, in the field of construction, increasing the temperature and concentration of carbon dioxide in the atmosphere has a significant impact on the life of urban structures in concrete. Moreover, the structures in a marine environment are damaged mainly by chloride ions. The evaluation of their service life is based on the determination of the penetration of these ions in concrete, that is to say, by calculating their diffusion coefficient, with Fick's law.

7.4.1 Fick's Law

Considerable research is devoted to high performance Sportswear and even the Ready-to-Wear segments of textiles. The desire is to maintain the body dry by wicking water away as the body generates heat and perspires. To illustrate the concept of diffusion, consider a semi-permeable membrane laminated to a fabric. Assume the sample was placed in a cell separating a column of air from a column of saturated water vapor. The membrane is permeable for water vapor but not air. At time $t = 0$, thanks to Brownian motion, the water molecules will traverse the membrane and then diffuse through the fabric sample to the air. (Evidently, the fabric sample is usually next to the skin—this example is for illustration purposes only.) The molecular transport through the fabric is referred to as diffusion.

In this system, we designate water as species A and air is referred to as species B. The mass fractions are denoted by x_A and x_B and they are defined as the mass of the individual species divided by the sum of the masses of the species. The molar fractions y_A and y_B equal the moles of species A or B with respect to the total number of moles. Remember that the total molar density equals $\frac{P}{RT}$ for gases. Figure 7.12 demonstrates the evolution of the mass fraction of water as a function of time. The first vertical solid line represents the semi-permeable membrane. The fabric extends from the membrane to the dotted line. Initially, the mass fraction of water in the fabric equals zero. At the beginning of the experiment, the mass fraction of water at the membrane equals one and it is zero in the fabric. As time advances, the mass fraction in the fabric increases (third image from the top). With time, the concentration profile of species A eventually becomes a straight line, as shown in Figure 7.12.

The mass flux of species A in the direction of z is given by Fick's law:

$$j_A = -\rho D_{AB} \frac{dx_A}{dz}, \tag{7.20}$$

where D_{AB} is the mass diffusivity of species A in species B in m^2 s^{-1}.

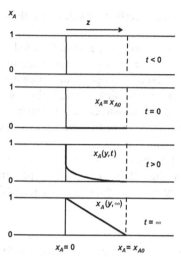

FIGURE 7.12 Diffusion of A Through a Semi-Permeable Membrane (Solid Line) Through a Fabric (Up to Dotted Line)

7.4.2 Schmidt Number

The Schmidt number is a dimensionless number equal to the ratio of kinematic viscosity to mass diffusivity. It is:

$$N_{Sc} = \frac{\nu}{D_{AB}} = \rho \frac{\eta}{D_{AB}}. \tag{7.21}$$

7.4.3 Measure of Diffusion

The device illustrated in Figure 7.13 is used to measure the diffusion coefficient of binary gas pairs. Two glass tubes with equal cross-sections are joined together at one end with a flexible tube that is pinched in the middle. Each tube is filled with a different gas and the clip is detached to initiate the experiment ($t = 0$).

The gases begin to diffuse due to the concentration gradient. After a prescribed time, the flexible tube is rejoined and the concentration of gas in each tube is determined. It would be sufficient to measure the concentration of one side to know what the concentration of the other side of the column should be. Good experimental practice would sample both columns to evaluate the measurement error.

The concentrations of both gases are given by the following equation:

$$F = \frac{N_{A1} - N_{A2}}{N_{A1} + N_{A2}} = \frac{8}{\pi^2} \sum_{k=0}^{\infty} \frac{1}{(2k+1)^2} \exp\left(-\frac{\pi^2 D_{AB} t (2k-1)^2}{4L^2}\right), \tag{7.22}$$

where 1 and 2 are the upper and lower tubes, t is the time of the experiment, L is the length of the tube, N_{A1} is the number of moles of A in the upper tube, and N_{A2} is the number of moles of A in the lower tube.

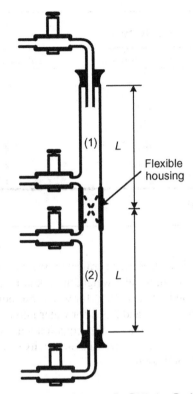

FIGURE 7.13 Loschmidt Instrument to Measure Gas Diffusion Coefficients (Holman, 2001)

The optimum time to run the experiment to determine D_{AB} is calculated from the following equation:

$$t_{opt} = \frac{4L^2}{\pi^2 D_{AB}}. \qquad (7.23)$$

The equation to calculate the factor F involves series terms. When the experiment is run up to t_{opt}, the higher-order terms may be neglected. Thus, the diffusion coefficient may be determined from the following equation:

$$D_{AB} = -\frac{4L^2}{\pi^2 t} \ln\left(\frac{\pi^2 F}{8}\right). \qquad (7.24)$$

Measuring diffusion rates is quite uncommon. Rather, the expected diffusion rates are calculated based on empirical relationships such as the following:

$$\frac{P D_{AB}}{(P_{cA} P_{cB})^{1/3} (T_{cA} T_{cB})^{5/12} \sqrt{\frac{1}{M_A} + \frac{1}{M_B}}} = a \left(\frac{T}{\sqrt{T_{cA} T_{cB}}}\right)^b, \qquad (7.25)$$

TABLE 7.7 Binary Diffusivity Pairs

Pairs A–B	Temperature (K)	Diffusivity, D_{AB} (cm^2 s^{-1})
CO_2–N_2	273.2	0.144
H_2–CH_4	298.2	0.726
H_2O–N_2	308	0.259
H_2O–O_2	352	0.357
He–Pyrex	283.2	4.5×10^{-10}
Hg–Pb	283.2	2.5×10^{-15}
Al–Cu	283.2	1.3×10^{-30}

Bird et al., 2006, p. 503.

where P is the pressure in atm, P_c is the critical pressure in atm, T_c is the critical temperature in K, M is the molecular weight in kg kmol^{-1}, a is 2.745×10^{-4} for nonpolar gas pairs and 3.640×10^{-4} for water into nonpolar gases, and b is 1.823 for nonpolar gas pairs) and 2.334 for water into nonpolar gases.

At pressures close to atmospheric, this equation agrees with experimental data to within 8%. Table 7.7 summarizes the diffusivity of various binary pairs-gas-gas, gas-solid and solid-solid.

7.4.3.1 Water Vapor Diffusion Through Membranes

Measuring water vapor transmission through membranes has been an active area of research for several decades now. Applications not only include textiles, but also housing, footwear, food packaging, and other industrial uses. The tests proposed in this section will be applied to measure the permeability of samples exceeding 32 mm thickness. There are two basic methods given by the ASTM:

1. *Method using hygroscopic powder (ASTM E96-00):*

   A film (membrane) is mounted on a sealed cup containing calcium chloride dried before-hand at 200 °C (or another hygroscopic powder). The anhydrous calcium chloride should have a particle size between 600 μm and 2.36 mm. The cup should be made of a light material that is impermeable to water vapor. The opening of the cup, over which the film is mounted, should be on the order of 3000 mm^2. The distance between the top of the anhydrous $CaCl_2$ and the film should be approximately 6 mm. (It is important to avoid touching the film with the powder.) The assembly is placed in an environmental chamber where the temperature and relative humidity should be checked frequently (temperature from 21 °C to 23 ± 1 °C and relative humidity at 50% ± 2%). Air must circulate continuously at a velocity of up to 0.3 m s^{-1}.

The cup assembly should be weighed at regular intervals until a constant value is achieved to determine the rate of absorption of water vapor by the $CaCl_2$. (For this reason, the cup should be made out of light—and impermeable—materials.) The weight scale should have a sensitivity well below 1% of the total weight of the cup plus the powder.

2. *Method using water (ASTM E96-00):*
The operating conditions of this test are similar to those of the test with the hygroscopic powder. However, in this case, the cup is filled with water and the rate of water transmission is calculated based on the drop in the weight of the cup and water assembly. Figure 7.14 illustrates the water bath into which the cups are placed to achieve a uniform temperature. Note that this instrument has room for 12 cups. In this way, many measurements can be made simultaneously and be either used to test many samples or to evaluate the uncertainty in the measurement of a single sample or even multiple samples.

Each cup should be filled with distilled water such that the space between the water level and the sample is from 13 mm to 23 mm. The level in the bottom of the cup should be on the order of 30 mm but the minimum specified depth is 3 mm.

The water vapor transmission rate, WVT, is calculated based on a mass balance:

$$WVT = \frac{\Delta W_{c+f}}{X_A t},\qquad(7.26)$$

where X_A is the cross-sectional area of the cup opening in m^2, ΔW_{c+f} is the weight change of the cup plus the film in g, and t is the time in h.

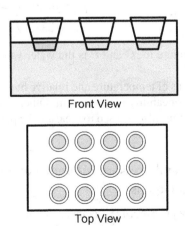

Front View

Top View

FIGURE 7.14 Device for Measuring the Permeability of Water Vapor

TABLE 7.8 WVP—Water Vapor Permeability of Selected Materials $(g\ m^{-2}\ d^{-1})$

Material	Permeability	Classification
Sports films (e.g. Gore)	>5000	
Sports wear (and film)	>2400	Very high
Suede	>1200	Very high (for leather)
Coated leather	600–1200	High
PU and fabric	240–600	Moderate
PVC and fabric	<120	Low

The water vapor permeability, WVP, is calculated based on the water vapor transmission rate:

$$WVP = \frac{WVT}{\Delta P} = \frac{WVT}{P_{sat}(RH_1 - RH_2)}, \qquad (7.27)$$

where ΔP is the pressure difference of water vapor (1 mmHg $= 1.333 \times 10^2$ Pa)), RH_1 is the relative humidity in the vapor space of the cup below the film, RH_2 is the relative humidity in the climate-controlled room, and P_{sat} is the saturation pressure of the water vapor at the temperature of the cup in mmHg.

Some values of permeability for common materials are summarized in Table 7.8.

The relative humidity in the hygroscopic powder is nominally 0%, while in distilled water, it is 100%.

The saturated water vapor pressure may be calculated according to:

$$P_s = 610.78 \exp\left(\frac{17.2694T}{T + 238.3}\right), \qquad (7.28)$$

where T is the temperature in °C and P is the water vapor pressure in Pa (see Figure 7.15).

Note: The conditions of temperature and relative humidity must be reported accurately for each permeability measurement. Other devices with infrared or colorimetric detectors may also be used to measure the permeability of the films to water vapor.

7.4.4 Temperature and Pressure Influence on the Diffusivity of Gases and Liquids

For a binary gas at low pressure, D_{AB} is inversely proportional to the pressure and increases with increasing temperature. Diffusivity is almost independent of the composition of the pair AB.

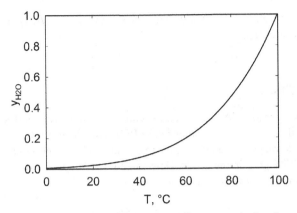

FIGURE 7.15 Mole Fraction of Water as a Function of Temperature (at 1 atm)

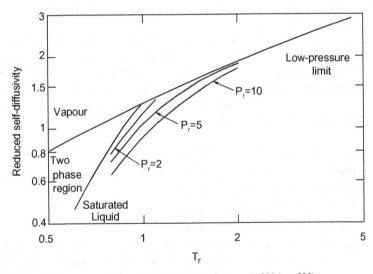

FIGURE 7.16 Diagram of Reduced Self-Diffusivity (Bird et al., 2006, p. 522)

At high pressure in liquids, the behavior of D_{AB} is more complex. However, it is easier to obtain experimental data for self-diffusivity (inter-diffusion of molecules within the same chemical species) of nonpolar solutions D_{AA*}.

Figure 7.16 shows that the self-diffusivity cD_{AA*} increases strongly with temperature, especially for liquids. For each temperature, it decreases to zero with increasing pressure. This diagram shows the reduced self-diffusivity, which is the ratio cD_{AA*} to pressure P and temperature T divided by the self-diffusivity reduced to the critical pressure P_c and the critical temperature T_c. This quantity is described in terms of the reduced temperature $T_r = T/T_c$ and reduced pressure $P_r = P/P_c$.

7.5 EXERCISES

7.1 The Governor of Syracuse chose you to replace Archimedes, who died after the siege by the Romans (212 BC). The Governor bought 20 beads of 22-karat gold, which are 1 cm in diameter. He suspects that there is less than 75% gold in the beads. Determine if the beads are at least 18 karats (assuming that the other metal is lead)

$$X = 24\frac{M_g}{M_m},$$

where X is the karat rating, and M_g and M_m are the mass of gold and the total mass, respectively. Also, $\rho_{Au} = 19\,200$ kg m^{-3} and $\rho_{Pb} = 14\,300$ kg m^{-3}.

The specific gravity of the oil in the drop ball viscometer is 0.8 and its viscosity equals $\mu = 10\,000 \pm 500$ cP. A persons pulse is used as a timer with a rate of 60 beats per minute and an uncertainty of 2 beats per minute. The falling ball drops a distance of 20 cm but the uncertainty of the measurement is ± 1 cm. The experimenta is repeated five times and the number of beats each time is: 21, 20, 23, 22, 22.

(a) What should be the speed of a pure gold bead?
(b) What is the uncertainty for the measure of speed of the pure gold?
(c) Calculate the density of a bead.
(d) Calculate the content of gold (in karats) in a bead.
(e) Is there less than 75% gold in the beads?
(f) The Governor wants to know with more certainty the amount of gold (or it is your skin that is in danger and not your brother-in-law's) and asks you to offer three ways to improve the measurement accuracy of the gold content with this device.

7.2 A Saybolt viscometer measures the viscosity of an oil with $\nu = 50$ cSt. What must the uncertainty in the measured time to drain be so that the uncertainty of ν is less than 1%.

7.3 You develop a new insulation (glass wool type) which has a thermal conductivity of about 0.05 W m^{-1} K^{-1}. The insulation is compressible— thus, while preparing the sample you must be careful not to move it. You place the sample, with a thickness of about 10 cm, and you apply a heat flux as the temperature difference across the sample is $(15.0 \pm 0.2)\,°C$.

(a) If the uncertainty of heat flow is 1%, what should the uncertainty in the measurement of the thickness of the material be, so that the thermal conductivity is less than $\pm 5\%$?

(b) If you double the heat flow, what would be the difference in temperature?

(c) If you notice a difference of 33 °C in (b), give two hypotheses explaining the difference between this and the one you calculated.

7.4 A cylindrical rod has a cross-section of 5.0 mm^2, which is made by joining a rod of 0.30 m silver to a rod of 0.12 m nickel. The silver side is maintained at a temperature of 290 K and the nickel side at 440 K. The thermal conductivity of silver is 0.42 kW m^{-1} K^{-1} and the thermal conductivity of nickel is 91 W m^{-1} K^{-1}. Calculate: *M. Hu*

(a) The steady-state temperature.

(b) The rate of conduction of heat down the rod (state any necessary hypotheses).

7.5 A rod is placed in an instrument to measure thermal conductivity, as shown in Figure 7.3. The thermal conductivity k_A of the calibration rod equals 0.5 W m^{-1} K^{-1}. Calculate k_B for the case where the lengths of both rods are equal and the temperatures at the heater, bath, and interface between the rods are $T_h = (40.00 \pm 0.02)$ °C, $T_b = (20.00 \pm 0.02)$ °C, and $T_{AB} = (22.50 \pm 0.05)$ °C. What is the uncertainty? *R. Boutrouka*

7.6 A gaseous mixture composed of CH$_4$ and O$_2$ is stored in a large room at 15 atm and -20 °C. Calculate the diffusion coefficient of oxygen in methane in the room. *É. Nguyen*

7.7 In the framework of a project on a process of fermentation, a chemical engineer focuses on heat transfer in a wall of his process which continuously circulates a refrigerant. The refrigerant is kept at a temperature of 25 °C and the temperature difference between coolant and the walls calculated by thermocouples indicates 52 °C. The sample used to estimate the heat transfer has a thickness of 15 mm and a diameter of 25 mm and the power applied to this sample is 7 kW. *É. Noiseux*

(a) Is it possible to determine the thermal conductivity?

(b) If the wall is insulated, determine the thermal conductivity.

7.8 What is the Reynold's number in a human capillary of 37 °C with a viscosity of 14×10^{-3} Pa s, a density of 1053 kg m^{-3}, and circulating 120 cm s^{-1}? A capillary has a radius of 4.5 μm. *A.L. Fakiris*

7.9 A Saybolt viscometer was used to calibrate the physical parameters of a falling ball viscometer. Methanol was chosen as the test fluid and its density equaled 792 kg m^{-3}. The drainage time in the Saybolt viscometer was 30.1 s, and it was 92.1 s in the falling ball viscometer. The diameter of the ball was 2 cm and the distance traveled was 0.3 m. The flow in the Saybolt viscometer was laminar: *L.M.K. Fondjo*

$$v = 0.00237t - \frac{1.93}{t},$$

where v is in 10^{-3} ft^2 s^{-1}.

(a) Determine the density of the ball.
(b) Explain (without calculating) how the students could determine the uncertainty of the calculated value in (a).

7.10 A parallel plate rheometer is used to characterize the viscosity of a molten polymer sample. The rheometer is made of two rectangular parallel plates of dimensions 60 cm × 10 cm that are separated by a 2 mm gap. The plates are located inside an oven to allow measurements at high temperature. The lower plate is stationary, while the upper plate is moving at a velocity $V = 10^{-3}$ m s^{-1}. For the sample investigated, the force required to maintain the upper plate velocity at 10^{-3} m s^{-1} is 100 N. *M.C. Heuzey*

(a) What is the shear rate $\dot{\gamma}$ applied on the polymer sample?
(b) What is the viscosity of the polymer sample?
(c) If the velocity of the upper plate is decreased by half, i.e. $V = 0.5 \times 10^{-3}$ m s^{-1}, the force decreases to 80 N. Is this behavior typical of molten polymers?

7.11 (a) Water and air are transported through two identical pipes at the same velocity. Which pipe operates at a lower Reynolds number?
(b) True or false, the advantages of an orifice compared to a venturi are:

(i) It takes less space.
(ii) It is easier to calibrate.
(iii) There is a permanent and smaller loss of charge.

(c) True or false:

(i) The loss of charge in a Coriolis is less than in a venturi.
(ii) The pitot tube can be used to measure the speed of a plane.
(iii) The vortex has the least charge loss, takes less space compared to most flow meters, and varies linearly with flow.

(d) Associate the terms in Table Q7.11 with their correct definition.
(e) True or false, the diffusion coefficient:

TABLE Q7.11 Terms and Definitions

Quantity	Definition
(1) N_{Re}	(A) Relationship between the diffusivity of the quantity of movement and the thermal diffusivity.
(2) N_{Pr}	(B) Relationship between the diffusivity of the quantity of movement and the mass diffusivity.
(3) N_{Sc}	(C) Relationship between the inertial forces and the viscous forces.
(4) N_{Ma}	(D) Ratio of the speed (fluid/object) to the speed of sound.

    (i) Increases with pressure.

    (ii) Increases with the square of temperature.

    (iii) Is higher for H_2O–N_2 than for H_2O–O_2.

    (iv) Varies more with temperature for NH_3 than oxygen.

  (f) Place the following materials in increasing order of thermal conductivity: gases, pure metals, alloys, liquids, and plastics.

  (g) True or false:

    (i) Helium is preferred to argon for window insulation.

    (ii) The heat flux equation by conduction is derived from Fourier's equation.

    (iii) Heat propagates in solids, fluids, and in void.

7.12 The thermal conductivity of a sample, shown in Figure Q7.12a, is sandwhiched between two blocks with known thermal conductivities. A constant heat flow of $Q = 10$ J s^{-1} is applied to the surface of plate A and the temperature of the outer surface of plate C is measured. The results of four experiments with the unknown sample.

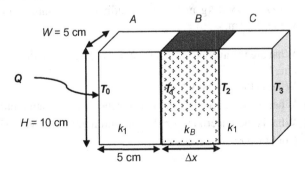

FIGURE Q7.12 Thermal Conductivity of Sample Setup

TABLE Q7.12a Thermal Conductivity of Sample B

Exp.	Δx (cm)	T_1 (°C)	T_3 (°C)
1	1	120	83
2	2	120	72
3	4	120	58
4	5	120	45

TABLE Q7.12b Thermal Conductivity and Interval of Confidence

α (%)	90	95	99	99.9
$t(\alpha, N), N = 4$	2.02	2.57	4.03	6.86
$K(\alpha)$	1.64	1.96	2.57	3.3

(a) Calculate T_2 in each of the experiments.

(b) Estimate k_B in each of the experiments.

(c) The uncertainty in the measurement of the temperature is 1 °C and that of the measurement of the thickness of plate B is 1 mm. What is the uncertainty in the value of the thermal conductivity of each sample B?

(d) Based on Table Q7.12b, calculate the interval of confidence of the estimated thermal conductivities when $\alpha = 95\%$. Comment on this value compared to those calculated in (c).

(e) Should we reject some of the experiments? Why?

7.13 You measure the viscosity of a new polymer in a falling ball viscometer. The density of the polymer is (970 ± 20) kg m^{-3}. The ball is made of titanium ($\rho = 4540$ kg m^{-3}) with a radius of 10.0 mm and it travels a distance of 0.40 m.

(a) What are the average and the standard deviation of the experimental data?

(b) What is the viscosity?

(c) What is the confidence interval for the time measurement?

(d) What is the uncertainty in the measurement of the viscosity?

(e) Could we reject a data point?

REFERENCES

Bird, R.B., Stewart, W.E., Lightfoot, E.N., 1960. Transport Phenomena. John Wiley & Sons.

Bird, R.B., Stewart, W.E., Lightfoot, E.N., 2006. Transport Phenomena. John Wiley & Sons.

McCoy, M., BASF hikes insulating material capacity, 2011. C&E News, 89 (46).

Morazain, J., L'avoin de 2025, 2011. Plan, Nov., 14–18.

Elsevier, 2007, December 18. Top 10 Advances in Materials Science Over Last 50 Years. ScienceDaily. Retrieved June 22, 2011, from: <http://www.sciencedaily.com/releases/2007/12/071218101208.htm>.

Holman, J.P., 2001. Experimental Methods for Engineers, seventh ed. McGraw-Hill, Inc., New York.

Karayianni, E., Coulston, G.W., Micka, T.A., 2008. Patent no. US2008176073-A1 and Patent no. US7665288-B2.

King, L.V., 1914. On the convection of heat from small cylinders in a stream of fluid, with applications to hot-wire anemometry. Philosphical Transactions of the Royal Society of London 214 (14), 373–433.

Marsh, K.N., Perkins, R.A., Ramires, L.V., 2002. Measurement and correlation of the thermal conductivity of propane from 86K to 600 K at pressure to 70MPa. Journal of Chemical and Engineering Data 47, 932–940.

McCabe, W.L., Smith, J.C., 1976. Unit Operations of Chemical Engineering, third ed. McGraw-Hill Chemical Engineering Series. McGraw Hill.

Patience, G.S., Hamdine, M., Senécal, K., Detuncq, B., 2011. Méthodes expérimentales et instrumentation en génie chimique, third ed. Presses Internationales Polytechnique.

Chapter 8

Gas and Liquid Concentration

8.1 OVERVIEW

Species concentration is a critical parameter for the control and safety of chemical processes, as well as dosing medication (pharmaceuticals) and mundane daily tasks like cooking; the most basic physiological function—breathing—relies on the carbon dioxide concentration in the blood. Air pollution is measured in terms of concentration of particulates or ozone. Mosquitoes seek their prey through a concentration gradient. The odor threshold of H_2S is 0.47 ppb. It irritates the eyes at 10–20 ppm and it is lethal at 800 ppm to 50% of the population when exposed for a period of 5 min. Properties of chemicals are listed on Material Safety Data Sheets (MSDS) and include physical characteristics—density, appearance, solubility, vapor pressure, boiling point, etc.—as well as toxicity and flammability characteristics. Toxicity is reported in terms of LC50 and LD50. LC50 represents the concentration at which 50% of the test specimens die after a period of time. The acute toxicity of isopropanol vapors (LC50) is $16\,000\ \mathrm{mg\,l^{-1}}$ in 8 h for rats. The LC50 for H_2S is 800 ppm after 5 min for humans. Another term used to express toxicity is LD50—the lethal dose at which 50% of the animal population dies. The acute oral toxicity (LD50) of isopropanol for mice is $3600\ \mathrm{mg\,kg^1}$ after a 4-h exposure. Its acute dermal toxicity (LD50) for rabbits is $12\,800\ \mathrm{mg\,kg^1}$ after a 4-h exposure.

In this chapter we discuss only a few of the more common instruments used to assess species concentration including chromatography, mass spectrometry, refractometry, spectroscopy, and X-ray.

8.2 CHROMATOGRAPHY

Co-authors: Danielle Béland, Cristian Neagoe

Experimental Methods and Instrumentation for Chemical Engineers. http://dx.doi.org/10.1016/B978-0-444-53804-8.00008-3

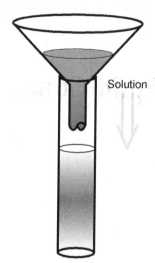

FIGURE 8.1 Example of Column Chromatography

Chromatography is a separation technique that was discovered by the Russian botanist Mikhail Semyonovich Tswett during the early 20th century (1903). He separated chlorophyll pigments dissolved in a mixture of petroleum ether and ethanol in a column of calcium carbonate (Figure 8.1). The different pigments with different degrees of affinity for calcium carbonate separated and eluted at different speeds. Chromatography is a contraction of two words from Greek: *chroma* (color) and *graphein* (to write).

Separating compounds in an unknown stream is achieved through their preferential adsorption (affinity), partition, on a column (stationary phase). The more affinity a compound has for the stationary phase, the more it will be retained, which results in a longer retention time. Compounds with low affinity elute first, followed sequentially by the compounds with a greater affinity. Retention time is the primary factor used to identify compounds: compounds eluting from the column are compared to a standard (using the same methodology). A further confirmation of the identity of the compound is to add a known quantity of the suspected compound together with the unknown stream to verify that the unknown peak increases. This verification is only required when matrix effects are suspected—the peaks may be shifted due to the presence of the other compounds in the stream. The third step of the process is to quantify the concentration, which is accomplished with external calibration, internal calibration, percent normalization, or standard addition.

Other types include column chromatography, paper chromatography, thin layer chromatography (TLC), high-performance liquid chromatography (HPLC), gas chromatography (GC), supercritical fluid chromatography (SFC)

TABLE 8.1 Different Forms of Chromatography

Stationary Phase	Mobile Phase	Type of Chromatography
Solid	Liquid	Liquid/Solid
Solid	Gas	Gas/Solid
Liquid	Liquid	Liquid/Liquid
Liquid	Gas	Gas/Liquid

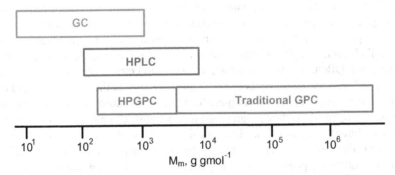

FIGURE 8.2 Applications of Chromatography

and size exclusion chromatography (SEC). The most common, which are treated in this chapter, are gas chromatography and high-performance liquid chromatography.

Chromatography is an analytical technique used for the separation, identification, and quantification of different compounds in a mixture. The separation is the result of the interaction between the different compounds with a stationary phase (liquid or solid) and the mobile gas phase (GC) or liquid phase (HPLC) (Table 8.1).

Applications of chromatography are given in Figure 8.2. Volatile and semi-volatile compounds are analyzed by GC and non-volatiles by HPLC. Sometimes the process of derivatization is used to increase the volatility of the compounds so they can be analyzed by GC. HPLC can also be used for thermolabile compounds and ionized products. SEC separates analytes based on physical dimensions (hydrodynamic volume or size), whereas the other techniques rely on chemical and physical interactions between the analyte and column. The distribution of molecular weight of polymers and protiens are determined by Gel Permeation Chromatography (GPC—a class of SEC). High Performance Gel Permeation Chromatography (HPGPC) is applicable for compounds with molecular weights on the same order as HPLC.

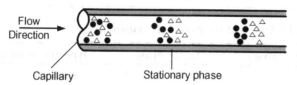

FIGURE 8.3 Separation of Two Compounds

The applications of chromatography are vast and with few restrictions. It plays an important role in the research and development of new molecules in the food industry and in the pharmaceutical industry—especially with respect to purity. It is also very useful for quality control in commercial processes (intermediates), product qualification, environmental monitoring, purification, and so on.

All chromatographs are equipped with an injection system, a column, and a detector. Often, multiple detectors, columns, and even injection systems are equipped with the instruments either as an obligation of the desired method or for multi-purpose applications. Figure 8.3 is a cartoon illustrating the separation of two compounds as they progress along a column. The mobile phase may be either a liquid or a gas and has little affinity for the stationary phase (and is not shown). The molecules represented by the open triangles have a lower affinity for the stationary phase. At the "beginning" of the column, the symbols are completely intermingled. In the middle of the figure, the molecules represented by the filled circles are further to the left. At the far right of the figure, the open triangles lead the filled circles entirely—perfect separation. With the help of a detector we obtain a chromatogram that is plot of an electrical signal (μV) as a function of time.

A chromatogram of the essential oil from the plant nepeta cataria as measured by a gas chromatograph is shown in Figure 8.4. The time interval from 22 min to 30 min is shown. The chromatogram consists of many minor peaks throughout the interval and three major peaks in the vicinity of 25–30 min. For the same class of molecules, the height of the peak generally correlates with concentration for the same detector. These three large peaks indicate that their concentrations are higher than the minor peaks. In HPLC, some detectors are deceiving: the detection efficiency is greater for double and triple bonds, by as much as a factor of 10! This is possible for GCs as well. For example, the sulfur response for a pulse flame photodetector is quadratic with concentration while it is linear for phosphorous. So, this must be taken into consideration when preparing samples for analysis because of the possibility to saturate the detector.

8.2.1 The Distribution Coefficient

In partition chromatography, the sample molecules are carried along with the mobile phase (liquid or gas) and adsorb, then desorb, as they proceed down

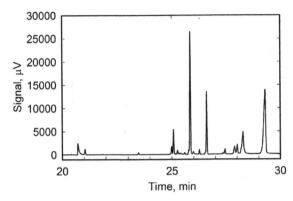

FIGURE 8.4 Chromatogram of the Essential Oil of *Nepeta cataria*

the column. The longer the molecule remains adsorbed on the stationary phase, the longer is its retention time. The equilibrium between the concentration in the mobile phase, C_m, and in the stationary phase, C_s, may be expressed as a ratio, which is known as the distribution coefficient, K_D:

$$C_m \longleftrightarrow C_s,$$
$$K_D = \frac{C_s}{C_m}. \tag{8.1}$$

The retention times of each solute can be determined for each column knowing the distribution coefficient. The concentration in the mobile phase is always much smaller than in the stationary phase $(C_m \ll C_s)$ so $K_D \gg 1$. Greater values of the distribution coefficient indicate greater solute retention and longer retention times. The value of K_D must be different for two compounds to be able to separate them.

The separation depends on several criteria:

1. Nature of the solute.
2. Nature of the stationary phase.
3. Temperature of the column.
4. Amount of stationary phase.
5. Linear speed of the mobile phase.
6. Column length.

8.2.2 The Capacity Factor

The capacity factor, or retention factor, k', expresses the speed of the passage of solute through the column. The k' factor determines the residence time of the solute in the stationary phase compared to the residence time in the

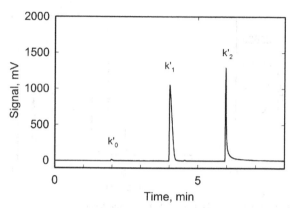

FIGURE 8.5 The Capacity Factor, k'

mobile phase. It characterizes the equilibrium of the solute between the stationary phase and the mobile phase and is assigned a value in relation to a reference (Patience et al., 2011). The retention factor of a component that is unretained is taken as the reference and denoted as $k' = 0$. A compound that exits the column at twice the time of the reference has a retention factor of $k' = 1$ and it is $k' = 2$ for molecules in which the time differential between the retained compound and the reference is twice the time of the reference. It is expressed by:

$$k' = \frac{t_R - t_0}{t_0}.$$
(8.2)

where t_R is the retention time for a known compound and t_0 is the retention time for an unretained compound (see Figure 8.5).

A low value of k' indicates that the solute is less retained in the column. Figure 8.6 illustrates a chromatogram in which the reference compound (unretained) elutes at 1.86 min (t_0). The first peak after that elutes at 2.16 min (t_R), thus the capacity factor is 0.3/1.86 = 0.16. The largest peak elutes at 2.60 min and thus its capacity factor equals 0.40. Different solutes will have different retention times depending on their polarity and the polarity of the stationary phase. Hydrocarbons (in the same family) will have increasing affinity with the stationary phase with increasing carbon number but permanent gases may have more or less affinity than methane, for example.

The velocity of the reference (and thus the mobile phase), u, is simply the length of the column divided by the retention time, t_0:

$$u = \frac{L}{t_0}.$$
(8.3)

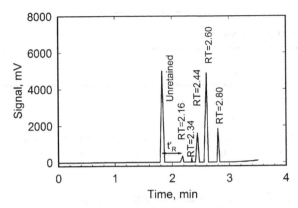

FIGURE 8.6 Reference (Unretained) Compound and Capacity Factor

The principal parameters that affect retention are:

1. Polarity of the stationary phase and the solute.
2. Temperature of the column (pertinent for gas chromatography).
3. Quantity of stationary phase.
4. Linear speed of the mobile phase.
5. Length of the column.

8.2.3 The Selectivity Factor

The degree of separation between successive peaks (of retained compounds) is known as the selectivity factor, α, or the relative retention. This is one of the most critical factors in chromatography: the selectivity factor should be great enough so that each peak is sufficiently resolved (i.e. there are no overlaps at the baseline). However, if the relative retention factors are too great, then the GC will operate inefficiently—the runs will be unnecessarily long. The selectivity factor is calculated as the ratio of the capacity factors and these are usually concerned with adjacent peaks:

$$\alpha_{n+a,n} = \frac{k'_{n+a}}{k'_n}. \tag{8.4}$$

The nomenclature concerning the subscripts is arbitrary except that the selectivity factor is always greater than 1: $\alpha_{4,3}$ would be the selectivity factor of the fourth peak versus the third peak and $\alpha_{6,3}$ would be that of the sixth peak versus the third peak.

Example 8.1. Calculate the selectivity factors for each successive peak in Figure 8.6. The retention time t_R for each peak starting from the leftmost peak is (in min): 2.16, 2.34, 2.44, 2.60, and 2.80.

TABLE E8.1 Selectivity Factors

$n+1,n$	2, 1	3, 2	4, 3	5, 4
$\alpha_{n+1,n}$	1.083	1.043	1.066	1.077

Solution 8.1. The peaks in Figure 8.6 are well resolved: there are no overlaps, and each one reaches the baseline. The selectivity factors are shown in Table E8.1.

The principal parameters that affect selectivity are:

1. Nature of the stationary phase.
2. Temperature of the column (in GC mostly).
3. Nature of the mobile phase (in HPLC).
4. Speed of the mobile phase.

8.2.4 The Number of Theoretical Plates

Theoretical plate theory was born from a static model that recognizes the similitude between the operation of a distillation column and a chromatographic column. Chromatographic efficiency, which depends on peak broadening, is expressed by the number of theoretical plates. This number characterizes the dispersion of all the molecules in the column. On each theoretical plate, equilibrium is instantaneous (distribution in fractions). The coefficient K_D is applied. In theory, each peak in the chromatogram represents the distribution of concentrations of a compound and is expressed in the form of a Gaussian curve. The calculations are statistical in nature and describe the shape of the peak. If we use the geometric characteristics of a Gaussian curve, we can deduce a column's number of theoretical plates for a given solute. In Figure 8.7, we calculate the number of theoretical plates of the tallest peak and the width (in min): The height of the peak is 5000 mV and the width, $w_{1/2}$, at 2500 mV is 0.01 min. The number of theoretical plates, N_{th}, is the square of the ratio of the retention time of the component and the width of the peak at half its height:

$$N_{th} = 5.54 \left(\frac{t_R}{w_{1/2}} \right)^2. \tag{8.5}$$

In the case of Figure 8.7, the number of theoretical plates is:

$$N_{th} = 5.54 \left(\frac{2.60 \ \text{min}}{0.01 \ \text{min}} \right)^2 = 3750.$$

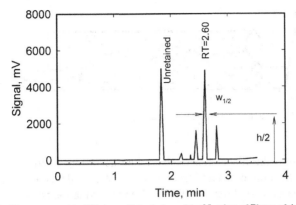

FIGURE 8.7 Chromatographic Efficiency Calculation of the Number of Plates of the Peak at Half of its Height

The principal parameters that affect efficiency are:

1. Geometry of the column—length, inside diameter.
2. Diffusion coefficient in the mobile phase.
3. Diffusion coefficient in the stationary phase.
4. Capacity factor.
5. Linear speed of the mobile phase.
6. Quantity of the stationary phase.

When we want to compare columns of different lengths, we use the height equivalent of a theoretical plate HETP (the plate number is generally expressed by μ):

$$\text{HETP} = \frac{L}{N_{\text{th}}}, \tag{8.6}$$

where L is the length of the column in cm and HETP is the height equivalent to a theoretical plate in cm.

The efficiency of a column is affected by a number of variables and many theories have been proposed to characterize the relationship between the variables and the height equivalent to a theoretical plate. The most widely used expression is that proposed by Van Deemter, which takes the following form:

$$\text{HETP} = A + \frac{B}{u} + Cu, \tag{8.7}$$

where A is the heterogeneous path length and represents eddy diffusion (absent in capillary columns), B is the term for longitudinal diffusion, C is the resistance to mass transfer, and u is the velocity of the mobile phase.

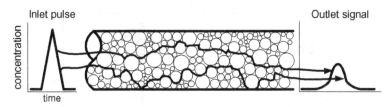

FIGURE 8.8 Dispersion of the Solute in the Column

It describes the dispersion of the solute according to the average linear velocity of the mobile phase. The optimum efficiency depends on an optimum velocity. The expanded form of the Van Deemter equation is:

$$\text{HETP} = 2\phi d_p + \frac{2\phi D_{\text{gas}}}{u} + \frac{8k'd_f^2}{\pi^2(1+k')^2 D_{\text{liq}}} u, \qquad (8.8)$$

where ϕ is the particle shape factor, d_p is the particle diameter, D_{gas} is the diffusion coefficient of the mobile phase, d_f is the film thickness, and D_{liq} is the diffusion coefficient of the stationary phase.

8.2.5 Eddy Diffusion

The eddy diffusion term, A, describes the effect of peak broadening caused by the presence of particles in the column. It exists only for packed columns. Because of the particles, the molecules travel different paths thus their elution is carried out at different times, as illustrated in Figure 8.8. It depends on the particle diameter, sphericity, and how the column is packed. Eddy diffusion is independent of the gas velocity vector, $\text{HETP} = A$. The initial peak as it enters the column is narrow and taller. As it exits the column the peak becomes much broader and the height decreases.

8.2.6 Longitudinal Diffusion

The term B relates to the diffusion of molecules in the mobile phase. The molecules have their own movement that is independent of the flow rate of the mobile phase and this movement is not restricted to one direction (Figure 8.9). The diffusion rate is determined by the type of molecule, the nature of the mobile phase, and the temperature. We can say that almost all of the same

FIGURE 8.9 Dispersion of the Solute According to its Diffusion Coefficient in the Mobile Phases

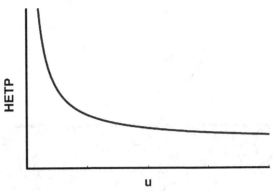

FIGURE 8.10 HETP $\propto \frac{B}{u}$

solute molecules leave the column at the same time. The higher the velocity of the mobile phase, the less apparent the effect is. This term is important when the mobile phase is a gas. B is influenced by the velocity of the mobile phase: HETP $\propto \frac{B}{u}$, as shown in Figure 8.10. This parameter only has a minor influence in HPLC.

8.2.7 Resistance to Mass Transfer

The term C is associated with the mass transfer of molecules between the mobile phase, C_m, and the stationary phase, C_s. The molecules are delayed in the column due to their interaction with the stationary phase. The term C expresses the resistance of the solute molecules between the fluid phase and the stationary phase—often this region through which the molecules diffuse is referred to as the boundary layer, as shown in Figure 8.11. This phase shift increases with increasing velocity: it is the result of the limitation of the kinetics of the adsorption-desorption process. The peak profile resulting from the resistance to mass transfer is directly proportional to the velocity of the mobile phase HETP $= Cu$ (Figure 8.12).

The curve resulting from the combination of parameters A, B, and C is hyperbolic and passes through a minimum corresponding to the optimum flow rate (Figure 8.13).

8.2.8 Resolution

When developing a chromatographic method, it is important to optimize all parameters in order to achieve optimum separation, while at the same time minimizing analysis time. The degree of separation between two adjacent peaks is characterized by its resolution, R, the ratio of the difference in retention times

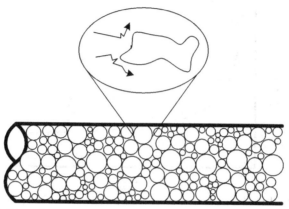

FIGURE 8.11 Dispersion of Solute between the Two Phases, Dynamic Equilibrium

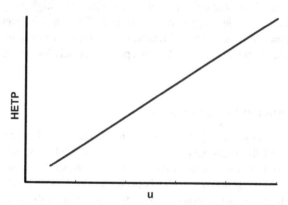

FIGURE 8.12 $HETP = Cu$

and the sum of the widths of the peaks at half the total height:

$$R = 1.177 \frac{t_{R2} - t_{R1}}{W_{1,1/2} + W_{2,1/2}}. \tag{8.9}$$

The resolution increases as the difference in retention times increases and it also increases for very narrow peaks. The following example demonstrates the relationship between the resolution, R, and the peak shape. At a resolution of 0.4, peaks A and B are entirely indistinguishable. At a value of 0.6, two shoulders appear at the top of the peak. The separation becomes more noticeable at 0.8 and 1. Finally, at approximately 1.25, the peaks are almost entirely resolved (see Figure 8.14).

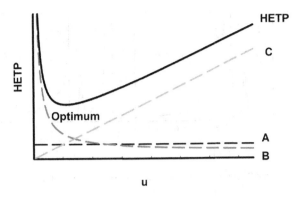

FIGURE 8.13 Curve Resulting from the Combination of Parameters A, B, and C

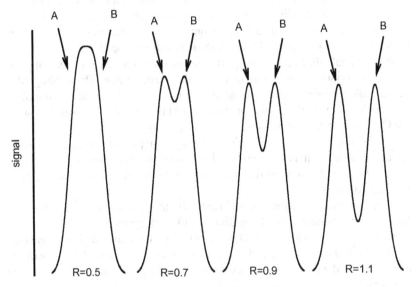

FIGURE 8.14 Value of the Resolution Related to the Degree of Separation

- If $R \leqslant 1.25$: no separation at the baseline.
- If $1.25 < R < 1.5$: separation depends on the symmetry of the peaks.
- If $R \geqslant 1.5$: separation at the baseline.

GC method development involves defining the most efficient columns for separation, length, and appropriate temperature and fluid velocity. Also, the injection method as well as the configuration of the columns if more than one

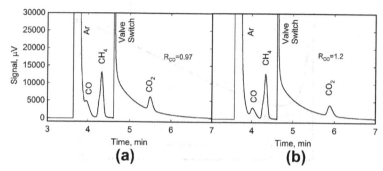

FIGURE 8.15   Column Performance (a) Before and (b) After Conditioning

is required. However, the peaks may initially be resolved but the resolution decreases with time; this could be due to a reversible deactivation of the column. A technique used to restore the selectivity and resolution of the column (or maintain it) is to "condition" the columns periodically. Conditioning involves heating the column to its maximum recommended temperature, which is often as much as 250 °C or more. Figure 8.15 demonstrates the change in resolution of the CO peak after conditioning. Before conditioning, the CO peak appears as a short, poorly defined shoulder after the valve switch. The leading edge of the peak approaches the baseline after conditioning. It would be difficult to accurately calculate the resolution with respect to methane (the peak following the CO) but it would be possible after conditioning.

Chromatography is an analytical technique that is now widespread in all areas of research. It is not only used by chemists, but also by biochemists, biologists, and chemical engineers. Many of them utilize chromatography in their line of work.

The previous sections introduced all the parameters that are the basis of chromatography. In addition, different types of chromatography were mentioned such as gas chromatography (GC) and high-performance liquid chromatography (HPLC). In this section GC and HPLC will be explained in depth to better understand their implications in research and production.

Together with the equation to calculate resolution, shown above, another expression is used for poorly resolved peaks that relies on the efficiency (number of plates, N_{th}), the capacity, and the selectivity:

$$R = \frac{\sqrt{N_{th}}}{4} \times \frac{k'}{k' + 1} \times \frac{\alpha - 1}{\alpha}. \qquad (8.10)$$

This expression is preferred over the one involving retention times and peak widths.

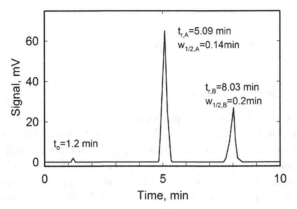

FIGURE E8.2 Graphical Analysis of the Complex Mixture

Example 8.2. Based on the chromatogram in Figure E8.2, calculate k', α, N_{th}, R, and HETP. The column length is 15 cm × 4.6 mm id.

Solution 8.2. k':

$$k' = \frac{t_R - t_0}{t_0},$$

$$k'_A = \frac{5.09 \text{ min} - 1.2 \text{ min}}{1.2 \text{ min}} = 3.24,$$

$$k'_B = \frac{8.03 \text{ min} - 1.2 \text{ min}}{1.2 \text{ min}} = 5.69.$$

α:

$$\alpha = \frac{k'_B}{k'_A}, \quad \alpha = \frac{5.69}{3.24} = 1.76.$$

N_{th}:

$$N_{th} = 5.54 \left(\frac{t_R}{W_{1/2}}\right)^2, \quad N_{th} = 5.54 \left(\frac{5.09 \text{ min}}{0.14 \text{ min}}\right)^2 = 7323.$$

R:

$$R = 1.177 \frac{t_{R2} - t_{R1}}{W_{1,1/2} + W_{2,1/2}}, \quad R = 1.177 \frac{8.03 \text{ min} - 5.09 \text{ min}}{0.14 \text{ min} + 0.20 \text{ min}} = 10.2.$$

HETP :

$$\text{HETP} = \frac{L}{N_{\text{th}}}, \quad \text{HETP} = \frac{15 \text{ cm}}{7323} = 0.002 \text{ cm}.$$

8.2.9 Gas Chromatography

Gas chromatography is useful in the pharmaceutical, food, environmental, and petrochemical fields as well as others. It is used for the separation of volatile and semi-volatile compounds. In gas chromatography, the compounds of interest are in a gaseous state or can be vaporized upon their introduction in the gas chromatograph. The mobile phase is a gas: the most commonly used are helium (He), hydrogen (H2), argon (Ar) and nitrogen (N2). We call the mobile phase the carrier gas. The choice of carrier gas depends on the detector (detectors will be discussed later), the separation efficiency, and the speed of analysis.

It is important to set the proper gas pressure or carrier gas velocity for the analysis. If the gas velocity in the column is too high, the compounds will have insufficient time to interact with the stationary phase, which will result in a poor resolution of the peaks. On the other hand, if the carrier gas velocity is too low, the separation will be maximized but the total time to perform the analysis will be prohibitively long. We need to select an appropriate carrier gas and velocity that will give the best separation in the least amount of time.

In gas chromatography, the parameters that must be optimized include physical and chemical properties, such as the temperature and the polarity of the molecule, column parameters and the polarity of the stationary phase. Finally, we must choose a detector and the appropriate injection system.

In capillary GC, the column length varies from 10 m to 100 m with an internal diameter as little as 1mm. The stationary phase can be liquid or solid. Generally there are three types of GC columns: WCOT (wall-coated open tubular), SCOT (support-coated open tubular), and PLOT (porous-layer open tubular). Figure 8.16 illustrates the three types.

These three column types illustrate possible configurations of the stationary phase. Together with the column configuration, we also choose the polarity of the stationary phase according to the type of molecules to be separated. For nonpolar compounds (hydrocarbons for example), the type of stationary phase we can use is a polymer called dimethyl polysiloxane or diphenyl polysiloxane. For compounds with a slight polarity, the type of stationary phase could be cyanopropylphenyl dimethyl polysiloxane. For polar compounds, two choices are possible: polyethylene glycol (often called Carbowax) or biscyanopropyl, cyanopropylphenyl polysiloxane. It is possible to choose the temperature at which a mixture of compounds will be analyzed. However, each column will have a different maximum operating temperature. When possible, it is best to use the column at 30 °C below its maximum allowable temperature in order to increase its lifetime. Using a column at its maximum operating temperature or even at a higher temperature will cause the column to deteriorate rapidly.

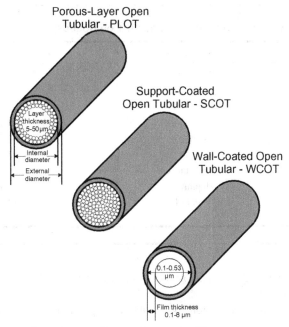

FIGURE 8.16 The Three Common Types of Columns for GC

The first compounds detected are often the most volatile as well as those of low polarity—the interaction with the stationary phase generally increases with polarity. It is often hard to resolve nonpolar peaks efficiently—Ar and N_2 is an example—when sampling air; unless specifically considered and designed for, the Ar peak will be masked by the nitrogen. (This could introduce a systematic error of 1% or more if not accounted for when the nitrogen is calibrated.) The order in which the compounds appear in the chromatogram is from the most volatile to the least volatile but there are exceptions. It is also important to consider the nature of the compounds and the type of stationary phase of the column. Take as an example the 10 compounds in Table 8.2.

It is possible to separate these compounds by GC. But the order in which the compounds are detected is not from most volatile to least volatile. If a nonpolar column (polydimethyl siloxane) is used, the order will be: 1, 2, 3, 4, 5, 6, 7, 8, 9, and 10.

If we use a column with a different polarity a polyethylene glycol column for example, then the order will change to: 2, 5, 8, 10, 1, 4, 7, 3, 6, and 9. So, it is important to be careful with the properties of the stationary phase and the polarity of the molecules to be separated.

The sample can be injected into the instrument in a liquid or gaseous form. In the case of a gas, we can directly introduce the sample into the column.

TABLE 8.2 Boiling Points of Ten Organic Compounds

	Substance	Boiling Point (°C)
1	Acetone	56
2	Pentane	36
3	Propanol	97
4	MEK	80
5	Hexane	69
6	Butanol	117
7	3-Pentanone	102
8	Heptane	98
9	Pentanol	136
10	Octane	126

In the case of a liquid sample, it is necessary to vaporize the liquid by using the proper injector type.

The three most common detectors for qualitative and quantitative analysis are:

1. Flame ionization detector (FID).
2. Flame photometric detector (FPD).
3. Thermal conductivity detector (TCD).

The flame ionization detector is one of the most commonly used detectors and is ideally suited for hydrocarbons and other flammable compounds. Although the flame may ionize inert compounds, the sensitivity is too low. Flame ionization is insensitive to H_2O, CO_2, CO, SO_2, CS_2, and NO_x and all noble gases. The chemical conductivity of the flame with the carrier gas (typically helium) and the detector gases—hydrogen and air (as the oxygen supply)—is essentially equal to zero. When a flammable compound enters the plasma produced by the hydrogen-air flame, the temperature is sufficiently high to pyrolyze the compound and produces electrons and positively charged ions resulting in a sharp increase in conductivity. The change in conductivity is measured using two electrodes: the nozzle head where the flame is produced is itself the cathode, and the anode (collector-plated) is positioned above the flame. The ions hit the collector plate inducing an electrical current, which is measured by a picoammeter. The sensitivity of the signal is approximately proportional to the reduced carbon atoms; therefore, the signal is proportional to mass and not concentration—the mass of the carbon ions. The FID detector has a high sensitivity and a wide linear range. However, the signal is lower for oxidized carbon compounds—functional groups such as alcohols, acids, carbonyls, amines, halogens, and others: the peak height of ethanol will be lower than that of ethane at the same concentration. Another disadvantage of FID detectors is that the sample is destroyed during its passage through the flame.

Flame photometric detectors are useful to analyze air and water pollutants and pesticides. It is specific and is mainly used for the analysis of sulfur- and phosphorus-containing compounds. When the effluent gas comes into contact with the flame, the phosphorus compounds are transformed into HPO and the sulfur compounds to S_2. Both HPO and S_2 will emit at different wavelengths. Depending on the filter used we will detect either the phosphorus or sulfur compounds. The light emitted will pass through a photomultiplier and the resulting current is then recorded by a data acquisition system. Depending on the detector types it can also analyze compounds containing nitrogen and metals like chrome or tin.

The thermal conductivity detector is a universal detector as it can be applied to organic and inorganic species. It operates by comparing the electrical conductivity of the effluent from the GC columns with that of a reference gas—the carrier gas. Because of their high thermal conductivity, either helium or hydrogen is used as a carrier gas. The carrier gas passes over an electrically heated filament made of platinum, tungsten, or nickel and operating at a specific temperature. Simultaneously, the effluent from the GC column passes over another electrically heated filament. As the compounds from the GC column elute, the thermal conductivity of the gas stream decreases, thus changing the thermal conductivity and the resistance of the filament. The lower conductivity causes the filament temperature to rise since less heat is carried away—hydrogen and helium have very high thermal conductivities compared to other gases. The change in temperature, and thus resistance, is sensed by a Wheatstone bridge circuit resulting in a voltage recorded in μV. Many organic compounds have a similar thermal conductivity, thus peak areas are comparable and the concentrations can be estimated by the relative ratios of the peak area. The TCD is less sensitive than the FID detector and has a lower dynamic range; however, one important advantage is that it is a non-destructive technique so the effluent may be analyzed with other instruments after the TCD. A mass spectrometer may be installed after the TCD—this instrument is referred to as a GC-MS.

8.2.10 High-Performance Liquid Chromatography (HPLC)

Gas chromatography is applicable for compounds with a reasonably high volatility or molecules that are insensitive to thermal decomposition at temperatures below 200–300 °C. HPLC instruments operate in the liquid phase and are particularly suited for compounds with poor thermal stability or low volatility. They may also be used for organic compounds that are volatile. HPLC is very useful for the pharmaceutical and biological industry to separate ionic compounds, polymers, and proteins. In this technique, the mobile phase and the compounds are liquids. To perform the separation of various compounds, we need to use high-pressure pumps and short columns packed with particles from 3 μm to 7 μm. The particles are silica based with different functional groups that allow different column polarities. The resolution increases with decreasing particle size. The column length varies from 5 cm to 30 cm and has an internal diameter from 1 mm to 5 mm. Compared to gas chromatography, liquid chromatography will show a lower decrease in efficiency when flow rate is increased. The number of theoretical plates for an HPLC column is calculated according to the following equation:

$$N_{th} = \frac{3500\,L}{d_p},\qquad(8.11)$$

where L is the column length in cm and d_p is the particle diameter in μm.

The number of theoretical plates is proportional to the column length and inversely proportional to the particle size. The advantage of using small particles is that they distribute flow more uniformly and, as a result, reduce the eddy diffusion, term A in the Van Deemter equation. However, the smaller particles increase the diffusional resistance of the solvent as well as the pressure drop (for a given flow rate). Choosing the flow rate is a critical parameter in developing an HPLC method. Low flow rates allow the analyte sufficient time to interact with the stationary phase and will affect both the B and C terms of the Van Deemter equation.

Compared with gas chromatography, the mobile phase is a liquid whose properties are defined by the analytes. The mobile phase consists of a single solvent or solvent mixture to adjust the polarity in order to optimize the analysis conditions.

8.2.11 Method Development

Approaching method development systematically and methodically is critical to successfully using chromatography. Chromatographs are quite robust when maintained properly but frequent maintenance is helpful to minimize experimental errors and achieve a high level of productivity. This section is very important because it outlines a systematic analytical method that can be applied to other instrumental techniques in chemistry.

First, determine the conditions of the preliminary analysis:

- Type of substance.
- Type of column.
- Type of detector.
- Type of injection.

After, you must perform the following steps:

1. With the chosen analytical instrument, determine the limit of detection (LOD): the lowest concentration injected into the instrument for which it is possible to define the presence of a compound in the sample. The signal-to-noise ratio should be greater than about 3. With modern instruments, this calculation is performed automatically by the software. Background noise is defined as the electronic noise of the various components of the instrument. In Figures 8.6 and 8.7, the signal of the highest peak is $5\,000\,000$ μV and the smallest peak is less than $100\,000$ μV. As long as the noise—the baseline fluctuations as measured in μV—is less than $30\,000$ μV, the signal is considered significant or quantifiable. The CO peak height in the chromatogram of Figure 8.15 is at approximately 4000 μV. The baseline fluctuations are indiscernible and so the signal-to-noise ratio is much greater than 100. Background noise should be minimized to achieve the lowest detection limit, otherwise higher concentrations of the solutes or larger volumes are required. Note that background noise is random and changes from week to week, from day to day, and even from hour to hour.

2. Determine the quantification limit which is the smallest concentration of a compound in the sample that gives a detectable signal reproducibly.

3. Calibrate the instrument.
   Calibration is essential for quantitative results, which relies on generating a calibration curve using standard solutions of known concentration. A typical curve will plot the concentration versus the area or height obtained, which is used to determine the linearity of the detector. It is preferable to perform a five-point calibration curve repeated two or three times. If necessary, it can be reduced to three points if the signal is linear in the range of interest.
   The limit of linearity is the concentration at which the behavior of the detector becomes nonlinear—successive increases in concentration result in a lower than proportional increase in the signal. If the concentrations of the samples or standards are too high, then they must be diluted by a factor of between 10 and 1000 or more to achieve a linear detector response. It is important to calculate the error on each point and the overall error of the curve.

4. Perform the analysis of the different samples.
   When analyzing a series of samples with known concentrations, the order of analysis should go from the least concentrated to the most concentrated. For unknown samples, a blank test (injection of solvent only) should be run after each successive sample to ensure that there is no carryover from the previous injection.

5. Derive the concentration and the analytical error from the calibration curve.

6. Document the results.

8.3 MASS SPECTROMETRY

Co-author: Patrice Perrault

Mass spectrometry is a powerful technique to identify atomic masses of compounds, elemental composition as well as chemical structures. The compound is first ionized to generate charged particles. The particles are separated in an applied electromagnetic field based on their mass-to-charge ratio (m/z) and they are subsequently detected. Ions may be generated from atoms, molecules, clusters, radicals, and/or zwitterions by high energy impacting electrons formed by an electrically heated filament—electron ionization—or by chemical-ion reactions—chemical ionization. Among several other ionization techniques, inductively coupled plasma—ICP—is commonly applied to generate cations. Electrically neutral plasma strips atoms of their outer electrons after the compound has been atomized by the high temperatures. Chemical ionization is a complex phenomenon in which positive and negative radical ions can be generated, and in which rearrangements of the resulting radicals can also occur. Ions can also be subject to isomerisation, dissociation, and various other pairings.

Analyte separation follows classical electromagnetic laws. As shown in Figure 8.17, a magnetic field B is applied perpendicularly to a charged particle, q, with an initial velocity, v. The force deflects the moving ions and the magnitude of the deflection depends on the mass-to-charge ratio.

The major mechanical components of a mass spectrometer (MS) include a sample inlet, an ion source, a mass analyser, and a detector. While the sample inlet is typically open to atmosphere, all other constituents are operated under high vacuum (on the order of 10^{-6} mbar), generated by a turbomolecular pump. The high vacuum is necessary to avoid bimolecular interactions between ions.

The sample inlet is constituted of a heated fused silica capillary, which is maintained at approximately 200 °C and is encased in a flexible tube. The ion source, in the case of electronic ionization, is composed of electrically heated metallic filaments. Mass analyzers, separating the analytes, include: time-of-flight (TOF), linear quadrupole (Q), linear quadrupole ion trap (LIT), quadrupole ion trap (QIT), Fourier transform ion cyclotron resonance (FT-ICR), etc. These detectors differ in their capacity to treat ion beams in a continuous or pulsed (TOF). Quadrupole mass analyzers stabilize and destabilize the ion paths with an oscillating electrical field. A "triple quad" is more recent technology and consists of three quadrupole stages. Quadrupole ion traps will sequentially eject ions that have been trapped in a ring electrode between two endcap electrodes.

MS is used for both quantitative and qualitative analysis (principally identification). Figure 8.18 is an analysis of air. The most abundant species is detected at an amu (atomic mass unit) equal to 28 and its value was 2.87×10^{-9}, which represents nitrogen. The second most abundant peak is at amu 32 at a value of 0.63×10^{-9}—oxygen. These values are representative of partial

FIGURE 8.17   Particle Trajectory Deflection in an Applied Magnetic Field

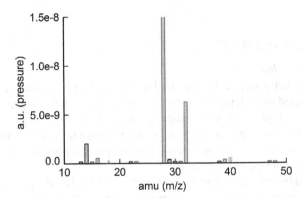

FIGURE 8.18   Mass Spectrum of Air

pressure (but require a calibration factor). Smaller peaks are recorded at amu 40 (argon) and 18 (water) and 44 (CO_2). Other peaks that represent fragments of molecules or atoms are also evident—amu 14 (nitrogen), amu 16 (oxygen), as well as amu 29, 30, 31, 38, and 39.

Quantitative on-line analysis is difficult without a precise calibration procedure. Furthermore, frequent calibration may be required to correct for drift. Calibration is achieved by analyzing a mixture of a known composition comparable to the similar sample of interest from which the instrument's sensitivity for each compound is evaluated. The sensitivity is a measure of the overall response of the instrument for a species when operated under well-defined conditions. As such, the sensitivity depends on species concentration, as well as on the other analytes' concentrations. For pure species, the sensitivity is an asymptotic function of the species' concentration. Process applications will often use mass spectrometry particularly to monitor gas phase compositions that are susceptible to the explosion: for selective hydrocarbon oxidation reactions, operating close to the explosion envelope may be desirable to achieve high production rates. Sampling the gas phase at a high frequency ($> 2\,Hz$) can minimize the hazard when the signal is automatically fed to the control system. When the concentration exceeds a threshold value (or even drops below a value), the interlock will trigger a pre-designated sequence of responses that may include reducing or shutting off the hydrocarbon flow (or oxygen), purging with inert, etc.

Some practical difficulties arise when analyzing mixtures of CO and N_2, both of which have the same amu. The intensity at the amu 28 is a contribution from both and differentiating one from the other is difficult. Secondary fragments may be used to quantify CO but it is more convenient, at the experimental scale, to use argon rather than nitrogen as an inert. For mixtures of compounds a GC installed upstream is useful such that the MS analyzes one compound at a time and thus overlapping is minimized.

8.4 REFRACTOMETRY

Co-author: Amina Benamer

Refractometry is a technique used to detect the concentration of binary mixtures based on differences in their refractive index. Besides concentration, it is also a simple method to quantify purity. Salinity of brine or sucrose concentrations in grapes or syrup are two typical applications. Urine-specific gravity is measured for drug diagnostics and plasma protein is assessed by refractometry in veterinary medicine.

Ernst Abbe first described the technique in 1874. His device consisted of placing a sample between two glass prisms. Incident light (monochromatic—typically 589 nm) is introduced from the bottom of the "illuminating prism." It passes through the sample and then is refracted before entering the measuring or refracting prism. When light passes from one object to another (for which the densities are unequal), it changes direction: the speed of light decreases with increasing density and this causes it to bend (Klvana et al., 2010, Florentin, 2004). The index of refraction, n, is defined as the ratio of the speed of light in vacuum, C, to the speed of light in the medium, v:

$$n = C/v.$$

Figure 8.19 demonstrates the change in angle as an incident ray travels from medium 1, with a refractive index n_1, to a denser medium 2 and a higher refractive index, n_2 (Klvana et al., 2010). The beam bends upwards with a smaller angle of incidence closer to the normal of the plane. If medium 1 were denser than medium 2, the angle of incidence would have been greater and the light beam would bend away from the normal plane. The relationship between the angle of incidence and of refraction is based on the Snell-Descartes law:

$$n_1 \sin \theta_1 = n_p \sin \theta_p,$$

where n_1 is the refractive index of the medium of interest and n_p is the refractive index of the prism. Generally, flint glass is chosen as the reference medium (prism) with a refractive index, n_p, equal to 1.7.

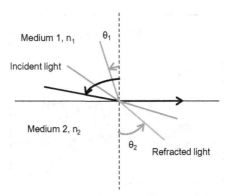

FIGURE 8.19 Angle of Incident Light

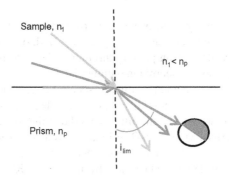

FIGURE 8.20 The Limiting Angle of Refracted Light

For the case in which the incident angle from the sample is at the limit ($\theta_1 = 90°$), the refracted angle equals $\arcsin\left(\frac{n_1}{n_p}\right)$. The operating principle of refractometers is based on identifying this, as shown in Figure 8.20. When looking through the telescope of a refractometer, the field is divided into two regions: a darker region, which represents light that is totally reflected and a lighter region for which the incidencent light enters at an angle lower than the critical angle (Figure 8.21). For example, the critical angle of a sugar solution with a refractive index of 1.4338 equals 81.7 °.

Refractometry is effective for liquids with a refractive index smaller than that of flint glass and typically between 1.3 and 1.7. The accuracy is one to two units to the fourth decimal place.

8.5 SPECTROSCOPY

Co-author: Cristian Neagoe

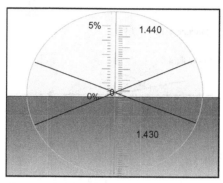

FIGURE 8.21   View of Internal Scale and Reflection

8.5.1 Historical

In the seventeenth century, Isaac Newton discovered that light decomposes into the seven colors of the electromagnetic spectrum—red, orange, yellow, green, blue, indigo, and violet—upon passing through a glass prism. This discovery was the beginning of spectroscopy. One century later, Joseph Fraunhofer noticed that light is formed from a great number of spectral lines and he measured the wavelengths of several Fraunhofer lines. Considerable progress occurred in the mid-nineteenth century when Foucault observed that the sodium flame emitted by line D also absorbs the radiation D emitted by an electrical arc placed nearby. For the first time, a correlation was made between radiation and the atomic structure of matter. In 1859, the German physician G. R. Kirchhoff discovered that the relationship between emissive power and absorptivity for a fixed wavelength is constant for all bodies at the same temperature. A few years later, Swedish Anders J. Ångström used glass gratings and succeeded in measuring with very good precision several wavelengths of solar radiation. He introduced the Ångström unit (1 Å $= 10\times10^{-10}$ m). In 1883, A. A. Michaelson used an interferometer and measured three Cadmium wavelengths to eight significant figures. The red cadmium line became a spectrometric standard. By the start of the 20th century, the analytical importance of spectrometric techniques was clear and the first commercial instruments were manufactured (Williams, 1976).

In spectroscopy, there are two types of spectra: emission spectrum and absorption spectrum.

8.5.2 Fundamentals

When an atom or molecule is bombarded by thermal, electrical discharge or electromagnetic radiation, it absorbs energy and enters into an excited state. The atom or molecule returns to its normal state by emitting a spectrum of frequencies of electromagnetic radiation. The emission spectrum is the signature of each atom and this property is used in industry for qualitative analysis.

The spectrum is composed of bands of light corresponding to the characteristic wavelengths.

When electromagnetic radiation (light) passes through a transparent medium, the amount absorbed is a function of the wavelength or frequency. The electromagnetic spectrum of the radiation exiting the medium is composed of dark lines that correspond to the wavelengths absorbed. The emission and absorption spectra are complementary: the wavelengths in the emission spectra correspond to the wavelengths absorbed in the absorption spectra.

The information provided by molecular spectroscopy, properly construed, is extremely useful in the qualitative and quantitative research on molecular composition. Spectroscopy has the advantage of being very fast and using only a small amount of material for analysis.

The energy difference between the final state (E_f) and the initial (E_i) is:

$$h\nu = h\frac{c}{\lambda} = E_f - E_i, \tag{8.12}$$

where h is $6.626\,069\,57 \times 10^{-34}$ J s (Planck's constant), ν is the frequency of the radiation emitted or absorbed, λ is the wavelength of the radiation emitted or absorbed, and c is the speed of light.

A molecule can have multiple modes of motion. In this sense, there can be movements of vibration in which the atomic bond varies in length (longitudinal vibrations) or in which the atoms oscillate in a direction perpendicular to the interatomic bond (transverse vibrations). Another mode of motion is the rotation of a group of atoms around a node, as shown in Figure 8.22.

The energy absorbed or emitted by the molecules has only quantum fixed values—energy is quantized and is distributed on the energy levels corresponding to vibrational, rotational, or electronic levels. Electronic transitions occur between two molecular electronic levels. They are usually accompanied by rotational and vibrational transitions.

The principle of spectroscopy is very simple: a sample is exposed to a varying frequency of radiation. The outgoing radiation is captured by a detector. Until the frequency reaches a quantum value, the radiation flux remains constant. When the frequency corresponds to a transition, described by Eq. 8.12, between two energy levels, a quantum of energy is absorbed. The absorption results in the

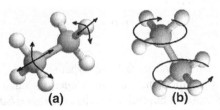

FIGURE 8.22 (a) Vibration Modes in the Molecule of Ethane; (b) Rotation in the Molecule of Ethane

detection of the absence of the respective frequency in the outgoing radiation of the sample.

8.5.3 IR Spectroscopy

The spectral range of IR spectroscopy is from 2500 nm to 16000 nm (Hart, 2000), which corresponds to the frequency range from 1.9×10^{13} Hz to 1.2×10^{14} Hz. This energy is too small to generate an electronic transition; IR spectra only characterize modes of molecular vibrations. To facilitate reading the spectra, values are reported as wavenumber, cm^{-1}, which is the reciprocal of wavelength, $\sigma = \frac{1}{\lambda}(cm^{-1})$.

The main application of IR spectroscopy is the identification of functional groups in organic molecules. Other uses are:

1. Determination of the molecular composition of surfaces.
2. Identification of chromatographic effluents.
3. Determination of molecular conformation (structural isomers) and stereochemistry (geometric isomers).
4. Determination of the molecular orientation in polymers and in solutions.
5. Determination of impurities in the test substance.

A great advantage of this method is that it does not destroy the sample analyzed. However, there are several limitations:

1. Information on the atomic composition of the molecule is still limited and often requires the use of complementary analytical techniques as nuclear magnetic resonance, mass spectroscopy, or Raman spectroscopy.
2. It is important that the solvent does not absorb in the spectral part of the molecule studied.
3. Several molecules are inactive in the infrared spectral part (Sherman Hsu, 1997).

The absorption IR spectra are presented in graphical form with the wavelength of the x-axis and the absorption intensity (A) or percent transmittance (%) on the y-axis. The transmittance is defined as the ratio between the intensity of radiation after passing through the sample (I) and the intensity of the incident radiation (I_0):

$$A = \log \frac{100}{T} = -\log \frac{I}{I_0},$$

Each group of atoms and each bond, whether single, double, or triple, is characterized by a vibrational transition energy. Values are available in several public databases (http://riodb01.ibase.aist.go.jp/sdbs/cgi-bin/direct_frame_top.cgi, http://webbook.nist.gov/chemistry/vib-ser.html, http://www.ir-spectra.com/indexes/index_d.htm) and some of them are given in Table 8.3.

TABLE 8.3 Characteristic Values of the Wave Number for Different Types of Bonds (Byrd, 1998)

Bond	Compound Type	Frequency Range (cm^{-1})
	Alkanes	2960–2850(s) stretch
		1470–1350(v) scissoring and bending
	CH$_3$ umbrella deformation	1380(m,w)—Doublet—isopropyl, t-butyl
	Alkenes	3080–3020(m) stretch
CH single bond		1000–675(s) bend
	Aromatic rings	3100–3000(m) stretch
	Phenyl ring substitution bands	870–675(s) bend
	Phenyl ring substitution overtones	2000–1600(w)—fingerprint region
	Alkynes	3333–3267(s) stretch
		700–610(b) bend
CC double bond	Alkenes	1680–1640(m,w) stretch
	Aromatic rings	1600–1500(w) stretch
CC triple bond	Alkynes	2260–2100(w,sh) stretch
CO single bond	Alcohols, ethers, carboxylic acids, esters	1260–1000(s) stretch
CO double bond	Aldehydes, ketones, carboxylic acids, esters	1760–1670(s) stretch
OH single bond	Monomeric—alcohols, phenols	3640–3160(s,br) stretch
	Hydrogen-bonded—alcohols, phenols	3600–3200(b) stretch
	Carboxylic acids	3000–2500(b) stretch
NH single bond	Amines	3500–3300(m) stretch
		1650–1580(m) bend
CN single bond	Amines	1340–1020(m) stretch
CN triple bond	Nitriles	2260–2220(v) stretch
NO$_2$	Nitro compounds	1660–1500(s) asymmetrical stretch
		1390–1260(s) symmetrical stretch

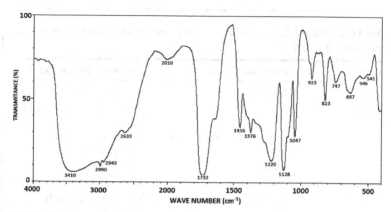

FIGURE 8.23 IR Spectrum of Lactic Acid (SDBSWeb, retrieved 2012)

TABLE 8.4 Wave Numbers and Bonds

Wave Number (cm^{-1})	Assigned Group
3410	OH single bond (alcohol)
2990, 2943	CH single bond stretch
2633	OH single bond stretch in carboxylic acids
1732	CO double bond stretch in carboxylic acids
1456	Umbrella effect in methyl, bending
1376	OH single bond bending in carboxyl group
1220	CO single bond stretch in CHO group
747	CO single bond bending in CHO group

Figure 8.23 is an example of a resolved IR spectrum of the lactic acid molecule. The data wave number and associated bonds are given in Table 8.4.

8.5.4 Spectroscopy UV/Visible

The spectral range corresponding to the spectroscopic UV/Visible is from 100 nm to 800 nm, but a narrower interval is more common from 190 nm to 750 nm. The UV range lies between 190 nm and 380 nm and the visible component is from 380 nm to 750 nm. The energy in this spectral range is greater than that of the IR range and generates electronic transitions in addition to vibrational and rotational transitions. The distances between the rotational levels are smaller than those between vibrational levels while the higher energies are found in electronic transitions. Electronic transitions are accompanied by

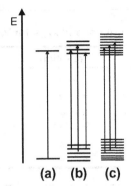

FIGURE 8.24 Transitions between Energy Levels: (a) Electronic; (b) Vibrational; and (c) Rotational

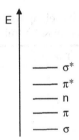

FIGURE 8.25 Orbital Energy Levels (Clark, 2007)

several vibrational transitions which, in turn, are associated with several small rotational transitions, as shown in Figure 8.24.

When a minimum threshold of UV or visible light radiation is absorbed by a molecule, electrons will pass to a higher energy state. There is a close relationship between the color of a substance and the energy change resulting from the transition. Three types of orbitals are involved: σ and π, which are bonding orbitals, and n, which is a non-bonding orbital. In addition, there are two antibonding orbitals designated as σ^* and π^*, which are at a higher energy state, as shown in Figure 8.25.

The energy between the electronic levels is determined by the types of groups of atoms rather than electrons and Table 8.5 demonstrates possible transitions of several compounds.

Atoms that absorb in the UV-Vis range can be further classified as chromophores and auxochromes. Chromophore groups are responsible for the color of the compound and absorb radiation at a specific wavelength. Examples are given in Table 8.6.

Auxochrome groups are groups of atoms which do not absorb in the 190 nm to 750 nm band but their presence in the molecule affects the absorption of chromophore groups. In this class there are OH, NH_2, CH_3, NO_2, Cl, OH, NH_2, CH_3, and NO_2.

TABLE 8.5 Electronic Transitions and λ_{max} for Various Substances

Compound	Transition	λ_{max} (nm)
Ethane	$\sigma \rightarrow \sigma^*$	135
Methanol	$\sigma \rightarrow \sigma^*$	150
	$n \rightarrow \sigma^*$	183
Ethylene	$\pi \rightarrow \pi^*$	175
Benzene	$\pi \rightarrow \pi^*$	254
Acetone	$n \rightarrow \pi^*$	290

TABLE 8.6 Characteristic Wavelengths of the Principal Chromophore Groups (Tables of Physical & Chemical Constants, 2005)

Group	Formula	λ_{max} (nm)
Nitrile	$-CN$	<180
Ethylene	$-C=C-$	<180
Sulphone	$-SO_2-$	180
Ether	$-O-$	185
Ethene	$-C=C-$	190
Thiol	$-SH$	195
Amine	$-NH_2$	195
Bromide	$-Br$	210
Iodide	$-I$	260
Ketone	$>C=O$	195, 275
Nitro	$-NO_2$	210
Nitrite	$-O-NO$	225
Nitrate	$-O-NO_2$	270
Azo	$-N=N-$	> 290
Nitroso	$-N=O$	300

The type of solvent used in the analysis may influence the absorption. In the case of a nonpolar substance in a nonpolar solvent, the influence of the solvent is negligible but in the case of polar molecules in polar solvents, there are fairly strong solute-solvent interactions, which usually lead to a decrease in resolution of the spectrum.

One of the most widespread applications in industry of the UV-Vis spectroscopy is the determination of the concentration of solutions. A wave passing through a transparent medium, a solution, for example, loses some of its energy. The Lambert-Beer law correlates intensity of absorbed incident radiation and the concentration of the solution. The energy absorbed, absorbance (A), or transmitted, transmittance (T), follows a logarithmic function of the absorption coefficient, ϵ, concentration, C, and the path length, l:

$$A = \log \frac{I_0}{I} = \log \frac{100}{T} = \epsilon C l.$$

UV-visible spectroscopy is used for quantitative analysis, studies of reaction rate and mixtures, identifying compounds and as detectors for HPLC. Quantitative analysis requires calibration curves to adequately characterize the variation of concentration and absorbance (or transmittance). Reaction rates may be derived by following the variation of the concentration of a compound in a vessel with time-on-stream. The total absorbance is the sum of the individual absorbances.

Spectroscopy has the advantage of being non-destructive and it is available at an affordable price. It can also be used for a wide variety of substances with good precision, sensitivity, and short analysis time. One major limitation is that the spectra are difficult to interpret for mixtures.

8.6 X-RAYS

Co-author: Amina Benamer

Wilhelm Conrad Röntgen discovered X-rays on the night of November 8, 1895 while he was conducting tests on cathode rays with Crookes tubes (Demerdijian, 2007). The "X" represents an unknown as was taken from algebra where it represents an unknown variable. This discovery earned him a Nobel Prize in physics in 1901. In December 1895, he produced an image of his wife's hand with an X-ray. As shown in Figure 8.26, it clearly shows bones as well as her wedding ring (Bellis, 2012).

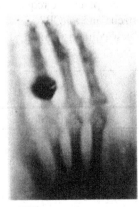

FIGURE 8.26 First X-ray Image, 1895

X-rays have different characteristics and unique properties, such as being able to traverse many transparent and opaque materials such as glass, tissue, and some metals, including lead. These rays are invisible and propagate in vacuum at the speed of light in a straight line.

This astounding discovery is widely used today and has taken an important place in society. Indeed, X-rays are now used in medicine to detect abnormalities in the human body (radiography, CT, etc.). X-rays are also used to remove cancer cells. Obviously, this practice can also be harmful to healthy cells if it is not used prudently. Also at airports, baggage is scanned by X-ray to identify liquids are other potentially hazardous objects.

Both X-ray fluorescence and X-ray diffraction are techniques used for chemical analysis. In X-ray fluorescence, gamma rays and high energy X-rays bombard a sample and secondary/fluorescent rays are emitted and detected. In X-ray diffraction, a X-ray beam is focused on a sample at various angles. The scattered intensity of the beam is detected. The nature of the scatter—refracted—beam reveals the crystal structure, chemical composition, and other physical properties.

8.7 EXERCISES

8.1 Para-nitrochlorobenzene (PNCB—$C_6H_4ClNO_2$) is an important inter-mediate and is used as an antioxidant in rubber. Parameters derived from a GC chromatogram of three compounds withdrawn from a reactor—chlorobenzene (C_6H_5Cl), toluene ($C_6H_5CH_3$), and PNCB—are shown in Table Q8.1. The GC column was 55 cm long. *M. Abdul-Halem*

(a) Calculate the number of theoretical plates for chlorobenzene.

(b) What is the height of an equivalent theoretical plate for chlorobenzene?

(c) What is the resolution between PNCB and toluene?

(d) How many theoretical plates are required to achieve a resolution of 1.8 for chlorobenzene and PNCB?

(e) What would the new column's length be?

TABLE Q8.1 GC Analysis

Compound	t_R (min)	$w_{1/2}$ (s)
Reference	1.00	-
Chlorobenzene	7.00	34.2
PNCB	8.175	36.9
Toluene	10.05	47.7

8.2 The chromatogram for a two-component mixture was separated on a 20 cm × 55 mm ID column. The reference peak was detected at 2 min and the retention time of the 2 compounds was 4 min and 8 min. The maximum peak heights of each peak are 50 mV (4 min) and 20 mV (8 min). Determine: *A. M. Bélanger*

(a) The capacity factor.

(b) The selectivity factor.

(c) The number of theoretical plates.

(d) The resolution.

(e) The equivalent height of a theoretical plate.

8.3 Natural gas is composed primarily of methane but may contain other hydrocarbons (ethane, propane, etc.), hydrogen sulfide, mercury, carbon dioxide, water, and nitrogen. *C. Mathieu*

(a) Five sample Hoke cylinders (originally containing air) were charged with shale. Complete the missing parameters in Table Q8.3.

(b) Among the components of the raw natural gas, some of them must be removed for reasons of safety, environmental hazards, and risk of corrosion in treatment facilities or during transportation. The gas is analyzed after a desulfurization step with a non-dispersive IR analyzer to evaluate the effectiveness of the desulfurization process installed:

(i) Determine the concentration of sulfur remaining in the natural gas stream coming out of the desulfurization unit.

(ii) If the maximum allowable concentration of H_2S in pipelines is 4 ppmv. Is the desulfurization unit efficient?

Note that $\epsilon = 4.2062 \times 10^3$ l mol^{-1} cm^{-1} for H_2S. The optical reference path is $b_0 = 50$ cm, the optical path of the sample is

TABLE Q8.3 Sample Data for Natural Gas Analysis

Exp	Sample Volume (ml)	Flow Rate (ml min^{-1})	Sampling Time (min)	Ratio c/c_i
1	3250	285	?	0.99
2	?	230	40	0.96
3	3300	?	55	0.975
4	3000	200	60	?

$b = 29.5$ cm. The concentration of reference natural gas in the analyzer is $c_0 = 15$ mg m^{-3}, and the conversion factor for the H$_2$S is $\frac{1\,mg}{1\,l\,air} = 717$ ppm (at 25 °C and 760 mmHg).

8.4 Parameters of a chromatogram to quantify the concentration of oxygen in air are: $t_0 = 0.6$ min, $t_{R,O_2} = 2.0$ min, $t_{R,N_2} = 3.2$ min. *S. Mestiri*

(a) Calculate the capacity factor of the first peak.

(b) Does changing the carrier gas change k'? Why?

(c) Calculate the selectivity. Are the two peaks well separated?

(d) Calculate the resolution knowing that $w_{O_2}, 1/2 = 0.1$ min and w_{N_2}, $1/2 = 0.3$ min.

(e) Are the peaks well resolved?

8.5 The concentration of iodine in solution is measured by UV/Vis spectrophotometry. The absorbance of the solution A equals 0.5 and $\epsilon(\lambda) = 250$. *N. Paulin*

(a) Using the Beer-Lambert law, find the iodine concentration in the sample.

(b) What would be the concentration of iodine if the readout for the transmittance was 56%?

Note that the cell used is cylindrical and has a volume capacity of 4 cm and $V = 5 \times 10^{-6}$ m^3.

8.6 Hydrazine reacts with benzaldehyde to form benzalazine. A chromatogram based on a HPLC analysis is shown in Figure Q8.6. *V. Messier* The column is 250 mm × 4.6 mm, phase KROMASIL 5 μm, the mobile phase is acetonitrile/acidified water 0.01%H$_2$SO$_4$ 80/20, the flow rate is 1.5 ml mm^{-1}, the volume injected is 25 μl, and the UV wavelength is 313 nm.

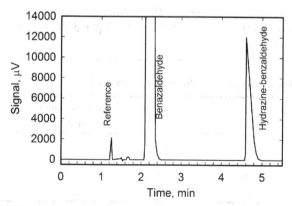

FIGURE Q8.6 Chromatogram of the Sample Collected During the Reaction

(a) For each component give the elution volume.

(b) What is the mobile phase velocity?

(c) If this column has a number of theoretical plates of 364, what is the height equivalent to one theoretical plate?

(d) Can we say that the chromatographic resolution depends on the peak symmetry?

http://www.inrs.fr/inrs-pub/inrs01.nsf/inrs01_metropol_view/ 42A5DA2 C5607DA9FC1256D5C0041C6D9/$File/052.pdf

8.7 Glycerol is a co-product of biodiesel manufactured from vegetable oils and fat. It will dehydrate catalytically to produce acrolein but may also produce undesirable compounds such as acetaldehyde (and coke, C^*).

$$C_3H_8O_3 \rightarrow C_3H_2O + H_2O,$$
$$C_3H_8O_3 \rightarrow CH_3CHO + 2H_2O + C^*.$$

Figure Q8.7 is a chromatogram of the effluent of a fluidized bed reactor (permanent gases). The widths at half maximum are 0.1 min for acetaldehyde, 0.25 min for acrolein, and 0.18 min for glycerol.

(a) From the chromatogram, calculate the retention factors for the three compounds.

(b) Calculate the number of theoretical plates for each compound.

(c) Calculate the resolution between acetaldehyde and acrolein. What is the minimum value to obtain a good separation? To reduced the resolution, how should the following operating parameters/conditions be modified?

(i) The length of the column.

(ii) The temperature.

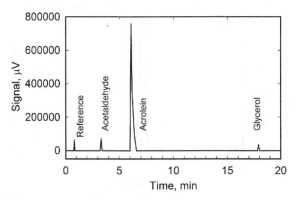

FIGURE Q8.7 Chromatogram of Glycerol to Acrolein

(iii) The linear velocity of the mobile phase.

(iv) The amount of stationary phase.

(d) The selectivity of the acrolein is calculated from the following equation:

$$S = \frac{[C_{acrolein}]}{[C_{acrolein}] + [C_{acetaldehyde}]}.$$

What is the uncertainty of selectivity for the following data (in μV, α concentration): for acrolein, 835 000, 825 000, 845 000, and 815 000; for acetaldehyde, 50 000, 80 000, 95 000 and 95 000, and $x_{acr} = 830\,000, s_{acr} = 13\,000, x_{ace} = 80\,000, s_{ace} = 21\,000$.

8.8 A gas chromatograph is used to examine the concentration of toxic chemicals from shale gas. The following compounds are suspected to be present with the methane: benzene, methylbenzene, parabenzene, and hydrogen sulfide.

(a) Which of the following three detectors would be appropriate for the analysis?

(i) FID (flame ionization detector).

(ii) TCD (thermal conductivity detector).

(iii) FPD (flame photometric detector).

(b) Figure Q8.8 is a chromatogram in which the peaks appear in order of increasing molecular weight. The reference peak appears at 55 s.

(i) What is the retention factor of each pollutant?

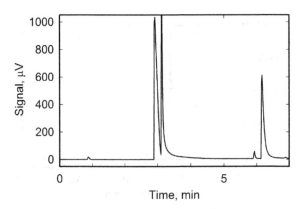

FIGURE Q8.8 Initial Analysis with Standards

(ii) What is the selectivity factor between hydrogen sulfide and parabenzene?

(iii) What is the resolution factor between benzene and methylbenzene?

(iv) Is the resolution sufficient?

(v) Identify two ways to increase the resolution between these two pollutants?

8.9 **(a)** What is the advantage of using a gas chromatograph with a temperature program?

(b) Why do we use hydrogen or helium as a mobile phase?

(c) Give three examples of solute that can be detected with the FID and three examples of non-detectable solutes.

(d) You measure the hydrocarbon concentration in a gas stream containing three solutes, including heptane and decane. The retention times of heptane and decane are respectively of 14.5 min and 22 min. The unknown solute leaves after 19 min. The residence time of tracer gas (non-retained) is 1 min. The peak widths at half-height of the solutes are 1.1 min for heptane, 1.4 min for decane, and 1.7 min for the unknown solute.

(i) If the unknown gas is an alkane, how many carbons could it have?

(ii) Calculate the capacity factor for each hydrocarbon.

(iii) Calculate the separation factor for both heptane and decane with respect to the unknown.

(iv) What are the parameters of resolution of heptane and decane with respect to the unknown? Are the peaks well resolved (separated)?

8.10 At the neolithic site of Kovacevo, Bulgaria, archeologists found several ceramic vases from 6200 to 5500 BC. For more than 20 yr, a debate has raged over the utility of these vases. A sample of black material was taken from inside one of the vases to determine its composition. An infrared spectrometric analysis (IRTF) was performed to determine the presence of organic material. To then identify the principal organic constituents, gas chromatography was used. The resulting chromatogram is shown in Figure Q8.10. The length of the column is 15 m and T_0 is 660 s. Table Q8.10 shows more data. *A. Bérard*

(a) Find the four constituents of the peaks of the black powder sample of vase KOV G264 1T. For ceramides, $k' = 2.08$, for cholesterol, the capacity factor is 0.995, for fatty acids, $N_{th} = 18754$, for glycerolipids, $HETP = 0.009$ cm, for triglycerine, the separation factor is 1.02, and for diglycerine, $R = 2.12$.

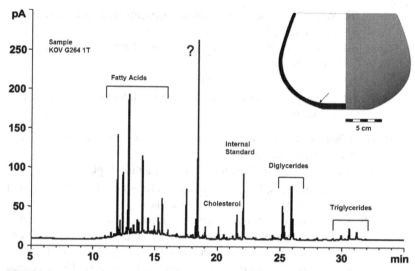

FIGURE Q8.10 Kovacevo Vase Chromatogram: (Vieugué et al., 2008)

TABLE Q8.10 Data for Kovacevo Vase Chromatogram

Peak	$W_{1/2}$ (s)
Fatty Acids	4, 6, 13, 12, and 10
Cholesterol	8
Diglycerides	16 and 14
Triglycerides	4, 7, 5

(b) What is the effect on the velocity of the mobile phase if the column length is decreased?

(c) Calculate the uncertainty in the resolution of diglyceride peak knowing that the uncertainty of the retention time is 0.1 min and that of $W_{1/2}$ is 1 s.

(d) Several samples of black powder were analyzed by chromatography. The internal standard was observed at several different times (in min): 21.3, 22.0, 24.2, 23.8, 20.4, 21.7, and 25.1. Can the internal standard peak, shown in Figure Q8.10, be considered as the internal standard (use the Chauvenet criterion).

8.11 Two peaks of a chromatogram appear near 7.5 min. Other conditions recorded were: $t_0 = 3$ min, $W_{1/2} = 0.35$ min, $HEPT = 0.005$ cm, and the selectivity is 0.98.

   (a) Calculate the number of theoretical plates. Determine the efficiency factor as well as the length of the column.

   (b) Calculate the capacity factor.

   (c) Calculate the resolution. What can you determine from this?

8.12 The chromatogram of a 17 cm × 4 mm column contains peaks at $t_0 = 0.3$ min, $t_{R,1} = 1.6$ min, and $t_{R,2} = 1.8$ min. In addition, $W_{1,1/2} = 0.1$ min and $W_{2,1/2} = 0.08$ min.

   (a) Find the separation factor α (selectivity).

   (b) What is the resolution?

   (c) Find the number of theoretical plates.

   (d) Determine the equivalent height of a theoretical plate.

   (e) What is the effect of increasing the temperature of the column on the retention time?

   (f) What is the effect of increasing the flow rate of the mobile phase on the retention time?

8.13 For each peak shown in Figure Q8.13, identify the corresponding compound and the bond.

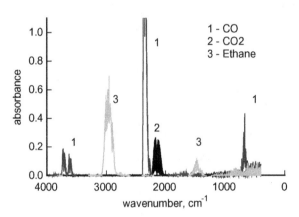

FIGURE Q8.13 IR Spectra

REFERENCES

Bellis, M., n.d. History of the X Ray. Retrieved July 2012, from About.com—Inventors: <http://inventors.about.com/od/xyzstartinventions/a/x-ray.htm>.

Byrd, J., 1998. Interpretation of Infrared Spectra, 1998. Retrieved July 20, 2012 from: <http://wwwchem.csustan.edu/tutorials/infrared.htm>.

Clark, J., 2007. UV-Visible Absorption Spectra. Retrieved July 2012, from UV-Visible: <http:// www.chemguide.co.uk/analysis/uvvisible/theory.html>.

Demirdjian, H., 2007, October 10. La radiographie (I)—Histoire de la découverte des rayons X et de leur application en médecine. Retrieved July 2012, from Rayons X: <http:// culturesciences.chimie.ens.fr/content/la-radiographie-i-histoire-de-la-decouverte-des-rayons-x-et-de-leur-application-en-medecine-1196>.

Florentin, E., 2004, May 10. Le réfractomètre. Retrieved July 2012, from ENS de Lyon-éduscol: <http://culturesciencesphysique.ens-lyon.fr/XML/db/csphysique/metadata/LOM_CSP_Refrac-tometre.xml>.

Hart, H., Conia, J.-M., 2000. Introduction à la chimie organique. Dunod, Paris.

Klvana, D., Farand, P., Poliquin, P.-O., 2010. Méthodes expérimentales et instrumentation—Cahier de laboratoire. École Polytechnique de Montréal, Montréal.

Kaye, G.W.C., Laby, T.H., 1911. Tables of Physical Constants and some Mathematical Functions, Longmans, Green and Co., New York. Tables of Physical & Chemical Constants (16th edition 1995). 3.8.7 UV-Visible Spectroscopy. Kaye & Laby Online. Version 1.0 (2005) www.kayelaby.npl.co.uk.

Patience, G.S., Hamdine, M., Senécal, K., Detuncq, B., 2011. Méthodes Expérimentales Et Instrumentation En génie Chimique, third ed. Department Chemical Engineering, Ecole Polytechnique de Montreal.

SDBSWeb. <http://riodb01.ibase.aist.go.jp/sdbs/>. National Institute of Advanced Industrial Science and Technology. Retrieved July 20, 2012.

Sherman Hsu, C.-P., 1997. Infrared spectroscopy. In: Settle, Frank (Ed.), Handbook of Instrumental Techniques for Analytical Chemistry. Retrieved July 20, 2012 from: <http://www.prenhall.com/settle/chapters/ch15.pdf>.

Vieugué, J., Mirabaud, S., Regert, M., 2008. Contribution méthodologique á l'analyse fonctionnelle des céramiques d'un habitat néolithique: l'exemple de Kovačevo (6 200-5 500 av. J.-C., Bulgarie), ArcheoSciences, 32, 99–113.

Williams, Dudley, 1976. Spectroscopy. Methods in Experimental Physics, vol. 13. Academic Press Inc., New York.

Chapter 9

Analysis of Solids and Powders

9.1 OVERVIEW

Particle technology and the analysis of solids involve the measurement of physical properties, the assessment of the impact of these properties on operations, and the identification of means to improve, develop, and design processes. Powder systems touch many chemical industries, and the majority of consumer products necessarily pass through the manipulation of particulates in their manufacturing process. The analysis and characterization of multi-phase systems not only includes gas-solids and liquid-solids systems but also two phase liquid systems, three phase systems, gas-liquids as well as micron-sized powders and nano-particles. Besides the chemical industry, other industries in which particle technology plays an important role include the pharmaceutical (pills, colloids, solutions, suspensions), agricultural (distribution of pesticides, herbicides, fertilizers, soil), food (flour, sugar, Cheerios, salt, etc.), construction (cement, concrete, asphalt), power industry (coal), mining, metallurgy, and the environment.

The combustion of coal to produce electricity is the greatest contributor to greenhouse gases and the manufacture of cement is third accounting for as much as 5% of the global CO_2. Concrete and asphalt are the two largest manufactured products and both are aggregates of particles. Annual production of concrete is approximately 2×10^9 kt yr^{-1} while the production of cement is on the order of 2×10^9 kt yr^{-1} (van Oss, 2011).

Cement manufacture is illustrative of unit operations involving solids—conveying (transport), grinding—milling—crushing (micronization), sieving (separation), and reaction. Mined limestone from a quarry passes through a primary jaw crusher (communition or grinding). The ore enters a vibrating screen to separate the ore into three fractions: particles greater than 120 µm in diameter are conveyed by a belt conveyor to a secondary crusher and then to

Experimental Methods and Instrumentation for Chemical Engineers. http://dx.doi.org/10.1016/B978-0-444-53804-8.00009-5

a hammer mill so that all the ore is less than 120 μm; particles from 50 μm to 120 μm are screened. After a secondary screening, particles are separated into three fractions—0–25 μm, 25–50 μm, and 50–120 μm and stockpiled in open pits. The next step is to micronize the particles to a fine powder less than 125 μm. This powder is fed to a rotary kiln in which the powder temperature reaches 1450 °C and the flame temperature reaches 2200 °C. Calcium carbonate (limestone) is calcined to produce lime and CO_2:

$$CaCO_3 \rightarrow CaO + CO_2.$$

In this step, 40% of the CO_2 is due to fossil fuel combustion to produce the flame, and 50% comes from the reaction. Because of the high temperatures, the powder agglomerates (forms clinkers). Before packaging and distribution, these particles are milled and sieved to the desired size.

Besides the processing steps mentioned for cement manufacture, other unit operations involving solids include: filtration, cyclonic separation, decantation (sedimentation), coating (polymer coatings of metals, for example), pneumatic conveying (of flour, grain, coal, alumina), crystallization (manufacture of catalysts and foodstuffs), mixing (pharmaceuticals—ensuring the active ingredient is uniformly distributed before making pills), drying, etc. In this chapter, we focus primarily on the physical characteristics of powders and solids—density, shape, and particle size.

9.2 DENSITY

The densities of gases are readily calculated through thermodynamic relationships and liquid densities are straightforward to measure compared to measuring solids and powder densities. Table 9.1 summarizes densities of many solids; they range from as low as $100 \, kg \, m^{-3}$ for nano-tubes to almost $20\,000 \, kg \, m^{-3}$ for tungsten. There is a significant variation between different types of wood. The highest density woods are lignum vitae ($1370 \, kg \, m^{-3}$) and ebony ($1120 \, kg \, m^{-3}$), while the lowest include balsa ($170 \, kg \, m^{-3}$), bamboo ($300 \, kg \, m^{-3}$), and red cedar ($380 \, kg \, m^{-3}$). The average density is $600 \, kg \, m^{-3}$ with a standard deviation of $200 \, kg \, m^{-3}$ (The Engineering Tool Box, retrieved 2011).

9.2.1 Bulk Density

Powders are characterized by three different measures of density: skeletal density, particle density, and bulk density. Bulk density, ρ_b, is the density of the powder that is typically packaged in a container (sack, box, silo, vessel). However, it generally increases with time and thus two other measures are often quoted to characterize the bulk density: pour density and tapped density. During shipping, or as the powder is moved from one place to another, the

TABLE 9.1 Densities of Typical Solids and Powders

Solid	Density (kg m^{-3})
Nano-tubes	100
Acrylic fibers	144
Colorant	144
Polypropylene	481
PVC chips	513
Coal dust	561
Wood	600 ± 200
Alumina	641
Polyethylene	689
Clay	802
Ice	917
Cerium oxide	994
Cement	1363
Phosphates	1443
FCC catalyst	1507
Sand	2651
Zinc powder	3367
Tungsten carbide	4008
Steel	7850
Tungsten	19 250

powder becomes more compacted. The pour density represents the lowest level of compaction—the conditions immediately after the powder is introduced into a container—this is the value generally reported as the "bulk" density. The highest level of compaction is referred to as the tapped density. The two are measured by first pouring the powder into a graduated cylinder to a certain height, h_b, (from which ρ_b is calculated) and then the cylinder is tapped gently on the side with a ruler until the solids level no longer drops, h_t. The ratio of the tapped density to the bulk density is referred to as the Hausner ratio:

$$H = \frac{\rho_t}{\rho_b} = \frac{h_b}{h_t}. \tag{9.1}$$

The Hausner ratio is a parameter that indicates how much a powder might compact with time, which correlates with how it will flow—flowability. Values greater than 1.25 indicate that the powder will have poor flowability characteristics. The Scott density is the standard test to measure bulk density or poured density.

FIGURE E9.1 Scott Density Measurement Apparatus

Example 9.1. The Scott density is measured using a standardized procedure in an apparatus shown in Figure E9.1. The powder is fed to a funnel at the top of the device and it falls through a series of baffles in a chute that redirect the flow and finally into a hollow cylinder (or cube) with a 1 cm^3 volume. When the cylinder at the bottom overflows, it is removed from underneath the chute and excess powder is leveled with a spatula (or ruler). The weight of the powder (g) divided by 1 cm^3 gives the Scott density—which is equivalent to bulk density.

    Example 9.1. Three students measured the mass of three powders—sand, fluid catalytic cracking catalyst (FCC), and $Ca_3(PO_4)_2$—10 times successively in a Scott Volumeter. The data is summarized in Table E9.1. Calculate the Scott density, uncertainty, and repeatability. The volume and uncertainty of the cylinder is 25.00 ± 0.03 cm^3.

Solution 9.1. The mean mass of catalyst, m_{cat}, collected, standard deviation and the uncertainty, Δ_{cat}, are first calculated. (Note that the uncertainty is defined as the 95% confidence interval and, in this case it is quite close to the standard deviation.)

    Since the uncertainty of the volume of the cylinder is less than five times the uncertainty in the mass of powder, it may be ignored. The uncertainty in the density is simply:

$$\Delta_\rho = \frac{\Delta_{m_{cat}}}{m_{cat}} \rho.$$

TABLE E9.1 Mass of Powder Collected by Three Students in Scott Volumeter

Sand

1	37.93	38.20	38.10	38.38	38.31	38.62	38.50	38.10	38.31	38.00
2	38.26	38.09	37.72	38.03	38.34	38.64	38.12	37.96	38.71	38.28
3	38.33	38.15	38.35	38.21	38.24	38.03	38.07	37.53	38.23	38.01

FCC

1	21.04	21.02	21.00	20.88	20.98	21.23	21.07	21.15	21.07	21.30
2	21.37	22.00	21.41	21.08	21.34	20.97	22.02	21.88	22.33	21.42
3	22.74	22.33	22.41	22.81	22.26	22.59	22.37	22.46	22.78	22.95

$Ca_3(PO_4)_2$

1	8.11	7.99	7.91	7.97	7.98	7.93	7.90	8.07	7.89	7.87
2	8.24	8.00	7.98	8.23	7.98	8.07	7.86	7.74	8.24	7.95
3	8.03	7.86	8.15	8.09	7.94	7.96	7.89	7.95	7.83	7.97

TABLE E9.1Sa Mean Mass and Statistics for the Scott Density Measured for Each Powder

	Sand			FCC			$Ca_3(PO_4)_2$		
Student	1	2	3	1	2	3	1	2	3
Mean mass	38.25	38.22	38.12	21.07	21.58	22.57	7.96	8.03	7.97
Standard deviation (s)	0.22	0.30	0.24	0.12	0.45	0.24	0.08	0.17	0.10
Δm	0.16	0.22	0.17	0.09	0.32	0.17	0.06	0.12	0.07

The agreement between the mean values of the students for the sand is better than for either the FCC or the calcium phosphate. The particle size was much larger for the sand and it was less cohesive. The repeatability variance, as described in Chapter 2, equals the sum of the variances of each sample divided by the number of samples subtracted by 1 (see Tables E9.1Sa and E9.1Sb):

$$s_r^2 = \frac{1}{n-1}\sum s_{r,i}^2.$$

The repeatability variance and standard deviation for the Scott density of each sample are shown in Table E9.1Sc.

If the samples were run in different laboratories, the reproducibility is (Table E9.1Sd):

$$s_R^2 = s_L^2 + s_r^2.$$

TABLE E9.1Sb Scott Density and Uncertainty Measured for Each Powder

Student	Sand			FCC			$Ca_3(PO_4)_2$		
	1	2	3	1	2	3	1	2	3
ρ	1.530	1.53	1.52	0.8430	0.8633	0.903	0.318	0.321	0.319
Standard deviation (s)	0.009	0.01	0.01	0.005	0.02	0.009	0.003	0.007	0.004
$\Delta\rho$		0.006	0.009	0.007	0.004	0.01	0.007	0.002	0.005	0.003
$\%\Delta\rho$	0.4	0.6	0.4	0.4	1.5	0.8	0.7	1.5	0.9

TABLE E9.1Sc Repeatability Variance and Standard Deviation for the Scott Density

	s_r^2	s_r (g cm^{-3})
Sand	0.007	0.09
FCC	0.008	0.09
$Ca_3(PO_4)_2$	0.003	0.06

TABLE E9.1Sd Scott Density Reproducibility

	s_R^2	s_R (g cm^{-3})
Sand	0.014	0.12
FCC	0.032	0.18
$Ca_3(PO_4)_2$	0.007	0.08

9.2.2 Particle Density

The particle density, ρ_b, represents the mass of a particle divided by the volume of its hydrodynamic envelope. It includes the void spaces of open pores that communicate with surrounding fluid but also internal pores:

$$\rho_p = \frac{m_p}{V_p}. \qquad (9.2)$$

As shown in Figure 9.1, the hydrodynamic envelope for a spherical particle is equal to the volume of the sphere. Most particles are non-spherical in shape

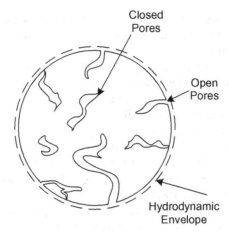

FIGURE 9.1 Particle Density

and thus defining the hydrodynamic envelope is problematic. Typically, the particle density is measured using pycnometry or by mercury porosimetry.

Either a graduated flask or a high precision pycnometer may be used to estimate the particle density, but these techniques are poor measures when the solids are soluble. Therefore, a liquid should be chosen for which the solids are insoluble. The test should be conducted rapidly to minimize either dissolution of the solids are the partial filling of open pores. To increase the precision, the graduated flask or pycnometer should first be calibrated with the fluid.

The flask or pycnometer is first filled halfway. A beaker containing the remaining portion of the liquid should also be prepared. The solid powder is weighed on a precision balance. The volume of the solids added to the flask should be less than about 25% of the volume, otherwise the powder might agglomerate. The powder should be poured steadily into the flask—slow enough to avoid agglomerates but quick enough to minimize dissolution—and then agitated gently to remove any gas bubbles—sonic pulses are an efficient means to remove bubbles. The fluid from the beaker is then poured into the flask until the level reaches the graduation line. The pycnometer with the solids and fluid is weighed. The particle density is then equal to:

$$\rho_p = \frac{m_p}{V_{\mathrm{pyc}} - \frac{m_t - m_p}{\rho_f}} = \frac{m_p}{V_{\mathrm{pyc}} - (V_{f,T} - \Delta V_f)}, \qquad (9.3)$$

where V_{pyc} is the pycnometer volume in cm^3, m_p is the mass of powder in g, m_t is the total mass of fluid and powder in g, $V_{f,T}$ is the volume of fluid measured out before the test in cm^3, and ΔV_f is the remaining fluid in the beaker (or burette) in cm^3.

An alternative (or supplement) is to subtract the total weight of fluid prepared to conduct the measurement from that remaining in the beaker after the pycnometer is filled.

The accuracy of pycnometry depends on the size and type of pores as well as the wettability of the solids—how well does the fluid envelope the solid and at what time frame. Mercury as a fluid minimizes these limitations and many instruments are available commercially.

Although mercury porosimetry is principally concerned with the measure of the pore size, pore volume, and distribution, it also measures density—bulk density, particle density as well as skeletal density. Unlike other fluids, it does not spontaneously fill pores but enters them only when a pressure is applied. The size of the pores that mercury enters is inversely proportional to the pressure (equilibrated): larger pores require less pressure compared to smaller pores.

Mercury is an ideal fluid for porosimetry because it is non-wetting and has a high surface tension. Surface tension is responsible for the shape that liquid droplets form on surfaces as well as their maximum size. The units are force per unit length or energy per unit area. The meniscus of water in a capillary tube is concave while it is convex for mercury. However, in a copper tube, the meniscus for both water and mercury is concave. Because of the adhesion between the wall of the capillary and the liquid, the liquid may be drawn up to a height h. The height depends on the liquid-air surface tension, γ_{la}, the contact angle, θ, the density of the fluid, ρ, and the radius of the capillary, r:

$$h = \frac{2\gamma_{la}\cos\theta}{\rho g r}. \tag{9.4}$$

For mercury to enter a pore, a force is required to overcome the resistance that arises from the surface tension:

$$f = 2\pi r \gamma_{Hg-a}\cos\theta. \tag{9.5}$$

The applied force to overcome the resistance from the surface tension equals the pressure acting over the surface area of the pore. At equilibrium, just before the mercury enters the pore, the force due to pressure equals that arising from surface tension:

$$2\pi r \gamma_{Hg-a}\cos\theta = \pi r^2 P. \tag{9.6}$$

The diameter of the pore at which mercury begins to penetrate is:

$$D = \frac{4\gamma_{Hg-a}\cos\theta}{P}. \tag{9.7}$$

Figure 9.2 demonstrates the volume of mercury intruded into a sample as a function of pore size. In the first part of the curve, the mercury fills the spaces between the particles—inter-particle void. The particle density equals the ratio

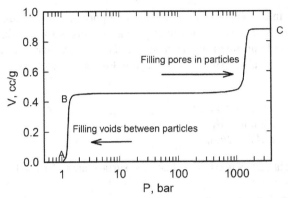

FIGURE 9.2 Change in Intruded Volume of Mercury with Decreasing Pore Size (Increasing Pressure)

of the particle mass to the difference between the total volume occupied by the solid sample, V_T, and the volume of mercury intruded, V_{AB}:

$$\rho_p = \frac{m_p}{V_T - V_{AB}}. \tag{9.8}$$

The bulk density is the product of the particle density and the solids fraction, which equals one minus the void fraction ϵ:

$$\rho_b = \rho_p(1 - \epsilon). \tag{9.9}$$

Note that the void fraction may take "v" as a subscript, ϵ_v, but often the subscript is omitted. After filling the inter-particle void (V_{AB}), the change in volume with pressure reaches a plateau—the volume remains constant as pressure increases. Finally, the volume of mercury intruded increases with increasing pressure as the small pores become filled with mercury.

The calculation of the skeletal density, ρ_s, is similar to that of the particle density. It is the ratio of the mass of particle to the difference between the total volume occupied by the solid sample, V_T, and the sum of volume of mercury intruded, $V_{AB} + V_{BC}$:

$$\rho_s = \frac{m_p}{V_T - V_{AB} - V_{BC}}. \tag{9.10}$$

The particle density is the product of the skeletal density and the solids fraction of the particle, which equals one minus the skeletal void fraction, ϵ_s:

$$\rho_p = \rho_s(1 - \epsilon_s). \tag{9.11}$$

The bulk density equals the product of the skeletal density and the solids fraction of the bulk solids and the solids fraction of the particles:

$$\rho_b = \rho_s(1 - \epsilon)(1 - \epsilon_s). \tag{9.12}$$

Mercury intrusion porosimetry is suited for pores as low as 2 nm and greater than 350 μm. N_2 intrusion porosimetry measures pores smaller than 2 nm but this technique is unsuited for determining the inter-particle void fraction: its upper range is smaller than for Hg intrusion. Pore sizes are classified into three ranges:

- *Macropores:* $d_p > 50$ nm;
- *Mesopores:* 2 nm $\leqslant d_p \leqslant 50$ nm;
- *Micropores:* $d_p < 2$ nm.

N_2 porosimetry is used to evaluate the surface area of particles and will be discussed further on.

Example 9.2. The data in Figure 9.2 represents the Hg intruded into a 0.5 g sample of an unknown powder. Based on the bulk characteristics given in Example 9.1, identify the powder. What is its particle density and skeletal density?

TABLE E9.2 Powder Properties from Example 9.1

	ρ	v	ϵ
Sand	1.528	0.655	0.69
FCC	0.870	1.15	0.39
$Ca_3(PO_4)_2$	0.319	3.13	0.14

Solution 9.2. The volume intruded between points A and B represents the inter-particle void fraction, ϵ, and is about 0.45 cm^3 g^{-1}. The volume between A and C represents the sum of the inter-particle and intra-particle void fraction and equals 0.88 cm^3 g^{-1}. The specific volume, v, of the powder is the inverse of the density and the inter-particle void fraction, ϵ, is the ratio of the specific volume and volume intruded between points A and B:

$$\epsilon = \frac{V_{AB}}{v}.$$

Note that the bulk void fraction, ϵ, of a powder may vary from 0.38 (if it is packed or has been tapped) to 0.42. Higher void fractions are possible for powders that are cohesive but values lower than 0.38 are uncommon.

The calculated void fraction based on the specific volume of mercury intruded is too high for the sand and much too low for the $Ca_3(PO_4)_2$. Therefore, it is reasonable to assume that the sample is FCC with an ϵ equal to 0.39. The particle density equals:

$$\rho_p = \frac{\rho_b}{1-\epsilon} = 1 - \frac{0.87}{1-0.39} = 1.43 \text{ g cm}^{-3}.$$

The intra-particle void fraction is calculated based on the volume intruded between points B and C—V_{BC} in Figure 9.2, which equals 0.43 cm^3 g^{-1}—and the specific volume of the particle, $v_p = 1/\rho_p$:

$$\epsilon_{sk} = \frac{V_{BC}}{v_p} = \frac{0.43}{0.70} = 0.61.$$

The skeletal density equals:

$$\rho_{sk} = \frac{\rho_p}{1 - \epsilon_{intra}} = \frac{1.43}{1 - 0.61} = 3.7 \text{ g cm}^{-3}.$$

Note that the skeletal void fraction may cover the entire range from zero for non-porous materials to greater than 0.95 for extremely porous materials.

9.3 DIAMETER AND SHAPE

Most engineering calculations (and correlations) related to powder technology are based on spherical particles for which the characteristic linear dimension is diameter. However, particles are rarely spherical or equal in size and thus the difficulty in choosing a characteristic linear dimension is related to choosing the dimension that characterizes the shape as well as the average length of the ensemble of particles. The term particle is generally applied to a body with a diameter less than 1 cm. It may be a solid, liquid (droplet), or gas (bubble). The characteristic dimension of a sphere is its diameter; it is the length of the side for a cube; the radius or height is the characteristic dimension of a cylinder or cone.

Particle shapes found in nature and used in commercial applications vary tremendously. They can have a major impact on physical properties, including: flow, tendency to agglomerate; activity and reactivity (in catalysis or combustion, for example); pharmaceutical activity (drug dissolution); gas absorption (chromatography); bulk density; optics; interactions with fluids; packing—particularly for fixed bed reactors; and pneumatic conveying—minimum gas velocity for transport. Fiber shape is used in the textile industry to achieve specific affects and fabric properties—prismatic, circular, trilobal, and combinations of shapes not only change appearance—luster, iridescence, sparkle—but also surface texture—softness and drape. In the paint industry, the particle size affects the opacity (light scattering behavior) of white pigments: the quantity of light scattered increases with decreasing particle size. Thixotropic agents—nanosized SiO_2, for example—are added to increase the viscosity of paints and reduce the tendency for pigments to settle (Patience, 2011).

Although spheres are characterized with a single dimension—diameter— a collection of spheres do not necessarily have the same diameter. Thus, the choice of a single dimension lies in assigning the value that best represents the application. This is discussed in greater detail later.

Acicular particles, as demonstrated in Figure 9.3.b, are formed of needles or with pointy shaped rods protruding from a nuclear mass. Among the dimensions

FIGURE 9.3 Particle Shapes: (a) Spherical - Calcite; (b) Acicular - Calcite with Stibnite; (c) Fibrous - Actinolite; (d) Dendritic - Gold; (e) Flakes - Abhurite; (f) Polyhedron - Apophyllite; (g) Cylindrical Particles (For color version of this figure, please refer the color plate section at the end of the book.)

that represent this shape are the length and diameter of each rod. However, for most practical applications the characteristic dimension would be the projected area or the length across the bulk of the needles.

Asbestos is an example of a fibrous mineral that may cause malignant cancer, mesothelioma, and asbestosis if inhaled over extended periods of time. Their form may be regular or irregular and at least two dimensions would be necessary to fully represent the characteristic dimension—both length and diameter. Figure 9.3.c is an example of a fibrous mineral called actinolite.

Snowflakes are good examples of dendritic particles that contain branches or crystal structures emanating from a central point. Two dimensions are necessary to adequately characterize this particle type. Figure 9.3.d demonstrates a dendritic form of gold.

Mica is an example of a material that forms very thin flakes. The characteristic dimensions are similar to the dendritic particles—thickness and width across the flake. Figure 9.3.e shows plate like structures of abhurite.

Figure 9.3.f demonstrates angular and prismatic shapes of the mineral apophyllite. Cubes—a specialized form of a polyhedron—like spheres may be characterized by a single parameter or dimension (e.g. sugar). Otherwise, for elongated angular particles, the characteristic dimension will be decided based on the application.

Cylindrical particles together with spherical are probably the most common shape. Many chemical reactors use cylinders (hollow cylinders) as the preferred shape for catalysts in fixed bed reactors (See Figure 9.3.g.). Rice has a cylindrical shape as do some bacteria (e.g. bacilli).

Geometric shape descriptors, two- and three-dimensional, are required to represent the dimensional properties of real particles. Three representations are often used to represent the non-standard particle forms—characteristic diameter, shape factor, and sphericity.

9.3.1 Engineering Applications

Many unit operations of chemical engineering deal with particles for which consideration of the particle diameter—particle characteristic dimension—is a fundamental design parameter. Table 9.2 lists several unit operations and the corresponding design parameter for which the particle diameter is required.

TABLE 9.2 Unit Operations Involving Powders

Unit Operation	Property Related, d_p
Transport	
Pneumatic	Particle terminal velocity
Bin flow	Cohesion, internal friction
Reactor design, performance	
Fixed bed	Pressure drop, heat transfer, kinetics
Fluidized bed	Flow regime
Spray drying	Heat transfer, mass transfer
Crystallization	Rate
Separation	
Entrainment	Particle terminal velocity
Decantation (settling)	Particle terminal velocity
Filtration	Pressure drop
Sieving	Shape
Centrifugation	Particle terminal velocity
Cyclone	Particle terminal velocity
Mixing (pharmaceutical)	Power consumption
Size reduction	Power consumption

Correlations, dimensionless numbers, and regime maps have been developed for many of these applications. The Reynolds number was introduced in Chapter 5 to differentiate flow regimes in pipes—laminar versus transition versus turbulent. In Chapter 6, the notion of particle Reynolds number was mentioned with respect to drag force and rotameters.

9.3.2 Particle Terminal Velocity

The Stokes flow regime applies to spheres at particle Reynolds number ($N_{Re,p} = \frac{\rho u d_p}{\mu}$) below 2; the intermediate range lies between 2 and 500; the Newton's law range extends beyond 500. Particle terminal velocity correlations have been derived for each of the flow regimes. In the case of cylinders and disks, other correlations apply. Correlations for other shapes are lacking, so estimating their particle terminal velocity requires approximating their characteristic particle dimension to that of a sphere. With this value, the Reynolds number is calculated (to identify the regime) and then the particle terminal velocity, u_t, as given by:

$$u_t = \left(\frac{g d_p^{1+n} (\rho_p - \rho_f)}{3 b \mu^n \rho_f^{1-n}} \right)^{1/(2-n)}, \qquad (9.13)$$

where ρ_f is the fluid density in kg m^{-3}, ρ_p is the particle density in kg m^{-3}, n equals 1, 0.6, and 0 for the Stokes' law, intermediate, and Newton's law regimes, respectively, and b equals 24, 18.5, and 0.44 for the Stokes' law, intermediate, and Newton's law regimes, respectively (McCabe and Smith, 1976).

Example 9.3. Calculate the velocity a water balloon 10 cm in diameter would achieve if dropped from the top of St. Paul's cathedral.

Solution 9.3. Besides experiments conducted by Galileo from the Leaning Tower of Pisa, Newton measured particle terminal velocities by dropping hog bladders from St. Paul's Cathedral. Galileo used his heart beat as a timer! In this example, the water balloon is assumed to be spherical (which is probably inaccurate because the balloon will deform to achieve the lowest drag). The density of air is approximately 1.18 kg m^{-3} ($\sim$1.2 kg m^{-3}) while it is 1000 kg m^{-3} for the water balloon. First, we determine the flow regime and from that calculate the terminal velocity. However, the flow regime depends on the Reynolds number, which depends on the terminal velocity. Assuming, that the terminal velocity equals 20 m s^{-1}, the Reynolds number is:

$$N_{Re,p} = \frac{\rho u d_p}{\mu} = \frac{1.2 \cdot 20 \cdot 0.1}{0.000018} = 130\,000.$$

The Newton's law regime applies to very high Reynolds numbers so:

$$u_t = \frac{g d_p (\rho_p - \rho_f)}{3 \cdot 0.44 \cdot \rho_f} = \frac{9.81 \cdot 0.1 \cdot 999}{3 \cdot 0.44 \cdot 1.2} = 25 \text{ m s}^{-1}.$$

9.3.3 Equivalent Diameter

The examples above demonstrate the need to define an equivalency between the characteristic length of an irregular shaped particle and a sphere. The correlations are derived for spherical particles but how would the pressure drop in a fixed bed reactor change if it were charged with cylindrical particles or prismatic? Shapes are measured coupling a microscope with an image analyzer. The characteristic length based on this technique is based on a two-dimensional image perpendicular to the lens. The equivalent diameter is the length of a line bisecting the particle and projecting a circle around this line (Figure 9.4). Optimally, the surface area of the circle could be equal to the surface area of the image of the project particle (Figure 9.5).

Other techniques of deducing shape (or size) include using sieve analysis or sedimentation. The equivalent diameter for a sieve analysis is the mesh size of the upper sieve through which particles pass. (A more precise definition of the sieve diameter is the mean between the mesh size through which the particles pass and on which the particles are retained.) The characteristic diameter of a sedimentation technique would be the diameter of a sphere that has the same settling velocity.

Other simple definitions include: a sphere having the same diameter as the largest width of the particle; a sphere having the same diameter as the minimum width; a sphere with the same weight. Table 9.3 summarizes several of the most common definitions of equivalent particle diameter.

9.3.4 Shape Factors—Sphericity

Besides equivalent diameter, shape factors are commonly used to express the characteristic diameter of an object as a function of an equivalent sphere and it is summarized in the ISO standard 9276-6. Sphericity, ϕ, is often used and it is defined as the ratio of the area of a sphere with the same volume of the particle

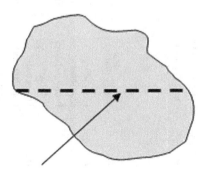

FIGURE 9.4 Equivalent Diameter

TABLE 9.3 Main Definitions of Equivalent Diameter

Symbol	Designation	Definition	Formula
d_v	Volume diameter	Diameter of the sphere having the same volume as the particle	$d_v = \sqrt[3]{\dfrac{6V_p}{\pi}}$
d_s	Surface diameter	Diameter of the sphere having the same surface as the particle	$d_s = \sqrt{\dfrac{S_p}{\pi}}$
d_{sv}	Surface-volume (Sauter) diameter	Diameter of the sphere with the same outer surface area to volume ratio	$d_{sv} = \dfrac{d_s^3}{d_s^2}$
d_d	Drag diameter	Diameter of the sphere having the same resistance to motion as the particle in a fluid of similar viscosity and at the same speed (d_d approaches d_s when $N_{Re,p}$ is small)	$F_D = C_D A \rho_f \dfrac{v^2}{2}$ $C_D A = f(d_d)$ $F_D = 3\pi d_d \eta v$ $N_{Re,p} < 2$
d_{St}	Stokes diameter	Diameter of a particle in free fall in the laminar flow regime ($N_{Re,p} < 2$)	$d_{St} = \sqrt{\dfrac{18\mu u_t}{(\rho_p - \rho_f)g}}$
d_f	Free-falling diameter	Diameter of the sphere having the same density and terminal velocity as the particle in a fluid of equal density and viscosity	$d_f = d_{St} N_{Re,p} < 2$
d_a	Projected area diameter	Diameter of the circle having the same area as the projected area of the particle in the steady state	$A = \dfrac{\pi}{4} d_a^2$
d_{ap}	Projected diameter surface	Diameter of the circle having the same area as the projected area of the particle in a random orientation	
d_c	Perimeter diameter	Diameter of the circle having the same circumference as the outer perimeter of the projected particle	
d_A	Sieve diameter	Width of the square mesh through which the particle will pass	
d_F	Feret diameter	Average value of the distance between two parallel tangents to the outer periphery of the projected particle	
d_M	Martin diameter	Average length of a chord dividing the outer periphery of the particle projected into two equal areas	

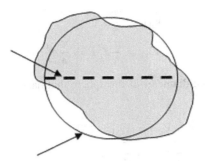

FIGURE 9.5 Martin's Diameter (Yang, 2003)

TABLE 9.4 Sphericity of Certain Particles

Particles	Sphericity, ϕ
Crushed coal	0.75
Crushed sandstone	0.8–0.9
Sand (round)	0.92–0.98
Crushed glass	0.65
Mica	0.28
Sillimanite	0.75
Salt	0.84

to the area of the particle, S_p:

$$\phi = \frac{\pi d_{sp}^2|_{v_{sp}=v_p}}{S_p}. \tag{9.14}$$

Some examples of values of sphericity, ϕ, for non-spherical particles are given in Table 9.4. Mica, which is very thin compared to surface area, has a sphericity of 0.28, while rounded sand has a sphericity between 0.92 and 0.98.

Example 9.4. Calculate the sphericity of a cylinder whose diameter equals its length.

Solution 9.4. In this case, the characteristic dimension of the cylinder is its diameter. If the length were different than the diameter, the characteristic dimension must be clearly specified. The volume of a sphere, V_{sph}, and cylinder, V_{cyl}, is:

$$V_{sph} = \frac{\pi}{6} d_s^3,$$

$$V_{cyl} = \frac{\pi}{4} d_p^2 L = \frac{\pi}{4} d_p^3.$$

We now calculate the diameter of the sphere that has the same volume as the cylinder:

$$V_{sph} = V_{cyl},$$

$$\frac{\pi}{6} d_{sph}^3 = \frac{\pi}{4} d_{cyl}^3,$$

$$d_{sph} = \sqrt[3]{\frac{3}{2} d_{cyl}^3} = 1.145 d_{cyl}.$$

The surface area of the sphere having this diameter is:

$$S_{sph} = \pi d_{sph}^2 = \pi (1.145 d_{cyl})^2,$$

$$S_{cyl} = \frac{\pi}{2} d_{cyl}^2 + \pi d_{cyl}^2 = \frac{3}{2} \pi d_{cyl}^2.$$

Finally, the sphericity is the ratio of the two surface areas:

$$\phi = \frac{S_{sph}}{S_{cyl}} = \frac{\pi (1.145 d_{cyl})^2}{\frac{3}{2} \pi d_{cyl}^2} = 0.874.$$

Zou and Yu (1996) have related the Hausner ratio and sphericity by the following equation:

$$H_r = 1.478 \cdot 10^{-0.136\phi}. \tag{9.15}$$

Many models have been proposed to correlate the particle terminal velocity with physical properties, including sphericity. The equation of Haider and Levenspiel (1989) depends on the dimensionless number known as the Archimedes number, N_{Ar}:

$$u_t = \sqrt[3]{\frac{\mu(\rho_p - \rho_f)g}{\rho_f^2}} \left(\frac{18}{N_{Ar}^{2/3}} + \frac{2.335 - 1.744\phi}{N_{Ar}^{1/6}} \right)^{-1}, \quad 0.5 < \phi < 1, \tag{9.16}$$

where

$$N_{Ar} = \frac{d_p^3 \rho_f (\rho_p - \rho_f)g}{\mu^2}. \tag{9.17}$$

9.3.5 Reactor Pressure Drop (Fixed/Packed Beds)

Multi-phase reactors—gas-solids, liquid-solids (slurry), gas-liquid-solids—are the heart of most chemical plants. Their design and operation determines the profitability of any process. Figure 9.6 demonstrates the commercial reactor

FIGURE 9.6 Formaldehyde Multi-Tubular Fixed Bed Reactor (Image courtesy of Haldor-Topsøe)

to produce formaldehyde by the partial oxidation of methanol. The reactor is about 2 m in diameter and it contains 10 000 tubes each with a diameter of 22 mm. Catalyst pellets are charged to each tube. In the early 1960s and 1970s, spherical pellets of about 3 mm were used. Better productivity was achieved with cylindrical pellets. The cylindrical pellets were modified by boring a hole through the end to make an "elongated ring." The motivation for changing the dimensions was twofold: (a) increase productivity and (b) decrease pressure drop. The pressure drop, $\Delta P / \Delta Z$, across tubular reactors for spherical particles is calculated based on Ergun's equation:

$$\frac{\Delta P}{\Delta Z} = \frac{U_g}{\phi d_p} \frac{1 - \epsilon_v}{\epsilon_v^3} \left(150(1 - \epsilon_v) \frac{\mu}{\phi d_p} + 1.75 \rho_g U_g \right), \qquad (9.18)$$

where U_g is the superficial gas velocity (velocity assuming the tube is absent of particles) in m s^{-1}, ϵ_v is the void fraction, d_p is the average particle diameter in m, μ is the fluid viscosity in Pa s, ρ_g is the gas density in kg m^{-3}, and ϕ is the sphericity.

By maximizing the particle diameter, the pressure drop is minimized. However, the rate of diffusion of heat and mass in and out of the particle is poorer and therefore productivity suffers.

The Thiele modulus, ϕ_s, is a dimensionless number that relates the reaction rate to the diffusion rate. When the reaction is fast compared to diffusion, the Thiele modulus is high and the effectiveness factor, η, equals one—all of the catalyst is active. When the Thiele modulus is low, and the diffusion rate (mass transfer) is the controlling factor, the effectiveness factor of the catalyst is below 1. The Thiele modulus is derived based on a mass balance assuming the reaction is first order (proportional to the concentration of the reacting species) and isothermal conditions. As shown, in the equation, ϕ_s is directly proportional to the particle diameter, decreases with the square root of diffusivity, D_E (m^2 s^{-1}), and increases with the first-order rate constant, k (s^{-1}):

$$\phi_s = \frac{d_p}{6}\sqrt{\frac{k}{D_E}} = \frac{\text{reaction rate}}{\text{diffusion rate}}. \tag{9.19}$$

The effectiveness factor is calculated according to Equation (9.20). For a Thiele modulus of 2, the effectiveness factor is approximately 0.5 and it equals 0.1 for a value of ϕ_s of 10:

$$\eta = \frac{1}{\phi_s}\left(\frac{1}{\tanh(3\phi_s)} - \frac{1}{3\phi_s}\right). \tag{9.20}$$

Example 9.5. A plant produces formaldehyde from methanol in a multi-tubular reactor. The reactor contains 10 000 tubes. Each tube is 1.2 m long with an inside diameter of 25 mm. The catalyst particles are spherical with an average diameter of 3 mm and the inter-particle void fraction equals 0.45. The reactor operates with 6.0 vol% MeOH in air at a temperature of 337 °C, a pressure of 1.5 bara, and a superficial gas velocity of 2 m s^{-1}:

(a) Calculate the pressure drop (assuming the volume change with reaction is negligible).
(b) Calculate the effectiveness factor for a rate constant of 1.0 s^{-1} and a diffusivity of 0.007 cm^2 s^{-1}.
(c) What would the pressure drop and effectiveness factor be for a particle 2 mm in diameter?

Solution 9.5a. The gas density, $\rho_f = MP/RT$, equals 0.86 kg m^{-3}. Since 94% of the gas is air, we can assume that its viscosity is equal to that of air at the reactor operating temperature—0.000 033 Pa s. All other operating parameters were defined in the problem. The pressure drop is calculated from Ergun's equation:

$$\Delta P = \frac{U_g}{d_p}\frac{1-\epsilon_v}{\epsilon_v^3}\left(\frac{150(1-\epsilon_v)\mu}{d_p} + 1.75\rho_g U_g\right)\Delta Z$$

$$= \frac{2}{0.003}\frac{1-0.45}{0.45^3}\left(\frac{150(1-0.45)\cdot 0.000033}{0.003} + 1.75\cdot 0.86\cdot 2.0\right)\cdot 1.2$$

$$= 19\text{ kPa}.$$

Solution 9.5b. The Thiele modulus is:

$$\phi_s = \frac{d_p}{6}\sqrt{\frac{k}{D_E}} = \frac{0.003\text{ m}}{6}\sqrt{\frac{1\text{ s}^{-1}}{0.007\text{ cm}^2\text{ s}^{-1}\cdot(0.01\text{ m cm}^{-1})^2}} = 0.60.$$

FIGURE 9.7 Pellet Morphology in Chemical Reactors (Image courtesy of Clariant.) (For color version of this figure, please refer the color plate section at the end of the book.)

The effectiveness factor equals:

$$\eta = \frac{1}{\phi_s}\left(\frac{1}{\tanh(3\phi_s)} - \frac{1}{3\phi_s}\right) = \frac{1}{0.60}\left(\frac{1}{\tanh(3 \cdot 0.6)} - \frac{1}{3 \cdot 0.60}\right) = 0.83.$$

Solution 9.5c. For particles with a diameter of 2 mm, the pressure drop is 32 kPa and the effectiveness factor becomes 0.92.

Whereas Figure 9.3 shows many different shapes that are common to engineering practice, Figure 9.7 shows the different particle shapes that are used in chemical reactors.

9.3.6 Fluidization

While the particle diameter of fixed bed reactors is on the order of 1–5 mm, the particle size used in another type of reactor—fluidized bed reactors—is on the order of 50–200 μm. Because of the small diameters, the effectiveness factor is close to 1 in most cases. The Ergun equation characterizes the pressure drop across a bed of solids—at low gas velocities (relative to the particle terminal velocity). When the drag force of the upward moving fluid exceeds the weight of the particles, the particles become fluidized—they begin to move up and down and the solids bed itself behaves like a fluid: objects that are denser than the bed will fall through the bed while objects that are less dense will be remain at the top. Based on a force balance, the pressure drop across the bed, $\Delta P/L$, will be equal to the head of solids (neglecting frictional forces):

$$\frac{\Delta P}{L_{mf}} = (\rho_{mf} - \rho_f)g, \tag{9.21}$$

where U_{mf} is the minimum velocity at which the solids become fluidized, L_{mf} is the height of the bed at minimum fluidization, and ρ_f is the fluid density in kg m^{-3}.

Considering that the mass of solids charged to the bed, W, is the product of the density and volume ($V = L_{mf} X_A$):

$$W = \rho_{mf} V = \rho_{mf} L_{mf} X_A. \tag{9.22}$$

The pressure drop in the bed may be estimated based solely on the cross-sectional area and mass:

$$\Delta P = \frac{g W}{X_A}. \tag{9.23}$$

For the case of very small particles, when the viscous forces dominate ($N_{Re,mf} < 20$), the minimum fluidization velocity, u_{mf}, is given by:

$$u_{mf} = \frac{d_p^2 (\rho_p - \rho_f) g}{150 \mu} \frac{\epsilon_{mf}^2 \phi^2}{1 - \epsilon_{mf}}, \tag{9.24}$$

where

$$N_{Re,mf} = \frac{\rho_f u_{mf} d_p}{\mu}. \tag{9.25}$$

When inertial forces dominate and viscous forces are negligible ($N_{Re,mf} > 1000$):

$$u_{mf}^2 = \frac{d_p (\rho_p - \rho_f) g}{1.75 \rho_f} \epsilon_{mf}^3 \phi. \tag{9.26}$$

When neither viscous nor inertial forces dominate, the equation relating the operating conditions and particle properties is:

$$\frac{1.75}{\epsilon_{mf}^2 \phi} N_{Re,mf}^2 + \frac{150(1 - \epsilon_{mf})}{\epsilon_{mf}^3 \phi^2} N_{Re,mf} = N_{Ar}. \tag{9.27}$$

When ϵ_{mf} is unknown and for irregular shaped particles (unknown ϕ), Equation (9.27) may be written as (Kunii and Levenspiel, 1991):

$$K_1 N_{Re,mf}^2 + K_2 N_{Re,mf} = N_{Ar}. \tag{9.28}$$

For a large number of particles, Grace 1982 recommends that $\frac{1}{K_1} = 0.0408$ and $\frac{K_2}{2K_1} = 27.2$. The minimum fluidization velocity is a parameter that is used in many correlations to predict mass transfer rates, heat transfer and other hydrodynamic characteristics. Particle diameter and shape are critical parameters affecting u_{mf}.

Geldart (1973) studied the fluidization characteristics of many sorts of powders and then classified their behavior into four categories, as shown in Figure 9.8:

- Group A powders are considered aeratable, i.e. easily fluidizable. Typically, these particles have a diameter between 50 μm and 200 μm with a high fraction of fines (as much as 30%)—particles with a diameter between 20 μm and 40 μm.

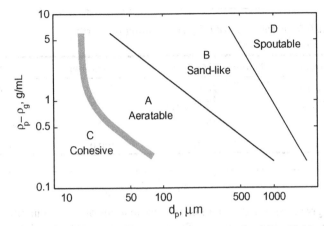

FIGURE 9.8 Powder Classification of Geldart in Air at Ambient Conditions (Geldart, 1973)

- Group B powders have a higher density and/or particle size compared to Group A powders. Sand is a typical example of a Group B powder.
- Group C powders are cohesive and fluidize with great difficulty. They generally have a lower particle density and/or particle size compared to Group A powders.
- Group D powders are difficult to fluidize—gas will channel through the bed and so shallow beds are common to minimize flow maldistribution.

9.4 PARTICLE SIZE DISTRIBUTION

The particle shape is only one factor in characterizing the average size of a sample population. The second critical parameter is the particle size distribution. Examples of particle sizes that are reasonably uniform include basmati rice and grain. For small particles however, the average population distribution may range from tens of μm to several hundred μm. This is particularly true for fluidized bed catalysts. (Fixed bed catalyst pellets—cylinders, for example—are generally considered to be monodispersed.) For polydispersed powders, selecting the appropriate diameter to represent the population is as difficult as selecting the appropriate shape for non-spherical particles.

9.4.1 Population of Particles

Algebraic expressions are used to reduce a population of particles with a wide size distribution to a single value. The expression relies on a physical characteristic of the population or a combination of characteristics—number, length (or some linear dimension representing the particle) area, mass, or volume, assuming that the particle density is invariant with volume.

TABLE 9.5 Distribution of Objects Orbiting the Earth (Wood-Kaczmar 1991)

Size (mm)	Number of Objects	% by Number	% by Mass
100–10 000	7000	0.2	99.96
10–100	17 500	0.5	0.03
1–10	3 500 000	99.3	0.01
Total	3 524 500	100.0	100.00

For a given particle distribution, the distributions of sizes in number, mass, and surface may differ significantly. For example, consider objects in orbit around the earth as summarized in Table 9.5. The majority of particles, some 3 500 000, are from 1 mm to 10 mm in diameter—these particles represent 99.3% of the total number of particles. However, they only account for 0.01% of the mass of particles orbiting earth. There are 7000 objects from 0.1 m to 10 m and these objects account for 99.96% of the total mass. If we consider mass, the average particle diameter is clearly between 0.10 m and 10 m, while if we consider the total number of objects, the average size lies between 1 mm and 10 mm. The question is, which distribution is most appropriate to calculate the characteristic diameter? The answer is that the diameter most representative of the distribution depends on the application: how much (in terms of mass) versus how many. Clean rooms for electronic applications are concerned with how many while catalytic reactors are concerned with how much.

Figure 9.9 illustrates the relationship between the size distribution function $F(d_p)$ and the particle size, d_p, in which the function represents the fraction of particles between d_p and $d_p + \Delta d_p$. The distribution of particles based on number is concentrated around the low particle sizes while the distribution based on volume (or mass) has a distribution with a weighting toward the larger particle size.

Several standard expressions of a population that are used include mode, median, and mean differences (arithmetic, geometric, square, harmonic, etc.), as shown in Figure 9.10.

The mode is the diameter that occurs most frequently and this value is the smallest compared to all other representation of a population of particles normally distributed. It is infrequently used. The median is most often used and represents the diameter for which 50% of the particles are smaller than that value (in terms of weight). It is also known as the d50. A d95 represents a diameter for which 95% of the particles are smaller than this value.

Besides the standard statistical representations of a population—mean, median, and mode—many other averaging techniques are used for powders. Consider a "population" of three particles with a diameter of 1 mm, 2 mm,

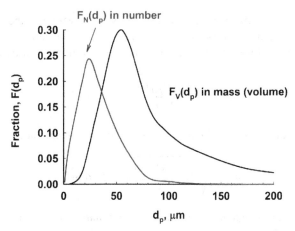

FIGURE 9.9   Distribution of Particles in Mass and Number

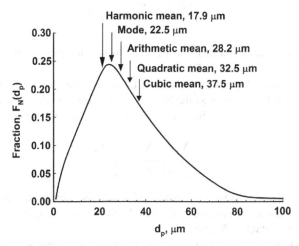

FIGURE 9.10   Different Options to Characterize the Size of Particles (According to Rhodes)

and 3 mm. The number length mean, d_{nl}, is the arithmetic mean of the three particles:

$$d_{nl} = \frac{\sum d_{p,i} N_i}{\sum N_i} = D_N[1,0].$$ (9.29)

This value is also written as $D_N[1,0]$ and in the case of the three particles, it will equal 2 mm. The first value in the expression $D_N[1,0]$—one—refers to the exponent with respect to diameter in the numerator. The second value—zero—refers to the exponent with respect to diameter in the denominator.

The number surface mean diameter considers averaging the population based on their surface area—it is the square root of the ratio of the sum of the surface area of the entire population and the total number of particles:

$$d_{ns} = \sqrt{\frac{\sum d_{p,i}^2 N_i}{\sum N_i}} = D_N[2,0]. \qquad (9.30)$$

For the example of the three particles, the surface mean diameter equals:

$$d_{ns} = \sqrt{\frac{1^2 + 2^2 + 3^2}{3}} = 2.16 \text{ mm.}$$

The volume mean diameter averages the population with respect to the total volume of each particle. It equals the cube root of the ratio of the sum of the volume of each particle and the number of particles:

$$d_{nv} = \sqrt[3]{\frac{\sum d_{p,i}^3 N_i}{\sum N_i}} = D_N[3,0]. \qquad (9.31)$$

The volume mean diameter is higher than the number mean and surface mean diameters and equals 2.29 mm.

In fluidization, the most common particle size distribution is the surface mean diameter. It is known as the Sauter mean diameter (d_{sv}) and represents the diameter of a sphere that has the same surface-to-volume ratio as the actual particle:

$$d_{sv} = \frac{\sum d_{p,i}^3 N_i}{\sum d_{p,i}^2 N_i} = \frac{\sum x_i}{\sum \frac{x_i}{d_{p,i}}}. \qquad (9.32)$$

This expression is most often expressed as a mass fraction, x_i, rather than in terms of the number of particles (N_i). For the three-particle system, the Sauter mean diameter equals 2.57 mm.

Table 9.6 summarizes many definitions of mean diameter. Note that the definition for number can be substituted for mass. The number length mean becomes the mass length mean—the symbol would be $D_m[1,0]$. Most often, data is reported as mass fraction and, therefore, to apply the number definitions requires a transformation from the mass fractions:

$$n_i = \frac{N_i}{\sum N_i} = \frac{\frac{x_i}{d_{p,i}^3}}{\sum \frac{x_i}{d_{p,i}^3}}, \qquad (9.33)$$

where x_i is the mass fraction (or percent) of each reported particle diameter, n_i is the number fraction (or percent) for each reported particle diameter, and N_i is the total number of particles in a given interval.

TABLE 9.6 Definitions of Number Mean Diameters (Allen 1990)

Designation	Definition and Symbol
Number length mean diameter	$D_N[1,0]: d_{nl} = \dfrac{\sum d_{p,i} N_i}{\sum N_i}$
Number surface mean diameter	$D_N[2,0]: d_{ns} = \sqrt{\dfrac{\sum d_{p,i}^2 N_i}{\sum N_i}}$
Number volume mean	$D_N[3,0]: d_{nv} = \sqrt[3]{\dfrac{\sum d_{p,i}^3 N_i}{\sum N_i}}$
Length surface mean diameter	$D_N[2,1]: d_{ls} = \dfrac{\sum d_{p,i}^2 N_i}{\sum d_{p,i} N_i}$
Length volume mean diameter	$D_N[3,1]: d_{lv} = \sqrt{\dfrac{\sum d_{p,i}^3 N_i}{\sum d_{p,i} N_i}}$
Surface-volume mean diameter	$D_N[3,2]: d_{sv} = \dfrac{\sum d_{p,i}^3 N_i}{\sum d_{p,i}^2 N_i} = \dfrac{\sum x_i}{\sum \frac{x_i}{d_{p,i}}}$
(Sauter mean diameter—SMD)	
Volume moment mean diameter	$D_N[4,3]: d_{lv} = \sqrt{\dfrac{\sum d_{p,i}^4 N_i}{\sum d_{p,i}^3 N_i}}$

The mass numbers for our example with three particles are much closer than the number distribution: the mass length mean diameter, $D_m[1,0]$, equals 2.72 mm. The mass surface mean diameter, $D_m[2,0]$, is 2.77 mm and the mass volume mean diameter, $D_m[3,0]$, is 2.80 mm.

Example 9.6. Table E9.6 summarizes the particle size distribution of commercial vanadium pyrophosphate catalyst to produce maleic anhydride from butane in a circulating fluidized bed reactor. Calculate $D_N[1,0], D_N[2,0], D_N[3,0], D_N[3,2]$, and $D_N[4,3]$.

Solution 9.6. Figure E9.6S is a plot of the wt% of each fraction as a function of the particle size and show a pseudo-bimodal distribution. There is a high fraction of fines and the highest fraction has a diameter of 55 μm. Because of the granularity of the particle size (small intervals between each particle size), a simple arithmetic mean of the upper and lower values is sufficient, i.e. the average particle diameter of the fraction between 32 μm and 44 μm is 38 μm. The Sauter mean diameter may be calculated directly from the data, since it is defined by a number fraction as well as a mass fraction in Table 9.6. Note that the sum of the weight fraction is not 1 but 0.994! Table E9.6S lists the average particle diameter for each segment of the wt% as well as the percentage based

TABLE E9.6 PSD for Commercial Vanadium Pyrophosphate Catalyst

d_p	wt%
$d_p < 10$	0
$10 < d_p < 20$	1.3
$20 < d_p < 25$	4.7
$25 < d_p < 32$	11.5
$32 < d_p < 44$	22
$44 < d_p < 66$	32.8
$66 < d_p < 88$	1.3
$88 < d_p < 100$	4.6
$100 < d_p < 125$	7.8
$125 < d_p < 150$	5.5
$150 < d_p < 200$	6.2
$200 < d_p < 250$	1.7

on the number, $N\%$. This value is calculated from the mass fraction as:

$$\frac{N_i}{\sum N_i} = \frac{\frac{x_i}{d_{p,i}^3}}{\sum \frac{x_i}{d_{p,i}^3}}.$$

$D[1,0] = 30 \ \mu m, D[2,0] = 33 \ \mu m, D[3,0] = 37 \ \mu m, D[3,2] = 48 \ \mu m,$
$D[4,3] = 68 \ \mu m.$

9.5 SAMPLING

The physical and chemical characteristics of several tons of material are often assumed based on a sample that may be no more than a couple of grams. The probability of withdrawing a representative sample is remote even when standardized procedures are followed with the same equipment and the same operator. Atmospheric monitoring is an example in which sampling is a critical factor that is mandated by government legislation. It is conducted to monitor health hazards and the EPA in the USA stipulates that the particulate matter should be inferior to 75 $\mu g \ m^{-3}$ on an average annual basis and that the instantaneous maximum concentration should not exceed 260 $\mu g \ m^{-3}$ (EPA, 2011). Manufacturers must monitor and control their particulate discharge rates incurring costs for cyclones, filters, scrubbers, electrostatic precipitators, etc.

Sampling can be broken into primary and secondary sampling. Primary sampling relates to removing a representative sample from the bulk—how to get a sample from a train car, drum, etc. ISO and other standards (ISO 13320 (3),

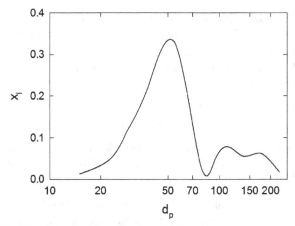

FIGURE E9.6S PSD of Vanadium Pyrophosphate Catalyst

TABLE E9.6S PSD for Commercial Vanadium Pyrophosphate Catalyst Solution

$d_{p,i}$ (μ m)	wt_i (%)	N_i (%)
15	1.3	20.17
22.5	4.7	21.60
28.5	11.5	26.01
38	22	20.99
55	32.8	10.32
77	1.3	0.15
94	4.6	0.29
112.5	7.8	0.29
137.5	5.5	0.11
175	6.2	0.06
225	1.7	0.01

USP 429) stress the importance of sampling and advise that a representative sample be prepared using a sample splitting technique. The golden rules of sampling suggested by Allen 1990 are:

1. Sample a moving stream;
2. Sample in increments the entire population instead of taking a single sample at one point.

Allen (1990) has defined the minimum mass of a sample to be collected in terms of the particle size distribution, particle density, ρ_p (kg m^{-3}) and mass fraction of the coarsest size class, w_λ. For a size range less than $\sqrt{2}$: 1 (120 μm:85 μm, for example) and where the mass fraction of the coarsest size class is less than 50 wt%:

$$m_s = \frac{1}{2} \frac{\rho_p}{\sigma^2} \left(\frac{1}{w_\lambda} - 2 \right) d_{p,\lambda}^3 \cdot 10^6, \tag{9.34}$$

where m_s is the minimum sampling mass in kg, $d_{p,\lambda}$ is the arithmetic mean of the coarsest fraction in mm, and σ^2 is the variance of the sampling error (generally 5%).

Example 9.7. Approximately 40% of the oil produced in the world is cracked catalytically to smaller molecules with zeolite catalysts—known as FCC (fluid catalytic cracking). The catalyst has an average diameter around 70 μm and it becomes coarser with time as the fine fraction of the powder is lost in the cyclones. For a FCC unit containing 200 t of catalyst, what is the smallest sample size required to achieve a sampling error less than 5% if the coarsest size range is from 177 μm to 210 μm. The particle density of FCC is 1200 kg m^{-3}.

Solution 9.7. The arithmetic mean of the coarsest fraction is:

$$d_{p,\lambda} = \frac{1}{2}(177 + 210) = 194 \, \mu m.$$

Assuming the coarsest sample size represents 5% of the total mass, the sample size to be collected, m_s, is:

$$m_s = \frac{1}{2} \frac{1200 \text{ kg m}^{-3}}{0.05} \left(\frac{1}{0.05} - 2 \right) (0.194 \text{ mm}^3)^3 \cdot 10^{-6} = 0.016 \text{ kg}.$$

The required sample size increases considerably as the coarsest fraction decreases.

Secondary sampling is how we place a representative sample into the instrument from the container that arrives in the laboratory. Care must be taken with sampling when dispersing a powder into a liquid. Many methods call for pre-dispersing the powder in a beaker and then pipetting the sample into the analyzer. When following this approach it is better to mix a concentrated paste in the beaker in order to minimize sampling bias during the transfer to the analyzer. If this is not practical for whichever reason, then the beaker pre-dispersion should be continuously stirred and the sample should be extracted halfway between the center of the beaker and the wall and also halfway between the liquid surface and the bottom of the beaker.

Suspensions and emulsions can sometimes be easily measured using the continuous phase of the original sample as the diluent in the analyzer. When powders are dispersed in liquid, the solvent must meet the following criteria:

- Negligible reactivity with powder;
- Does not swell or shrink particles by more than 5% in diameter;
- Solubility must be less than 5 g powder in 1 kg liquid;
- Have a refractive index (RI) different than the sample;
- Be free from bubbles and particles;
- Have suitable viscosity to enable recirculation; and
- Be chemically compatible with materials in the analyzer.

Dispersing powders in liquid can often present challenges. The ISO 14887 standard provides useful insight into this realm. Among the suggestions in ISO 14877 is to prepare the sample on a slide and look at it under a microscope. Determine if you are looking at individual particles or clumps. See if exposing the sample to an external ultrasonic probe eliminates the clumps.

Surfactants are often required to wet the powder for proper dispersion. ISO 14887 provides a comprehensive listing of commercially available dispersing agents. Some instrument manufacturers (Horiba, for example) recommend surfactants including Micro 90 solution (also good for cleaning the instrument), Triton X-100, Igepal CA-630, Tween 80, and lecithin.

Once a powder is dispersed, it sometimes helps to add a stabilizer (or admixture) to the sample, such as sodium hexametaphosphate. The stabilizer alters the charge on the surface of the particles, preventing re-agglomeration.

9.5.1 Stability Testing

After the dispersing liquid or mixture has been chosen, test the system for stability by collecting multiple measurements as a function of time. Measuring the recirculating sample should generate extremely reproducible results. The sample should be measured at least three times over a time frame of several minutes. A particle size distribution which steadily shifts to a finer particle side together with an increase in light transmission may indicate dissolution. An increase in particle size may indicate agglomeration or swelling. An increase in transmission alongside the disappearance of the coarsest particles may indicate settling. Random variations are more difficult to interpret but could arise from thermal fluctuations or poor mixing.

9.6 PSD ANALYTICAL TECHNIQUES

Techniques for analyzing the size and shape of particles and powders are numerous. Some are based on the light, others on ultrasound, an electric field, or gravity. They can be classified according to the scope and scale of observation (Webb, 2011):

- *Visual methods:* Microscopy (optical, electronic, and electronic scanning) and image analysis;
- *Separation methods:* Screening, classification, impact and electrostatic differential mobility, sedimentation;

- *Continuous scanning methods:* Electrical resistance, optical detection zones;
- *Field scan techniques:* Laser diffraction, sound attenuation, the photon correlation spectroscopy;
- *Surface techniques:* Permeability and adsorption.

It is important to define the characteristic dimension of each analytical technique, some of which are reported in Table 9.7. Electro-zone techniques will typically report a volume average, $D_m[3,0]$ and a number average $D_N[3,0]$. Most optical techniques report a surface dimension, $D_N[2,0]$, while microscopy will report a length dimension, $D_N[1,0]$ (although, if combined with image analysis, they also report a surface dimension). Sedimentation techniques separate particles based on the Stokes diameter. Sieve analysis separates powders based on the narrowest projected surface area but sphericity can also play a role for particles elongated in one direction. The mass of the particles on each sieve is weighed, so the defining particle diameter would be $D_m[2,0]$. When the distribution is not explicitly mentioned, it can be deduced by examining the population distribution curve. If the population of particles at the low end is low, then most likely mass fractions are reported. When the fraction of the smallest particles is high, it is most likely that the particles are counted.

Figure 9.11 shows the characteristic operating ranges of a selected number of instruments. Sieve analysis is used for larger particles from as low as about $10 \mu m$ to $1000 \mu m$. Microscopy is adequate from below $1 \mu m$ to as much as $1000 \mu m$. For fine particles, laser diffraction is the most broadly accepted technology and coincidentally can measure the widest range of particle sizes—it can measure particles as small as low as 40 nm.

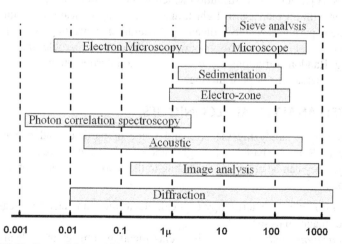

FIGURE 9.11 Range of Selected Measurement Techniques

TABLE 9.7 Analytical Technique and Characteristic Dimension

Analytical Technique	Size Range (μm)	Representative Dimension
Sieving		
Dry	>10	Combination of D_m [2,0] and ϕ
Wet	2–500	
Image analysis		
Optical	0.2–100	D_N [2,0]
Electron microscopy	0.01–500	D_N [1,0] and D_N [2,0]
Radiation scattering		
Laser diffraction (Horiba)	0.01–3000	D_m [4.3], D_m [3,2]
Electrical zone sensing		
(Coulter counter)	0.6–1200	D_N [3,0], D_m [3,0]
	1–800	
Entrainment, elutriation		
Laminar flow	3–75	Stokes diameter, d_t
Cyclone	8–50	
Gravity Sedimentation		
Pipette	1–100	
Photo-extinction	0.05–100	Stokes diameter
X-ray	0.1–130	
Centrifugal classification	0.5–50	Stokes diameter

Agreement between analytical techniques is often poor. Moreover, reproducibility of measured distributions with the same technique but different instruments may be in disagreement by over 10%. Typically, when monitoring a process, or for quality control of the manufacture of a powder, the same instrument should be used and, whenever possible, the same operator. Between technologies for non-spherical particles, on average laser diffraction (Horiba) will give a lower value compared to sieve analysis and an electrical zone sensing (Coulter counter):

$$(d_{sv})_{\text{laser}} = 1.35(d_a)_{\text{sieve}} = 1.2(d_v)_{\text{electrosensing}}. \qquad (9.35)$$

9.6.1 Sieve Analysis

Screening is one of the simplest, cheapest, and most commonly used techniques to classify particles according to their size. It was one of the first separation techniques used by ancient civilizations for the preparation of foodstuffs,

TABLE 9.8 Standard US Mesh Sizes (McCabe and Smith, 1976)

Mesh	μm	Mesh	μm	Mesh	μm
4	4760	20	840	100	149
5	4000	25	710	120	125
6	3360	30	590	140	105
7	2830	35	500	170	88
8	2380	40	420	200	74
10	2000	45	350	230	62
12	1680	50	297	270	53
14	1410	60	250	325	44
16	1190	70	210	400	37
18	1000	80	177	500	31

including the Egyptians. They were made of woven fabric and some were made by punching holes in plates. Agricola illustrated woven wire sieves in 1556. In 1867, Rittinger suggested a standardized progression of aperture size $\sqrt{2}$ with 75 mm as the reference point. Between successive screens in a series, the open area is double. Modern standards use a $\sqrt[4]{2}$ progression (1.189) (except for the French AFNOR standard).

Sieving separates powders solely based on their size—it is independent of density or surface properties (although the sieving time may be a function of density). Table 9.8 summarizes the sizes of the US Mesh from 31 μm to 4760 μm.

To separate a powder into particle fractions, a series of sieves are stacked one on top of the other starting with a collection pan (receiver) followed by the screen with the smallest aperture. The powder is weighed then poured on to the top screen. A lid is placed on the top sieve and the stack is then shaken by hand (very inefficient) or vibrated with a machine. National standards BS1796, ASTM 452, and ASTM C136A each recommend different criteria for the length of time to vibrate. The first vibration period should be on the order of 20 min. After this point, the sieves are weighed and vibrated an additional 10 min. If the change in weight from one period to another is greater than 0.5%, the stack should be vibrated an additional 10 min (ASTM 452). Compressed air sieving is recommended for particles up to 20 μm and wet sieving for particles that have a tendency to form agglomerates. Wet sieving is useful for particles down to 5 μm.

The result of a sieve analysis is a tabulation of the mass of powder (or mass fraction) of each screen increment. Two numbers are required to specify the particle size: the screen through which the particles pass and the screen on which they are retained. The data may be reported as 120/170—the particles were retained on the 170 mesh screen and passed through the 120 mess screen.

This would mean that the particles on the 170 mesh screen lie between 88 μm and 125 μm. This representation of the data is referred to as a "differential analysis." In a cumulative analysis, the particles retained on each screen are summed sequentially starting from either the receiver or the screen with the largest aperture.

Since the particles must pass through an opening between wires forming a square mesh, the most appropriate diameter to represent the size distribution would be $D[2,0]$.

Errors incurred during the sieving operation may include:

1. Insufficient duration;
2. Wear on the sieves (thus allowing larger particles through a given sieve);
3. Errors of sampling (the powder loaded to the stack is unrepresentative of the populations);
4. Measurement errors;
5. Operational errors (too much or too little sample, humidity causing agglomeration, or static);
6. Sensitivity of powder to vibration (resulting in breakage); and
7. Interpretation (particularly related to particle shape).

9.6.2 Laser Diffraction

Co-author: Mark Bumiller

Laser diffraction is the most popular, widespread modern sizing technology. It has been used successfully for an array of applications across many industries. The current state-of-the-art instruments have a measurement range from 10 nm to 3 mm, the ability to measure suspensions, emulsions, and dry powders within seconds, full push-button automation, and software. The instrument consists of at least one source of high intensity, monochromatic light, a sample handling system to control the interaction of particles and incident light, and an array of photodiodes to detect the scattered light over a wide range of angles.

The primary function of the array of photodiodes is to record the angle and intensity of scattered light. This information is then input into an algorithm that converts the scattered light data to a particle size distribution. The algorithm consists of one of two optical models—the Fraunhofer approximation and the Mie scattering theory—with the mathematical transformations necessary to generate particle size data from scattered light. The Fraunhofer approximation, which enjoyed popularity in older laser diffraction instruments, assumes that particles:

- are spherical;
- are opaque;
- scatter equivalently at wide angles and narrow angles;
- interact with light in a different manner than the medium.

These assumptions are reasonable for particles greater than 20 μm. The Mie scattering theory is a closed-form solution (not an approximation) to Maxwell's electromagnetic equations for light scattering from spheres. This solution is superior to the Fraunhofer approximation: it has a much greater sensitivity to smaller particles (which scatter at wider angles); it includes a wide range of particle opacity (i.e. light absorption); and, it is user-friendly since only the refractive index of the particle and dispersing medium is required. Accounting for the light that refracts through the particle allows for accurate measurement even in cases of significant transparency. The Mie theory makes certain assumptions:

- all particles are spherical;
- each particle is of the same material;
- the refractive indices of particle and dispersing medium are known.

The over-reporting of small particles is a typical error seen when using the Fraunhofer approximation.

9.6.3 Microscopy

Co-author: Milad Aghabararnejad
Optical microscopy is often used as an absolute method of analysis of particle size, since it is the only method that allows on the one hand, to examine the particles individually for shape and composition, and on the other hand, to measure the diameter with a resolution of 200 nm (for optical microscopes) and particle sizes as low as 5 μm. It can provide information on the size distribution in number and form. Different diameters can be measured: the equivalent diameter of Martin, d_M, Feret diameter, d_F, the diameter of the projected area, d_a, etc. New microscopes use cameras to output a digital image for analysis. Below 5 μm, the images are blurred due to diffraction effects.

For particles less than 5 μm, other electron microscopy can be used, such as transmission electron microscopy (TEM), which can measure particles up to 1 nm, and scanning electron microscopy (SEM), which can measure particles up to a few nanometers. In TEM, electrons penetrate the sample while for SEM an electron beam scans the surface resulting in secondary electron emission, backscattered electrons, light, and X-rays. SEM is faster than TEM and gives a greater three-dimensional depth. Magnifications reach as high as $100\,000\times$ with resolutions of 15 nm. Significant sample preparation is required for both microscopy techniques.

9.6.4 Electrical Sensing Instruments

Electrical sensing instruments measure changes in electrical conductivity when particles pass through an orifice together with a conductive fluid. The magnitude of the change in conductivity is representative of the particle size and the number of particles passing the aperture equals the number of pulses detected. The principle relies on suspending a powder in a liquid and circulating the suspension through an orifice across which a voltage is applied. The capacitance changes each time a particle passes the orifice.

The instrument is quite easy to use but there are certain limitations:

- The electrolyte must be compatible with the particle (no swelling, breaking, or dissolution);
- The method is slow relative to laser diffraction (but substantially faster the sieving);
- Dense or large particles may settle in the beaker and thus are not measured.

9.7 SURFACE AREA

Surface area, together with density and particle size, is a critical parameter in many chemical engineering applications. It is a determining factor in making catalyst; moreover, the loss of surface area often correlates with loss in activity.

The theory of physical adsorption of gas molecules on solid surfaces was derived by Brunauer, Emmett, and Teller (BET). The theory serves as the basis for the most widely used technique to assess specific surface area of powders and solids. It extends the Langmuir isotherm concept.

The BET theory accounts for monolayer and multilayer molecular adsorption according to the following hypotheses: (a) gas molecules physically adsorb on a solid in layers infinitely; (b) there is no interaction between each adsorption layer; and (c) the Langmuir theory can be applied to each layer. The resulting BET equation is expressed by:

$$\frac{1}{W\left(\frac{P_0}{P} - 1\right)} = \frac{c-1}{W_m c}\frac{P}{P_0} + \frac{1}{W_m c}, \tag{9.36}$$

where P and P_0 are the equilibrium and the saturation pressure of adsorbates at the temperature of adsorption, W is the adsorbed gas quantity, and W_m is the monolayer adsorbed gas quantity. c is the BET constant, which is expressed by:

$$c = \exp\left(\frac{E_1 - E_L}{RT}\right), \tag{9.37}$$

where E_1 is the heat of adsorption for the first layer, and E_L is that for the second and higher layers and is equal to the heat of liquefaction. Equation (9.36) is an adsorption isotherm and can be plotted as a straight line with $\frac{1}{W\left(\frac{P_0}{P}-1\right)}$ on the y-axis and $\phi = \frac{P}{P_0}$ on the x-axis.

The linear relationship of this equation is maintained only in the range of $0.05 < \frac{P}{P_0} < 0.35$. The value of the slope A and the y-intercept I of the line are used to calculate the monolayer adsorbed gas quantity W_m and the BET constant c. Consider I and A as intercept and slope of the BET plot:

$$W_m = \frac{1}{A + I},$$ (9.38)

$$c = 1 + \frac{A}{I}.$$ (9.39)

The BET method is widely used in surface science for the calculation of surface areas of solids by physical adsorption of gas molecules. The total surface area, S_{total}, is evaluated by the following equation:

$$S_{total} = \frac{W_m N A}{M},$$ (9.40)

where N is Avogadro's number (6.022×10^{23} molecule mol^{-1}), A is the area of the cross-section of adsorbate molecules ($16.2\ A^2\ mol^{-1}$ for nitrogen), and M is the molecular weight of the adsorbent gas in g mol^{-1}.

The measured surface area includes the entire surface accessible to the gas whether external or internal. Prior to the measurement, the sample is pre-treated at high temperature in vacuum in order to remove any contaminants. To cause sufficient gas to be adsorbed for surface area measurement, the solid must be cooled (normally to the boiling point of the gas). Most often, nitrogen is the adsorbate and the solid is cooled with liquid nitrogen. Adsorption continues until the amount of nitrogen adsorbed is in equilibrium with the concentration in the gas phase. This amount is close to that needed to cover the surface in a monolayer (see Table E9.2).

Example 9.8. An α-alumina powder with an average d_p of 130 μm and total surface area of 46 $m^2\ g^{-1}$ is impregnated by an 8 M solution of nickel nitrate

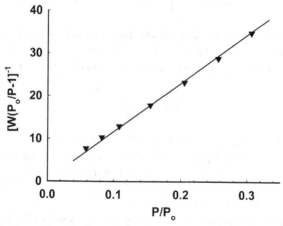

FIGURE E9.8S Seven Adsorption Point BET Plot

TABLE E9.8 BET Results (P in mmHg, Volume in cm^3)

| Run 1 ($m = 0.2413$ g) | | Run 2 ($m = 0.2251$ g) | | Run 3 ($m = 0.2482$ g) | |
P ($P_0 = 768$)	Adsorbed Volume N$_2$	P ($P_0 = 780$)	Adsorbed Volume N$_2$	P ($P_0 = 769$)	Adsorbed Volume N$_2$
45	1.6	41	1.3	45	1.6
63	1.7	64	1.4	63	1.7
83	1.8	85	1.5	83	1.8
118	2.0	123	1.6	119	2.0
157	2.2	162	1.7	157	2.1
196	2.3	200	1.8	197	2.3
235	2.4	239	1.9	234	2.5

hexahydrate Ni(NO$_3$)$_2 \cdot$ 6H$_2$O. A seven-point BET test was completed using nitrogen as adsorbate on the impregnated powders. The BET test was repeated three times and Table E9.8 represents the volume of adsorbed nitrogen at different pressures. Calculate the total surface area.

Solution 9.8. The first step is to plot $\dfrac{1}{W\left(\frac{P_0}{P}-1\right)}$ versus $\frac{P}{P_0}$. W is the mass of adsorbed nitrogen per mass of sample (in g). Table E9.8S shows the calculated value for the first run.

TABLE E9.8S First Run

$\dfrac{P}{P_0}$	W	$\dfrac{1}{W\left(\frac{P_0}{P}-1\right)}$
0.058	0.0081	7.7
0.082	0.0087	10.3
0.108	0.0093	12.9
0.154	0.0102	17.8
0.205	0.0111	23.2
0.256	0.0119	28.8
0.306	0.0127	34.8

Nitrogen is considered an ideal gas. Figure E9.8S shows the BET plot of the first run.

Using a linear regression, the slope is:

$$A = 108.7$$

the intercept is:

$$I = 1.2$$

and the correlation coefficient is:

$$r = 0.999,$$

$$W_m = \frac{1}{A + I} = \frac{1}{108.7 + 1.2} = 0.009 \, \frac{g \, N_2}{g \, sample},$$

$$S_{total} = \frac{W_m N A}{M} = \frac{\left(0.009 \, \frac{g \, N_2}{g \, sample}\right) \cdot 6.022 \times 10^{23} \cdot 16.2 \times 10^{-20} \, \frac{m^2}{mol \, N_2}}{28.0134 \, \frac{g \, N_2}{mol \, N_2}}$$

$$= 31.7 \, \frac{m^2}{g \, sample}.$$

The same procedure can be used for the second and third runs. The surface areas of the three runs are (in $m^2 \, g^{-1}$): 31.7, 26.7, and 31.4.

9.8 EXERCISES

9.1 Calculate the total void fraction of bulk solids if both the inter- and intra-void fractions equal 0.4.

9.2 To evaluate the particle density using a 100 ml graduated cylinder, it is first "calibrated" with water: the meniscus reached 98.3 ml when exactly 100.0 g of water is charged to it at 20 °C. The procedure to measure the particle density involves first drying the cylinder, adding 50 g of water, then measuring approximately 25 g of powder and pouring it into the cylinder, and then finally adding water until it reaches the 98.3 ml mark. Calculate the particle density and uncertainty for the following five measurements:

- *Initial mass of water:* 50.3, 50.9, 48.7, 50.2, and 49.5.
- *Mass of catalyst added:* 23.9, 25.2, 27.3, 24.8, and 26.3.
- *Mass of water added to 98.3 ml:* 37.0, 37.8, 36.9, 38.0, and 37.5.

9.3 Determine the Hausner ratio and its uncertainty for a 100.0 g sample of catalyst that is poured into a 100 ml graduated cylinder. Based on the Zou and Yu correlation, what is the sphericity of the catalyst particles:

- *Poured height:* 87.3, 86.9, 88.5, 87.0, and 85.8.
- *Tapped height:* 80.2, 80.9, 81.3, 80.3, and 79.8.

9.4 A Peregrine falcon can dive at a velocity of 300 km h^{-1}. They weigh as little as 0.91 kg and measure up to 0.59 m in length. Calculate the diameter of the falcon assuming that its shape approximates that of a cylinder.

9.5 Calculate the sphericity of a pyramid whose base equals the height.

9.6 Calculate the sphericity of a hollow cylinder 4 mm in diameter, 6 mm in length with a 2 mm hole.

9.7 What is the sphericity of a sphere with a cylindrical hole passing through it whose diameter equals one fourth the diameter of the particle?

9.8 What is the sphericity of a hexahedron (cube)?

9.9 What is the surface area of coal with a sphericity of 0.75 and a volume of 1.5 mm^3? Calculate the surface-volume diameter (*A. Benamer*).

9.10 The particle size of crushed sandstone was measured with a Coulter Counter, a Sedigraph, and by Permeability, and the diameters for each were $d_v = 48.2$ μm, $d_{st} = 45.6$ μm, and $d_{sv} = 38.2$ μm, respectively. Are these results consistent with expectation? Calculate the expected sphericity based on this data (*C. Ndaye*).

9.11 Ceramic is produced from powders including oxides, ferrites, and silicates. Using a transmission electron microscope (TEM) calibration standard of 1 pixel = 1 μm, coupled with image processing, the company obtains the data shown in Table Q9.11 (*A. Artin*):

   (a) Determine the mode diameters of the size distribution in μm.

   (b) Calculate the average diameter in numbers.

   (c) Calculate the Sauter mean diameter d_{sv}.

   (d) The Sauter mean diameter from a Coulter Counter analysis was reported to be 23 μm. Are the powders in the form of platelets?

TABLE Q9.11 PSD

Size of Particles (pixels)	Number of Particles
]0, 25]	1
]25, 50]	426
]50, 100]	2940
]100, 200]	6593
]200, 350]	6127
]350, 600]	3649
]600, 790]	2468
]790, 1500]	284
]1500, ...]	0

9.12 A 250 g powder sample is loaded into a 40 ml vial. Mercury is pumped into the vial and the volume $V_{AB} = 1.5$ cm^3 and $V_{BC} = 2.5$ cm^3.

Calculate the skeletal density, the particle density, and the bulk density (R. Tohmé).

9.13 The results of a particle size analysis are shown in Table Q9.13:

TABLE Q9.13 Salt Particle Size Analysis

Mesh	d_p (μm)	Weight of Retained Particles (g)
20	840	0
30	590	4.8
40	420	8.9
60	250	24
80	177	19
120	125	12
140	105	5.7

(a) What is the d_{50} for this powder.

(b) Calculate the $D_N[1,0]$ and $D_N[3,2]$ for the powder that passes through the 40 Mesh sieve.

9.14 By controlling the precipitation conditions, zinc oxide particles will form an octahedral shape with a narrow particle size distribution (R. Silverwood):

(a) Calculate the sphericity of an octahedral particle.

(b) Based on Table Q9.14, calculate the $D[3,2]$.

TABLE Q9.14 PSD of Zinc Oxide

d_p (μm)	Fraction
$5 < d_p \leqslant 10$	0.09
$10 < d_p \leqslant 15$	0.37
$15 < d_p \leqslant 20$	0.42
$20 < d_p \leqslant 25$	0.12

(c) What is the average particle diameter?

9.15 The measured diameters (d_p) of a group of powders are: 0.12, 0.13, 0.14, 0.14, 0.15, 0.13, 0.12, 0.12, 0.11, 0.14, 0.15, 0.15, 0.13, 0.11, 0.20, and 0.13. Calculate d_{nl}, d_{n-sa}, d_{n-v}, and d_{sv}. Based on Chauvenet's criterion, can you reject one of the measurements?

9.16 Derive the expression for the minimum fluidization velocity for the case where viscous forces dominate.

9.17 What is the particle terminal velocity in air of sand that is $150\,\mu m$ in diameter? $60\,\mu m$ in diameter? $30\,\mu m$ in diameter?

9.18 Calculate the Archimedes number and the terminal velocity of a grain of sand $250\,\mu m$ in diameter in water.

9.19 A new process to convert glycerol to acrolein uses a catalytic fluidized bed reactor:

$$C_3H_8O_3 \rightarrow C_3H_4O + 2H_2O.$$

The particle size distribution is given in Table Q9.19:

(a) Calculate the hydraulic diameter and the Sauter mean diameter.
(b) Of the four particle types given below, which has the greatest surface area?
(c) Which particle has the smallest surface area?
(d) Is the charge loss in a tubular reactor greater for particle (i) or (ii)?
(e) Is the charge loss in a tubular reactor greater for particle (ii) or (iii)?
(f) What is the equivalent diameter and the sphericity of particle (iv)?

Particles:

(i) Sphere of diameter d_p.
(ii) Cylinder of diameter and height d_p.
(iii) Hollow cylinder of inner diameter $d_p/2$, full diameter d_p, and height d_p.
(iv) Sphere of diameter d_p with a hole of diameter $d_p/4$ passing through it.

TABLE Q9.19 Catalyst Particle Size Distribution

$45 < d_p \leqslant 68$	4.7
$68 < d_p \leqslant 89$	16.6
$89 < d_p \leqslant 102$	15.5
$102 < d_p \leqslant 117$	16.7
$117 < d_p \leqslant 133$	15.3
$133 < d_p \leqslant 153$	11.9
$153 < d_p \leqslant 175$	8.1
$175 < d_p \leqslant 200$	5.0

9.20 Demonstrate the equivalence of the two expressions of the Sauter mean diameter:

$$D_N[3,2] = d_{sv} = \frac{\sum d_{p,i}^3 dN}{\sum d_{p,i}^2 dN}$$

and

$$d_{sv} = \frac{1}{\sum \frac{x_i}{d_{p,i}}}.$$

9.21 Methanol is produced in a tubular reactor with a Cu-Zr-Zn catalyst. The particle size distribution is obtained from sieving (see Table Q9.21).

$$\frac{dP}{dZ} = -\frac{U_g}{\phi d_p}\frac{1-\epsilon_v}{\epsilon_v}\left(\frac{150(1-\epsilon_v)}{\phi d_p}\mu + 1.75\rho_0 U_g\right),$$

where $\epsilon_v = 0.41 \pm 3\%$, $\mu = 0.000030$ Pa s $\pm 3\%$, $Z = 3.2$ m, $U_g = (2.0 \pm 0.2)$ m s^{-1}, $\rho_0 = 1.0$ kg m^{-3}, and $\phi = 0.95 \pm 0.05$:

(a) What definition of average particle is the most appropriate to characterize these particles? What characteristic diameter represents the flow rate of a tubular reactor?

(b) What is the average diameter and the charge loss across the bed?

(c) The charge loss across the bed is greater than predicted according to the calculations in (b). Identify three possible reasons for this.

(d) Calculate the uncertainty in the charge loss taking into account the uncertainties.

TABLE Q9.21 Particle Size Analysis by Sieving

Mesh	d_p (μm)	Mass %	Uncertainty (%)
4	4760		
6	3360	10	1
10	2000	20	2
12	1680	30	3
14	1410	20	2
18	1000	20	2

REFERENCES

Allen, T., 1990. Particle Size Measurement, fourth ed. Chapman and Hall.

ASTM 452, 2008. Standard Test Method for Sieve Analysis of Surfing for Asphalt. Roofing Products.

ASTM C136A, 1994. Field Sampling and Laboratory Testing Procedures Rejuvaseal Pavement Sealer.

BS1796, 1989. Test sieving. Methods using test sieves of woven wire cloth perforated metal plate.

Geldart, D., 1973. Types of gas fluidization. Powder Technology 7 (5), 285–292. doi: http://dx.doi.org/10.1016/0032-5910(73)80037-3.

Grace, J.R., 1982. In: Hetsroni, G. (Ed.), Handbook of Multiphase Systems, Hemisphere, p. 8-1.

Haider, A., Levenspiel, O., 1989. Drag coefficient and terminal velocity of spherical and non-spherical particles. Powder Technology 58, 63–70.

ISO 13320, 1999. Analyse granulométrique—Méthode par diffraction laser—Partie 1: Principes généraux.

ISO 14887, 2000. Sample preparation—dispersing procedures for powders in liquids.

ISO 9276-6, 2008. Representation of results of particle size analysis—Part 6: Descriptive and quantitative representation of particle shape and morphology.

Kunii, D., Levenspiel, O., 1991. Fluidization Engineering. Butterworth-Heinemann Series in Chemical Engineering, second ed.

McCabe, W.L., Smith, J.C., 1976. Unit Operations of Chemical Engineering. McGraw-Hill Chemical Engineering Series, third ed.

EPA, 2011. Particulate Matter Sampling. Retrieved 2011 from APTI435:ATMOSPHERIC SAMPLING COURSE: <http://www.epa.gov/apti/Materials/APT%20435%20student/Student%20Manual/Chapter_4_noTOC-cover_MRpf>.

Patience, G.S., Hamdine, M., Senécal, K., Detuncq, B., 2011. Méthodes expérimentales et instrumentation en génie chimique, third ed. Presses Internationales Polytechnique.

van Oss, H.G., 2011. United States Geological Survey, Mineral Program Cement Report, January.

USP 429. Light Diffraction Measurement of Particle Size.

Webb, P.A., 2011. Interpretation of Particle Size Reported by Different Analytical Techniques, Micromeritics Instrument Corp. Retrieved 2011 from Particle Size: <http://www.micromeritics.com/pdf/mas/interpretation%20of%20particle%20size%20by%20different%20techniques.pdf>.

Wood densities, n.d.. Retrieved 2011, from The Engineering ToolBox: <http://www.engineeringtoolbox.com/wood-density-d_40.html>.

Wood-Kaczmar, B., 1991 The junkyard in the sky: space is full of rubbish. From tiny flecks of paint to the broken remains of old satellites and rockets, the debris orbiting the Earth could mean the end of spaceflight within decades. New Scientist, October 13.

Yang, W.-C., 2003. Particle dharacterization and dynamics, Handbook of Fluidization and Fluid-Particle Systems. Marcel Dekker.

Zou, R.P., Yu, A.B., 1996. Evaluation of the packing characteristics of monosized nonspherical particles. Powder Technology 88, 71–79.

Solutions

CHAPTER 1

1.1 (a) 10.73 ft^3 psi lb-mol^{-1} °R^{-1} (b) 0.082056 atm l mol^{-1} K^{-1}

1.3 2.86 atm

1.5 0.8 kPa

1.7 (a) 74 μm (b) 246 μm

1.9 (a) Your mass on the Moon is the same as on Earth—72.6 kg. The scale would read a "weight" of one-sixth that: 12.1 kg. Force is the equivalent of weight in SI with the unit of N. Therefore, your "weight" on the Moon would be 119 N (b) 72.6 kg

CHAPTER 2

2.1 (a) 3.2×10^2 (b) 15.3 (c) 1530 (d) 1.26

2.3 $\mu = 28.5$, $\sigma = 9.9$, $\alpha = -1.87$, $P(Z < -1.88) = 0.470$, we can reject 10

2.5 0.081

2.7 0.04 mV

2.9 (a) $\mu = 112.7273$, $\sigma^2 = 1247.11$, $\sigma = 35.31$ (b) no

2.11 $P_{O_2} = 0.150$ atm, $W_{O_2} = 0.001$ atm

2.13 $t = 23.43$ min, $W_t = 1.73$

2.15 (a) $V_C = 50.9$ ml (b) $\Delta E = 0.0472$ (c) $\Delta I = 0.0103$ (d) Reading errors and not respecting significant figures.

2.17 (a) $\mu = 156.6$, $\sigma = 31.75$, $\sigma^2 = 1008.09$ (b) 0.1664 (c) We can reject 90

2.19 $\eta = 25.2127$ mPa s, $W_\eta = 1.0186$ mPa s

2.21 56 min 30 s

2.23 (a) $\mu = 20.28$, $\sigma^2 = 10.17$, $\sigma = 3.19$ (b) Yes, we can reject 12 because:

$$P\left(\frac{x_m - \mu}{\sigma}\right) = P\left(\frac{12 - 20.28}{3.19}\right) = P(-2.6) = 2(0.495) = 0.99$$

and

$$1 - 0.99 < \frac{1}{2(10)}$$

(c)

$$\mu = \frac{(7850 - 850)\frac{1}{3} \cdot 4\pi \left(\frac{1.5}{100}\right)^3 \cdot 9.81 \cdot 20.28}{6\pi \frac{1.5}{100} \cdot 0.4} = 174.08$$

$$W_{\rho - \rho_o} = \sqrt{W_\rho^2 + W_{\rho_o}^2} = 4 \text{ kg}$$

$$\frac{W_\mu}{\mu} = \sqrt{\left(\frac{W_{\rho - \rho_o}}{\rho - \rho_o}\right)^2 + \left(\frac{W_t}{t}\right)^2}$$

$$= \sqrt{\left(\frac{4}{7000}\right)^2 + \left(\frac{2 \cdot 3.19}{20.28}\right)^2} = 0.31$$

$$W_\mu = 0.31$$

CHAPTER 3

3.1 (a) Five (b) Although the model has different values for the exponents, these are not strictly fitted parameters since they are not allowed to vary. The equation is a nonlinear model with four fitted parameters.

3.3 (a)

$$H_0 : \mu_1 > \mu_2 + \delta$$
$$H_1 : \mu_1 \leqslant \mu_2 + \delta$$

(b) If we take a value of $\delta = 0.2$:

$$\mu_1 - \mu_2 = 0.2 \mu_1 = 1.4$$

$$\bar{X}_1 - \bar{X}_2 - t(\alpha, df)\sqrt{\frac{S_1^2}{n} + \frac{S_2^2}{n}} < \mu_1 - \mu_2 < \bar{X}_1 - \bar{X}_2$$

$$+ t(\alpha, df)\sqrt{\frac{S_1^2}{n} + \frac{S_2^2}{n}}$$

$$6.90 - 6.44 - 2.26\sqrt{\frac{0.03 + 0.06}{10}} < \delta \mu_1 < 6.90 - 6.44$$

$$+2.26\sqrt{\frac{0.03 + 0.06}{10}}$$
$$0.23 < 1.4 < 0.89$$

Thus we can accept the null hypothesis. If we had defined $\delta = 0.1$, then we would have rejected the null hypothesis and accepted the alternative that Basmati is longer than UB. (c) Discuss.

3.5 (a) See Table QS9.5a (b) See Table QS9.5b (c) There is a confounding factor for (a) with the second and third models. The columns corresponding to β_0 and $\beta_{11}X_1^2$ are parallel. There is also a confounding factor for (b) with the first and third models. The columns corresponding to $\beta_3 X_3$ and $\beta_{12}X_1 X_2$ are parallel (d) Part (a) with the third model and the base case presents no confounding. Part (b) with the third model and the base case does present confounding.

TABLE QS9.5a Eight Experiments Using Two-Level Full Factorial Design

−1	−1	−1
−1	−1	1
−1	1	−1
−1	1	1
1	−1	−1
1	−1	1
1	1	−1
1	1	1

TABLE QS9.5b Eight Experiments Using Three-Level Full Factorial Design

−1	−1	1
−1	0	0
−1	1	−1
0	−1	0
0	1	0
1	−1	−1
1	0	0
1	1	1

3.7 (a) There will be four parameters and the model will be of the following form:

$$E = a_1\epsilon + a_2 T + a_3 F + b$$

(b) A minimum of four experiments are required to obtain the four parameters. Eight experiments are needed for a full factorial design. It is outlined in Table QS9.7. (c) The new model will have six parameters and be of the following form:

$$E = a_1\epsilon + a_2T + a_3F + a_{12}\epsilon T + a_{13}\epsilon F + b$$

(d) A minimum of six experiments must be performed to obtain the six parameters. 2^5 experiments are required for a full factorial design, which is 32.

TABLE QS9.7 Full Factorial Design for the Four Parameters

Exp	E	T	F
1	1	1	1
2	1	1	−1
3	1	−1	1
4	1	−1	−1
5	−1	1	1
6	−1	1	−1
7	−1	−1	1
8	−1	−1	−1

3.9

$$P = 140 + 0.905L_t - 45.1C_t - 0.187L_tC_t$$
$$R^2 = 0.9889$$

3.11 (a) The resistance depends on the concentration. (b) 20% (c) No

3.13 (a) See Table QS9.13. (b)

$$\Phi = \begin{pmatrix} 1 & 1 & 1 & 1 \\ 1 & 1 & -1 & -1 \\ 1 & -1 & 1 & -1 \\ 1 & -1 & -1 & 1 \end{pmatrix}$$

(c)

$$Y = \Phi\theta$$

$$\begin{pmatrix} Y(1) \\ Y(2) \\ Y(3) \\ Y(4) \end{pmatrix} = \begin{pmatrix} 1 & 1 & 1 & 1 \\ 1 & 1 & -1 & -1 \\ 1 & -1 & 1 & -1 \\ 1 & -1 & -1 & 1 \end{pmatrix} \begin{pmatrix} \beta_0 \\ \beta_1 \\ \beta_2 \\ \beta_{12} \end{pmatrix}$$

> **TABLE QS9.13** Factorial Design for the Relationship between Temperature, Pressure, and Quantity of Solvent to Produce Pesticide
>
> | Exp | X_1 | X_2 |
> |-----|-------|-------|
> | 1 | 1 | 1 |
> | 2 | 1 | −1 |
> | 3 | −1 | 1 |
> | 4 | −1 | −1 |

3.15 (a) Null hypothesis: $H_o : \mu_1 = \mu_2 = \mu_3 = \mu_4$, and the alternative hypothesis is that the means differ (b) Yes (c) Yes

3.17 (a) $\alpha = 2.8$ (b) $R^2 = 0.993$

3.19

$$Emf = 1.53 + 0.0436T$$
$$R^2 = 0.972$$
$$Emf = 0.045T$$
$$R^2 = 0.94$$
$$Emf = 0.0365T + 0.0000416$$
$$R^2 = 0.9993$$

CHAPTER 4

4.1

$$P = \frac{1}{3}\tilde{n}Mv_{rms}^2 = \frac{1}{3}\frac{n}{V}Mv_{rms}^2$$
$$Mv_{rms}^2 = 3kNT$$
$$P = \frac{1}{3}\frac{n}{V} \cdot 3kNT$$
$$PV = nkNT = nRT$$
$$R = kN$$

4.3 $P = 0.32$ atm, $P_{O_2} = 0.067$ atm

4.5 126.6 kPa

4.7 (a) 771 mbar (b) 20.2 m (c) 0.16 m (d) 5.6 mm

4.9 $\Delta P = 1.1$ MPa, $\Delta P = 1800$ Pa

4.11 (a) $P_A = 0.637$ atm, $P_B = 2.81$ atm, $P_C = 0.877$ atm, $P_D = 16.0$ atm
(b) $W_{\Delta P} = 1.08$ atm

4.13 18.74 kPa

4.15 (a) 2 atm (b) The Nonette (c) 85 m (d) Same height

4.17 $t = 0.41$ mm

4.19 (a) $H = 1.96$ kPa, $H = 6.13$ kPa, $H = 10.3$ kPa (b) Glycerol

4.21 0.0326 m

4.23 (a) 0.007 mm, (b) 20. kPa, (c) 2.0 m, (d) 0.030 mm, (e) $0.06 - 6\%$ @ FS

CHAPTER 5

5.1 (a) $T = 45.4\,^\circ$C (b) $P_2 = 173.79$ kPa

5.3 (a) Toluene (b) Galinstan (c) Toluene, 10X

5.5 (a) No (b) 2.822 mV

5.7 (a) $r = \frac{2}{3}\frac{t}{(\alpha_2 - \alpha_1)(T - T_0)}$ (b) 2.43

5.9

$$\left(\frac{\Delta R}{R}\right)^2 = \left(\frac{1}{R_0}\Delta R_0\right)^2 + \left(\frac{a + 2bT}{1 + aT + bT^2}\Delta T\right)^2$$

5.11 $-0.1275 + 0.0415T + 1.226 \times 10^{-7}T^2, R^2 = 0.99994$

5.13 J: 0.663 mV, T: 0.513 mV, K: 0.519 mV, and J: 8 °C, T: 8.25 °C, K: 8.5 °C

5.15 (a) 12.183 mV (b) 205.7 °F (96.5 °C) (c) 13.145 mV

5.17 (a) T: $T = 24.01\Delta E + 25.03$, K: $T = 24.41\Delta E + 24.99$ (b) See Table QS5.17 (c) Assuming the experiment was well conducted, the thermocouples should have a greater error (w_T) than calculated in (b).

TABLE QS5.17 Absolute and Relative Errors for Each Device (T and w_T in °C, relative error in %)

	Exp. Therm.			Type T			Type K	
T	w_T	Rel. Err.	T	w_T	Rel. Err.	T	w_T	Rel. Err.
21	0.5	2	21.4	0.1	0.6	21.4	0.1	0.6
25	0.5	2	26.0	0.1	0.5	25.7	0.1	0.5
29	0.5	2	29.2	0.1	0.4	29.7	0.1	0.4
33	0.5	2	33.4	0.1	0.4	35.0	0.1	0.3
39	0.5	1	38.3	0.1	0.3	38.5	0.1	0.3
42	0.5	1	42.4	0.1	0.3	42.5	0.2	0.6
47	0.5	1	46.9	0.1	0.2	46.5	0.1	0.3
51	0.5	1	50.4	0.1	0.2	49.9	0.1	0.2

CHAPTER 6

6.1 (a) $\Delta_{\Delta P} = 21$ Pa (b) $Q = 5.40 \times 10^{-3}$ m^3 s^{-1} (c) $\Delta Q = 1.3\%$

6.3 Pipe 1: 105.9 l min^{-1} at STP, pipe 2: 139.3 l min^{-1} at STP. The flow rate is higher for the second pipe.

6.5 (a) 657 Pa (b) Turbulent (c) 1%

6.7 $Q = 596202.31$ ml s^{-1}, $\dot{m} = 24.03$ g s^{-1}

6.9 $u = 6$ m s^{-1}, $\dot{m} = 84$ kg s^{-1}

6.11 0.051 m^3 s^{-1}

6.13 (a) 27.4% of the height (b) Steel: 43% of the height, tantalum: 30% of the height

6.15 0.55% ~ 6%

6.17 $P_{dyn} = 0.45$ mbar, $\Delta u_1 = 0.0166$ m s^{-1}

6.19 14 m s^{-1}

CHAPTER 7

7.1 (a) 0.010 m s^{-1} (b) $\Delta_u = 0.0006$ m s^{-1} (c) 17 800 kg m^{-3} (d) 17.1 karats (e) Yes (f) Take more data points, prolong the time the falling ball falls, use a more viscous oil.

7.3 (a)

$$\frac{q}{A} = k\frac{\Delta T}{\Delta x}$$

$$\Delta x = \frac{k\Delta T}{q/A}$$

$$\frac{W_{\Delta x}}{\Delta x} = \sqrt{\left(\frac{W_k}{k}\right)^2 + \left(\frac{W_{\Delta T}}{\Delta T}\right)^2 + \left(\frac{W_{q/A}}{q/A}\right)^2}$$

$$W_{\Delta x} = \sqrt{0.05^2 + \left(\frac{0.2}{15}\right)^2 + 0.01^2} \cdot 10 \text{ cm} = 0.53 \text{ cm}$$

(b) $\Delta T_2 = 2\frac{q}{A}\frac{\Delta x}{k} = 2\Delta T_1 = 30\,°C$ (c) The uncertainty of the temperature measurement is incorrect, or the heat flux is 2.2 or 10% instead of 1%.

7.5 $k_2 = -\frac{T_m - T_1}{T_m - T_2}k_1$, so $k_2 = 3.5$ W m^{-1} K^{-1} and $W_{k_2} = 0.085$ W m^{-1} K^{-1}

7.7 (a) If the walls of the process in which the refrigerant circulates are not isolated, it is impossible to determine the thermal coefficient because there is heat loss to the environment (b) $k = 1.03$ k W m^{-1} °C^{-1}

7.9 (a) $\rho = 8784$ kg m^{-3}

7.11 (a) Air (b) T, F, F (c) T, T, T (d) 1C, 2B, 3A, 4D (e) F, T, F, F (f) gases, liquids, plastics, alloys, metals (g) F, T, F

7.13 (a) 13.3, 0.54 (b) $\mu = \frac{(\rho - \rho_0)Vg}{6\pi r L}t = \frac{(4540-970)\cdot 2 \cdot 0.01^2 \cdot 9.81}{9 \cdot 0.4} \cdot 13.3 = 25.9$ Pa s
(c) $\Delta = k(\alpha)\sigma = 2.57 \cdot 0.54 = 1.4$ s (d) $W_\mu/\mu = \sqrt{(W_\rho/\rho)^2 + (W_t/t)^2}$
$= \sqrt{(20/3570)^2 + (1.4/13.4)^2} = 0.10 = 10\%, W_\mu = 0.1 \cdot 25.9$
$= 2.6$ Pa s (e)

$$a = \frac{X - \mu}{\sigma} = \frac{14.1 - 13.3}{0.54} = 1.48$$
$$P(Z < 1.48) = 0.431$$

$5(1 - 2 \cdot 0.431) = 0.691 > 0.5$, therefore accept. t is 13.4 s, 12.9 s, 14.1 s, 12.7 s, and 13.2 s.

CHAPTER 8

8.1 (a) 836 plates (b) 0.066 mm (c) 1.57 (d) 1502 plates (e) 84.61 cm
8.3 (a) 2858 ml, 221.3 ml min^{-1}, 52.5 min, 0.982 (b) 18.23 ppm. The concentration is too high. The desulfurization unit must be revised to have a better process efficiency.
8.5 (a) $C = 0.0016$ mol l^{-1} (b) $C' = 0.0008$ mol l^{-1}
8.7 (a) $t_0 = 0.8$ min, $t_1 = 3.2$ min, $t_2 = 6$ min, $t_3 = 18$ min, $k_1 = 3, k_2 = 6.5, k_3 = 21.5$ (b) $N_{th} = 5.54(t_r/w_{1/2})^2$, so $N_{th,1} = 5700, N_{th,2} = 3200, N_{th,3} = 55\,400$ (c) $R = 9.4$, the minimum value is 1.25 although 1.5 is accepted, shorten the column, increase the temperature, increase the linear velocity of the mobile phase, and decrease the amount of stationary phase (d) $\frac{\Delta s}{s} = 5\%$
8.9 (a) Reduce the analysis time. (b) Solute diffusion is faster than in N_2, so the balance between the mobile phase and the stationary phase is reached more quickly. For DCT it is easier to detect higher resolution. Conductivity. (c) Detectable: alcane, organic; undetectable: oxygen, nitrogen, carbon dioxide, etc. (d) 8 or 9, k' is 13.5, 18, and 21, $\alpha_{1,2} = 1.33, \alpha_{1,3} = 1.56, \alpha_{2,3} = 1.17, R_{2,1} = 1.89 > 1.25$ so it is well resolved, $R_{3,2} = 1.13 < 1.25$ so it is not well resolved
8.11 (a) $N_{th} = 2544, \frac{N_{th}^2}{4} = 12.61, L = 13$ cm (b) $k' = 1.5, \frac{k'}{k'+1} = 0.6$ (c) $R = 7.41 > 1.5$, there is separation.
8.13 See Fig. QS8.13

CHAPTER 9

9.1 0.68
9.3 $H = 1.08, \Delta_H = 0.02, \phi = 1$

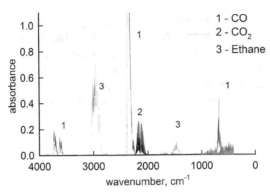

FIGURE QS8.13 IR Spectra Solution

9.5

$$V_s = \frac{\pi}{6} d_p^3$$

$$V_{py} = \frac{1}{3} h^3$$

$$S_{A,sp} = \pi d_p^2$$

$$S_{py} = h^2 + 4 \left(\frac{1}{2} h \sqrt{h^2 + (h/2)^2} \right) = h^2 (1 + \sqrt{5})$$

$$V_s = V_{py}$$

$$\frac{\pi}{6} d_p^3 = \frac{1}{3} h^3$$

$$d_p = \sqrt[3]{\frac{2}{\pi}} h$$

$$\phi = \frac{\pi d_{sp}^2 |_{v_{sp} = v_p}}{S_p} = \frac{\pi (2/\pi)^{2/3} h^2}{(1 + \sqrt{5}) h^2} = 0.72$$

9.7 One—sphericity refers to the outer surface only

9.9 1.69 mm^2, 0.847 mm

9.11 (a) $d_p = 13.62$ μm (b) $\bar{d}_{NL} = \frac{\sum_{i=1}^{9} d_{p,i} dN}{\sum_{i=1}^{9} dN} = 18.12$ μm (c) $\bar{d}_{SV} = \frac{\sum_{i=1}^{9} d_{p,i}^3 dN}{\sum_{i=1}^{9} d_{p,i}^2 dN} = 22.83$ μm (d) Yes

9.13 (a) $d_{50} = 250$ μm (b) $D_N[3,2] = 210$ μm, $D_N[1,0] = 161$ μm

9.15 $d_{nl} = 0.136$, $d_{ns} = 0.137$, $d_{nv} = 0.139$, $d_{sv} = 0.143$

9.17 0.242 m s^{-1}, 0.099 m s^{-1}, 0.506 m s^{-1}

9.19 (a) 105 μ m (b) iii (c) i (d) ii (e) ii (f) $d_{p,eq} = 0.968 d_{p,s}$, $\phi = 0.794$

9.21 (a) $D[2,0] = 2.24$ mm (b) $\Delta P = 122$ kPa (c) The definition of the mean diameter should be based on the drag. There are many small particles. The gas velocity is 2.0 m s^{-1} at the inlet, but the average speed is higher. The temperature gradient is important, therefore the gas velocity is higher. The uncertainty of the variables is greater than estimated (d) $\Delta_{\Delta P} = 0.24$

Index

A

Accuracy, Instrumentation concepts,
42–43
ADC resolution, 36
Airline instrumentation, 133
Airweight result, barometric pressure, 114
Alternative hypothesis, planning
experiments, 73
American Society for Testing and Materials, 11
Amontons' law, 104
Boyle's law and, 104
Charles law and, 104
Analysis of Variance (ANOVA), 72–73
Antoine equation, 151–152
Arithmetic function, uncertainty propagation,
31–32
Arrhenius equation, 88
Atom absorption, 295

B

Bacon, Sir Francis, 15
Bar charts, 56
histogram, 57
usages, 56
Barometric pressure
airline instrumentation, 133
airweight result, 114
changes, 113
elevation impact on, 114
pressure gauge, 110, 120–121
Bellows
deformation factors, 125–126
principle, 124–125
BET theory
in monolayer adsorption, 343
in multilayer adsorption, 343
in surface area, 343–344
temperature of adsorption, 343
Bimetallic thermometers
coil application, 164
curvature equation, 165–166
deflection, 166

thermal expansion coefficiency, 164–165
thermostats, use in, 164
Binary gas diffusion, 250
pressure effects, 256
Bourdon gauge
for pressure indication, 124
principle, 124
Boyle's law
Amontons' law vs, 104
description, 104
BSI (United Kingdom), 10–11

C

Calorie
defined, 9
Capsule Pressure Gauge, 128
Celsius scale, 8
Cement manufacturing, 307–308
fossil fuel combustion, 308
powder analysis, 307–308
processing steps, 307–308
solid analysis, 308
Central Limit Theorem and, 28
CGPM recommendations
standards, quantity, 5–7
cgs (centimeter-gram-second)
conversion, SI system, 7
length standard, 8
standard of mass, 7–8
time standard, 8
unit of force, 9
volume measurement in, 8
Charles law
Amonton's law, 104
description, 104
Chemical industry
multi-phase reactors, 324–325
particle technology, 307
Chemical literature
mixed units, 8
prefixes, 8
unit kmol, usage, 8

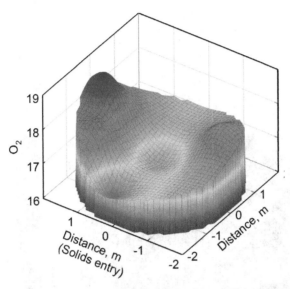

FIGURE 2.22 3D Plot of Oxygen Distribution Across a 4 m Diameter Regenerator

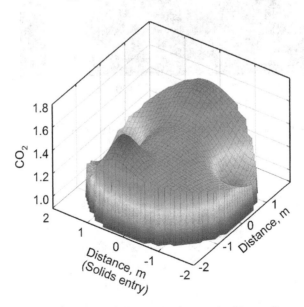

FIGURE 2.23 3D Plot of Carbon Dioxide Distribution Across a 4 m Diameter Regenerator

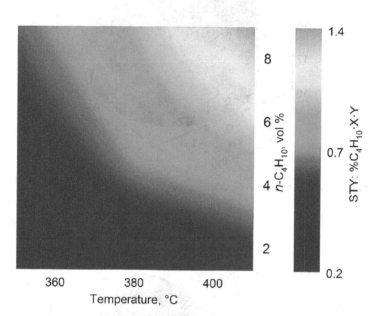

FIGURE 2.24 Contour Plot of STY Versus Temperature and vol% $n\text{-}C_4H_{10}$

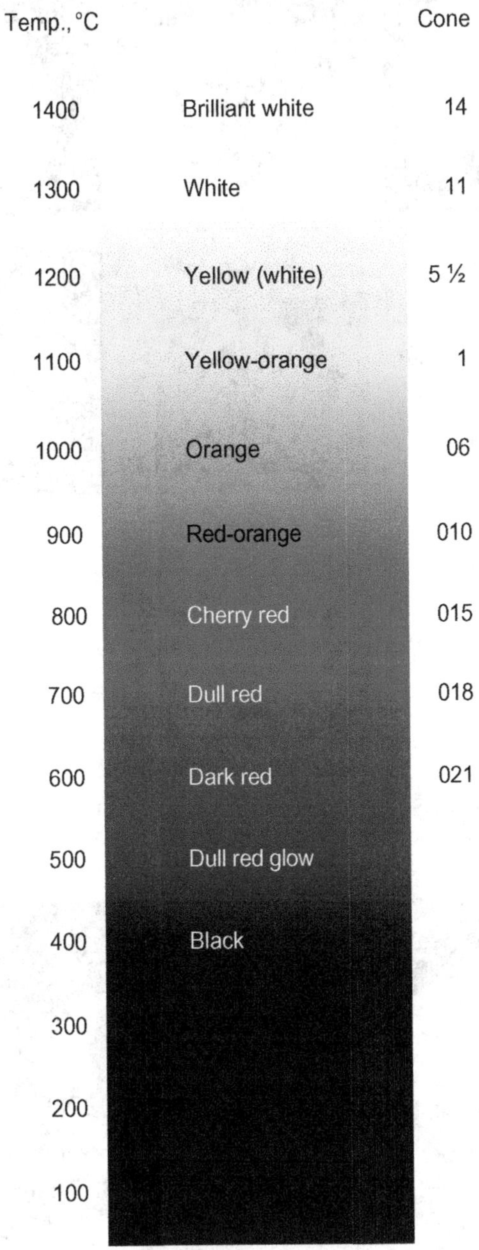

Temp., °C		Cone
1400	Brilliant white	14
1300	White	11
1200	Yellow (white)	5 ½
1100	Yellow-orange	1
1000	Orange	06
900	Red-orange	010
800	Cherry red	015
700	Dull red	018
600	Dark red	021
500	Dull red glow	
400	Black	
300		
200		
100		

FIGURE 5.2 Kiln Firing Chart

FIGURE 9.3 Particle Shapes: (a) Spherical - Calcite; (b) Acicular - Calcite with Stibnite; (c) Fibrous - Actinolite; (d) Dendritic - Gold; (e) Flakes - Abhurite; (f) Polyhedron - Apophyllite; (g) Cylindrical Particles

FIGURE 9.7 Pellet Morphology in Chemical Reactors (Image courtesy of Clariant.)

Edwards Brothers Malloy
Ann Arbor MI. USA
April 1, 2014